Cracked Rotors

A Survey on Static and Dynamic Behaviour Including Modelling and Diagnosis

Nicolò Bachschmid, Paolo Pennacchi, and
Ezio Tanzi

Cracked Rotors

A Survey on Static and Dynamic Behaviour Including Modelling and Diagnosis

Springer

Prof. Nicolò Bachschmid
Politecnico di Milano
Dept. of Mechanical Engineering
Via La Masa, 1
20156 Milano
Italy
E-mail: nicolo.bachschmid@polimi.it

Ezio Tanzi
Politecnico di Milano
Dept. of Mechanical Engineering
Via La Masa, 1
20156 Milano
Italy
E-mail: ezio.tanzi@mecc.polimi.it

Paolo Pennacchi
Politecnico di Milano
Dept. of Mechanical Engineering
Via La Masa, 1
20156 Milano
Italy
E-mail: paolo.pennacchi@polimi.it

ISBN 978-3-642-01484-0

e-ISBN 978-3-642-01485-7

DOI 10.1007/978-3-642-01485-7

Library of Congress Control Number: 2009941533

Typesetting: Data supplied by the authors

Production: Scientific Publishing Services Pvt. Ltd., Chennai, India

Cover Design: WMX Design, Heidelberg, Germany

Printed on acid-free paper

9 8 7 6 5 4 3 2 1

springer.com

Foreword

A very rich, but also in some way confusing, literature about cracked rotors has appeared in the last 30 years and is still developing. Since the authors have been involved in analyses of experimental data from power plants, in studies, in laboratory tests, in development of models and in numerical analyses of cracked rotors for more than 20 years, they felt that time was ready to publish a book about cracked rotors that should contain the main achievements obtained.

The focus of this book was intended on practical aspects related to industrial machinery and to numerical analyses aimed to represent their behaviour, rather than on theoretical investigations.

Since the background of the authors is mainly rotor-dynamics, some contributions of other experts have been asked for to cover some areas that are not strictly related to this field.

The book is devoted to all engineers or technicians that are in some way involved in the design, in the condition monitoring, in the maintenance of rotating machinery or in the management of any plant in which rotating machineries are installed, especially to those who are responsible of the safety of the plant, as well as to researchers or students that are interested in the topic of developing cracks in rotating shafts.

Chapter 1 is dedicated to the general introduction and to the overview of development and propagation of cracks in rotating shafts.

The typical experimental behaviour of cracked rotating shaft is described in chapter 2, as it has been measured in industrial machines.

Chapter 3 introduces the possible testing techniques that can be employed for detecting a crack in rotating shafts.

Chapter 4 is dedicated to provide a deeper insight into the breathing mechanism of a crack and into its thermal sensitivity, as it results from a series of experimental laboratory tests.

The modelling of the stiffness variation due to the presence of cracks in shafts, as proposed by different researchers, the modelling of breathing mechanism and related stiffness variation and finally the calculation of the dynamical response of a full size cracked rotor is described in chapter 5.

Chapter 6 is dedicated to a comparison of calculated results to experimental results obtained using both a medium size test rig and a full size shaft-line of a turbo-generator and to the sensitivity analysis performed with the most suitable models: how the position of the crack, how its shape and how its depth influence the system response.

Chapter 7 describes some second order effects, like: i) the excitation of torsional and axial vibrations, ii) the effect of a slightly helicoidal development of cracks, as it can occur in case of huge torsion loads, compared to the more common transverse cracks and finally iii) the comparison between the results obtained with linear models and those obtained with the fully non-linear approach, showing what could be the effect of very deep cracks on very light shafts loaded with rather high unbalances. These last effects are shown using the model of the shaft of a small machine, supported by oil-film bearings, that is anyhow more representative than the usual very simple Jeffcott / de Laval rotor.

Chapter 8 is entirely dedicated to the diagnosis of cracks in rotating shafts, assuming that only the usual measurements in correspondence of the bearings are available to detect the presence of a crack, in a possible early stage of its development. It is shown that not only the crack can be detected, but also its position and depth can be identified, using a model based method.

Contents

Acronyms

1X: once per revolution harmonic component
2X: twice per revolution harmonic component
3X: three-times per revolution harmonic component
AC: alternate current
BWR: boiling water reactor
CCL: crack closure line
CETIM: Centre Technique des Industries Mécaniques
CF: certainty factor
DC: direct current
d.o.f.s: degrees of freedom
EDF: Électricité de France
EDM: electrical discharge machining
EE: environmental effects
EFIT: elasto-dynamic finite integration technique
ET: eddy current testing
FCP: false-call probability
FE: finite element
FEM: finite element model
FES: far end scan
FN: false negative response
HF: Human factors
HP: High pressure
ID: inner diameter
IP: Intermediate pressure
LEFM: Linear Elastic Fracture Mechanics
LP: Low pressure
MPI: magnetic particles
NDE: non-destructive evaluation
NDT: non-destructive testing
NES: near end scan
OD: outer diameter
PA: phased array
POD: probability of detection
POND: probability of non detection
POR: probability of recognition
PT: dye penetrant testing

PWR: pressurized water reactor
RCP: Reactor cooling pump
RHS: right hand side
ROC: relative operating characteristic
RRP: Reactor recirculation pump
RT: x-ray testing
SERR: strain energy release rate
SIF: stress intensity factor
TOFD: time of flight diffraction
UT: ultrasonic testing
UV: ultraviolet
VT: visual testing

Chapter 1
Cracks in Rotating Shafts

1.1 Introduction

One of the most common incipient losses of structural integrity in mechanical structures is the development and propagation of cracks. A crack may propagate from some small imperfections on the surface of the body or inside of the material and it is most likely to appear in correspondence of high stress concentration. For example, figure 1.1 shows a crack developed in the stiffening plate of a generator casing.

Fig. 1.1. Crack developed in the stiffening plate of a generator (Courtesy of Dr. Carlo Stoisser, Generation, Thermal Engineering Dept. EDF – Électricité de France).

Fretting corrosion, in case of shrink fitted connections, may generate cracks, the development of which is also apparently favoured by wet and corrosive environments. Thermal stresses and thermal shocks are also responsible for generating high local stress intensity factors, which can cause the starting of a crack and its propagation. Figure 1.2 shows a crack developed in the blade root of a steam turbine.

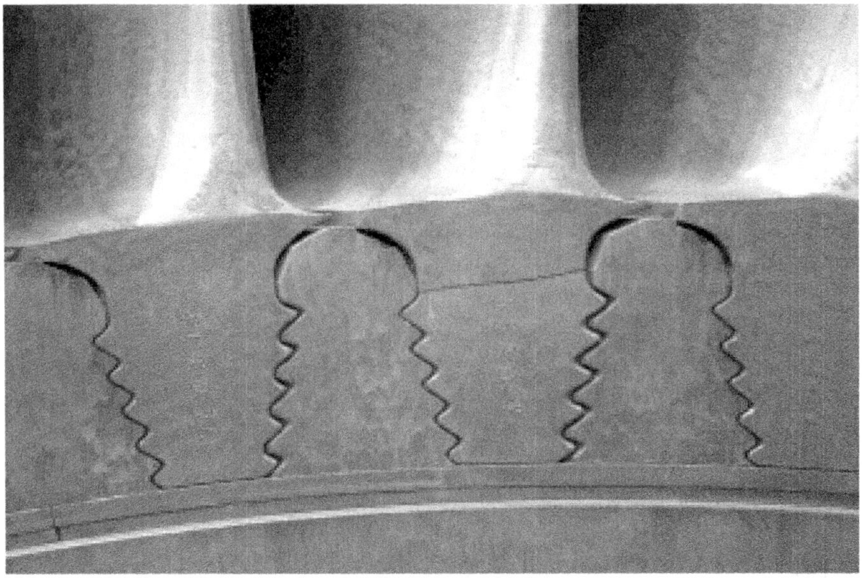

Fig. 1.2. Crack developed in the blade root of a steam turbine (Courtesy of Dr. Carlo Stoisser, Generation, Thermal Engineering Dept. EDF – Électricité de France).

Crack propagation, direction of propagation and propagation speed as a function of the stress level are problems to which many studies have been dedicated in the last 50 years. These studies form what is generally called *Fracture Mechanics*. There are scientific journals and international conferences that are dedicated exclusively to fracture mechanics and its basic achievements are taught generally in university lectures about strength of materials. Indeed cracks propagate mainly due to low or high cycle fatigue and crack initiation and propagation are strongly dependent on material properties.

Once a crack, although small, has appeared, high stress intensity factors will develop at the crack tip and allow the crack to propagate deeper, even if the external loads are not changing. Crack propagation path is generally determined by the direction of maximum stress or by the minimum material strength. Inspections with dye penetrant, ultrasonic tests or x-rays and similar non-destructive tests are commonly used for revealing the presence of cracks and defining their extensions. Dynamic tests, based on vibration measurements are also suggested for identifying presence and location of cracks in complex structures: the change in the overall dynamic behaviour might be caused by the local reduction of the stiffness due to a crack that has developed somewhere in the structure.

A very special case of damage in mechanical structures is the development and propagation of *transverse cracks* in rotating shafts.

Figure 1.3 shows an example of an hollow shaft, in which the crack starts from a keyhole.

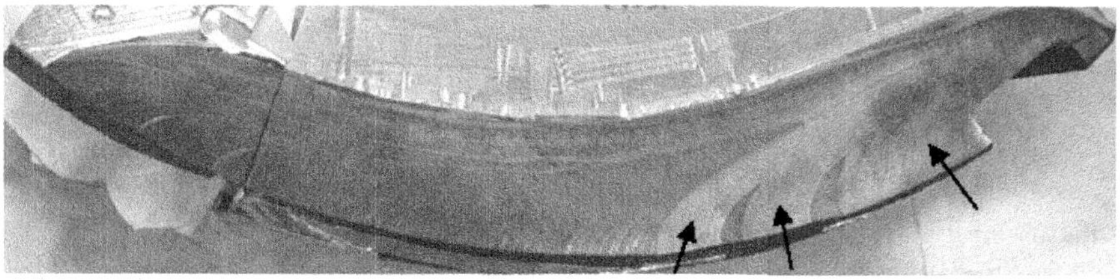

Fig. 1.3. Typical crack propagation in a shaft. The crack start from the keyhole on the right (Courtesy of Dr. Carlo Stoisser, Generation, Thermal Engineering Dept. EDF – Électricité de France).

For the analysis of the behaviour of cracked rotating shafts, the research field of *Rotor Dynamics* merges with that of Fracture Mechanics. The term transverse cracks indicate cracks in which the crack surface is orthogonal to the rotation axis of the shaft. Also few other types of cracks have been reported, like slant cracks, but transverse cracks are the most common cracks in rotating machinery. The obvious difficulty to inspect rotating shafts during the operation of the machines makes the detection of cracks much more difficult in these structures than in static (non-rotating) structures. Therefore suitable symptoms are needed that can easily be measured (typically vibrations) and that are able to indicate clearly the presence of a crack in a rotating shaft. If these symptoms arise in a machine, the machine may be stopped, the shaft may be uninstalled and inspected with standard procedures and the catastrophic failure of the complete unit can be avoided.

The propagation speed of a crack can be very low at the beginning, often periods of rest alternate with periods of propagation. When the crack has propagated to a relevant depth, then the propagation speed increases dramatically and the shaft may fail in few hours of operation. Therefore, the surveillance and the condition monitoring of the rotating machine should be able to identify all symptoms of a propagating crack in the early stage.

The accurate modelling of cracked shafts allows the simulation of its dynamic behaviour, as a function of different positions, different depths and shapes of transverse cracks. The comparison of calculated vibrations and measured vibrations allows the identification of not only the presence of a crack in a rotating shaft, but also of its location and depth, using a diagnostic procedure. This justifies the need of reliable and accurate models for representing cracks in rotating shafts and for simulating their dynamic behaviour.

1.2 The Development of Transverse Cracks in Rotating Shafts

As it was stressed in the introduction, cracks in rotating shafts are most likely to appear in correspondence of sharp changes of the diameter or of the geometry of the shaft (for instance due to the presence of holes, slots for keys, threads and so on) in regions of high stress concentration.

Also thermal stresses that develop as the consequence of the working fluid in thermal machines, like steam and gas turbines, and thermal shocks are responsible for generating high local stress intensity factors, which can cause the starting of a crack and its propagation.

In rotating shafts, the cracks propagate generally in a plane perpendicular to the shaft axis, when axial bending stresses are prevailing, generating a transverse crack. Anyway, also conical crack surfaces and helical crack surfaces, as well as longitudinal cracks, have been reported in literature.

The crack may have any orientation at its starting point, depending on local conditions, but when the crack propagates more deeply, its direction is mainly radial and the cracked surface, although not exactly, is roughly planar, with vanishing small curvatures.

The propagation velocity may change from case to case in rotating shafts. Propagation times of 74000 up to 101000 hours of operation are reported from detection, or suspected presence, to dangerous crack depth, but only 2500 hours of operation has been enough to deepen consistently the crack in other cases [1.1]. Very frequently the crack moves by steps, alternating growths to stops: both can be seen on the cracked surface pattern where the rest lines called *beach marks* are recognizable. These are evident in figure 1.4, in which a particular of a cracked surface of a generator is shown, or in figure 1.3 where they are highlighted by arrows.

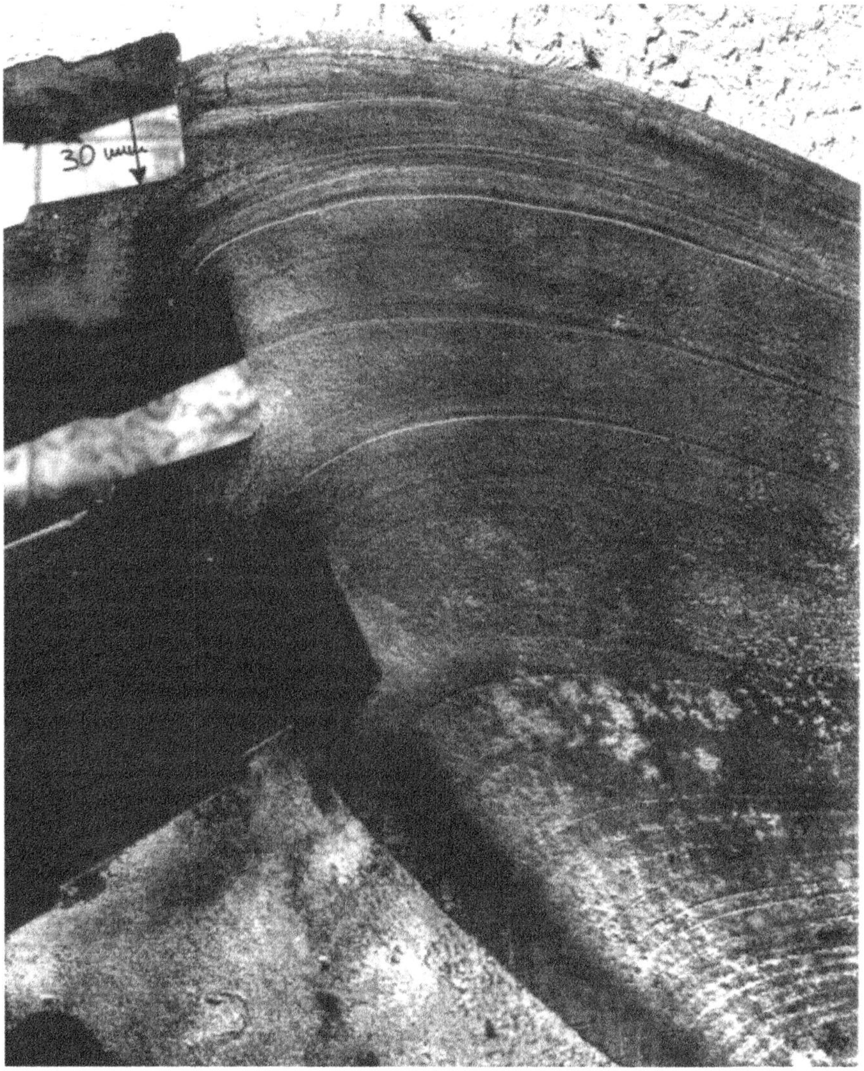

Fig. 1.4. Crack beach marks on a crack surface of a generator. Winding slots are partially visible on the left.

Generally, when the crack approaches a dangerous depth, it propagates more quickly, with a propagation velocity that increases almost exponentially. The final growth, up to a critical dangerous depth, takes sometimes only few days of operation. This is rather evident from the analysis of the condition monitoring data shown in figure 1.5, figure 1.6 and figure 1.7 that are relative to turbo generators of three different power plants, located in different countries.

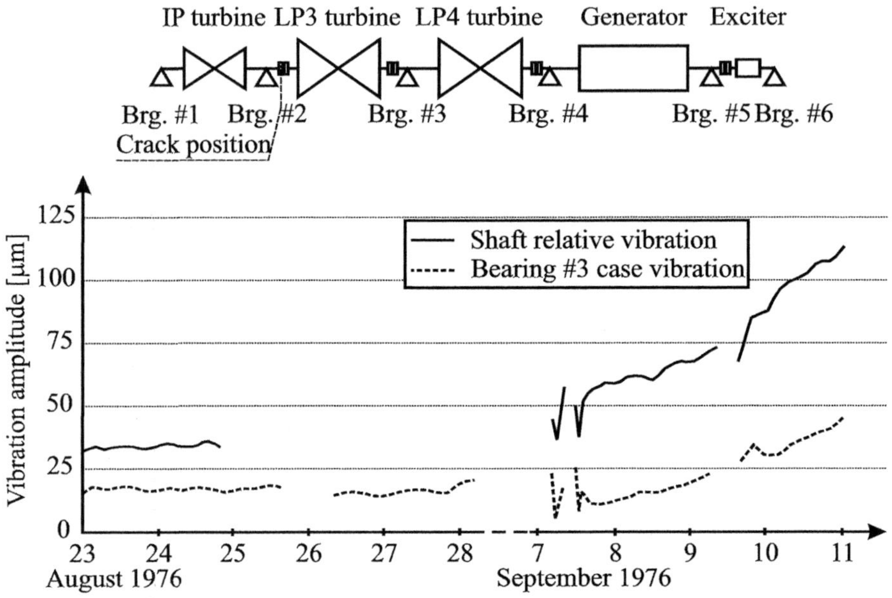

Fig. 1.5. Vibration trend of the intermediate pressure shaft of cross-compound unit of Cumberland power plant.

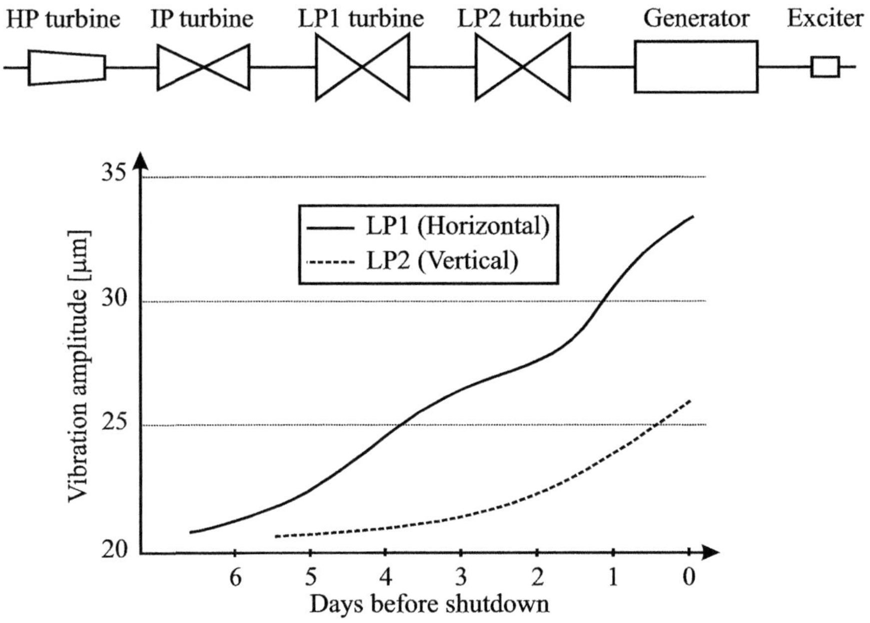

Fig. 1.6. Vibration trend of the steam turbine of Wuergassen power plant.

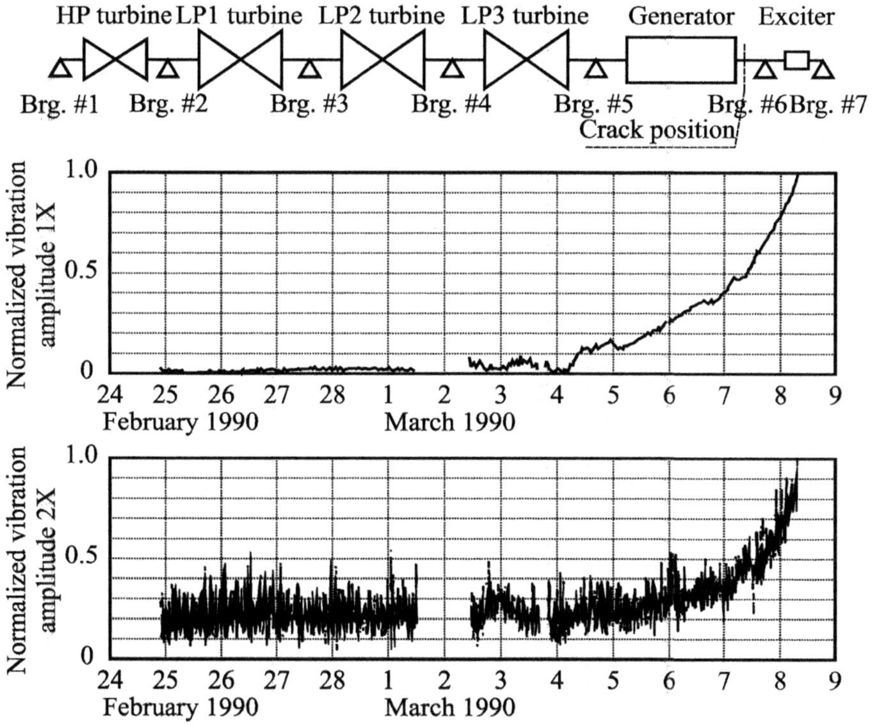

Fig. 1.7. 1X and 2X vibration trends in generator bearing #5 of Darlington power plant (vibration amplitudes are normalized to their maximum value).

The first example is relative to the 1.34 GW cross-compound unit in the Cumberland (USA) coal power plant. This steam turbo generator has two shaft lines and figure 1.5 shows the journal relative to bearing and the bearing case vibrations measured in bearing #3 of the intermediate pressure shaft. This shaft is composed of the intermediate pressure (IP) turbine, two low pressure turbines – LP3 and LP4 –, the generator and the exciter. Some data were missed by the condition monitoring system, but the final trend is rather clear.

The second example is shown in figure 1.6, where the vibrations measured in correspondence of the two LP turbines of the 670 MW steam turbo generator installed in the Wuergassen (Germany) nuclear power plant are displayed. Vibration amplitudes increase quickly in the last 5 days of operation before shutdown.

The third example in Figure 1.7 shows the 1X and 2X vibration trends in the bearing on the driven-end side of the generator of the 936 MW steam turbo generator installed in the Darlington (Canada) nuclear power plant. In this case both harmonic components show the same increasing trend. Also in this case the vibration amplitude increases quickly in the last days of operation.

Even if the turbo generators, the data of which are considered in figure 1.5, figure 1.6 and figure 1.7, are of different types and sizes, all of them had a very similar trend of the vibration level increase in the last 4-5 days of operation before the shutdown of the machine. The inspection revealed cracks that had developed up to dangerous depths in all these cases. In other cases documented in literature less than 50% of the cross section was left un-cracked.

Similar vibration trends are found also in reactor cooling pump (RCP) shafts just before failure even if the rate of the vibration increase is much slower: figure 1.8 shows the vibration trend of the unit 1 RCP of Palo Verde (USA) nuclear power plant. Another vibration time history of a cracked RCP shaft is shown in figure 2.5 of section 2.2.

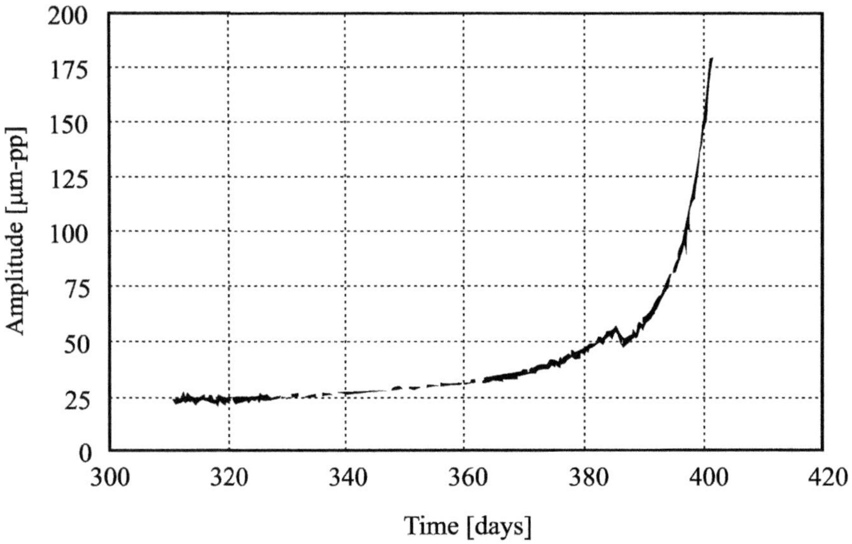

Fig. 1.8. Vibration trend of unit 1 RCP of Palo Verde power plant.

In many cracked shafts, it can be assumed that the growth takes place not continuously during normal operating conditions, in which millions of cycles are consumed in few hours of operation, but during some kind of transient condition, such as start-up or coast-down. In these conditions also some thermal stresses may sum up to the normal bending stresses and additional stresses are caused by high vibration amplitudes experienced when crossing critical speeds due to resonant conditions.

Cracked rotors can follow on to operate, if spare rotors are not available, providing that the crack propagation is controlled by accurate vibration monitoring, as was done for instance by a British electricity board [1.2] in the past.

Sometimes cracked rotors can be repaired by welding, but in most cases they are removed definitively from service. Various cracked rotors have been finally cut or broken, in order to allow a close inspection on the cracked surface and to study in detail the crack generation and propagation mechanism. In few circumstances the cracked rotor has been subject to test runs in test cells, in order to analyse accurately its dynamic behaviour and its thermal sensitivity.

Information about cracks and failures is generally kept confidential by the plant management and by the machine manufacturer; therefore not all the cases have been reported and analyzed in literature, especially in the recent years. Anyhow the available pieces of information prove that cracks in power and industrial plant machinery, like steam turbines, generators and pumps, have been discovered in Europe, in North America as well as in the Far East.

In 1987, Allianz, an industrial insurance company, promoted a Conference about preventing catastrophic failures in rotating machinery, in which 37 different crack cases were reported in Europe in the previous 20 years [1.1]. Also other 16 catastrophic failures were reported, but only very few could be attributed with certainty to the presence and propagation of transverse cracks. Sometimes it can be very difficult to identify the cause of the accidents after the complete bursting of the rotor system.

A typical example is represented by the accident occurred on March 1956, in Pittsburgh (USA). During an over speed test at 3920 rpm performed to set the trip, a 125 MW generator burst [1.3]. The machine went to service less than two years before and had a regular inspection and overhaul two months prior to the catastrophic failure. The generator body burst lengthwise approximately into two half cylinders like a log split down the middle. The two halves also broke into smaller pieces by following transverse cracks, see figure 1.9. The question to be solved was whether the transverse cracks or the longitudinal crack developed first. The investigation ascertained that the forged core of the generator presented an original relevant longitudinal flaw that was not discovered even with ultrasonic examination during the acceptance, bearing in mind that 50's technology was used. The longitudinal crack was the cause and several major transverse cracks, not caused by fatigue, were generated during the burst in this case.

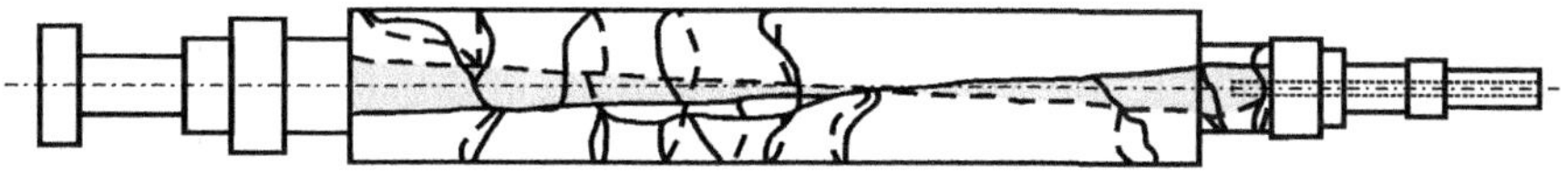

Fig. 1.9. Diagram of the generator showing the location of the primary longitudinal fracture (shaded area) and the major transverse cracks.

An overview about occurrence of cracks in industrial machines and some catastrophic failures caused by cracks is presented by Ishida in [1.4]. About 50 cases are reported between the 50's and the 90's, mainly in western countries and in Japan. The shafts affected by cracks were principally steam turbines of different sizes (from 60 MW up to 1300 MW), but also generators and RCP/RRP (reactor recirculation pump) of nuclear power plants were affected by cracks. Some interesting cases, reported chronologically in literature, are:

- 1953, Tanners Creek power plant (USA). The steam turbine unit, operating at 1800 rpm, increased suddenly its vibration level. After the shutdown a transverse crack with a circumferential extension of approximately 160° was discovered in a turbine [1.5].
- March 1954, Arizona (USA). During the balancing in the factory, a generator rotor operating at 3600 rpm burst. Also before bursting, several cracks were found in that generator [1.6].
- September 1954, Cromby (USA). A generator burst while running at 3780 rpm during an over-speed trip test [1.6].
- December 1954, Rigeland (USA). During a routine over-speed trip test, a LP steam turbine burst [1.7].
- 1956-1957, Castle Donnington power plant (UK). Three different cases of cracks were discovered in as many large turbine rotors. The causes were ascribed to fatigue under very low nominal gravitational bending stresses. Transverse cracks propagated up to 75% of the diameter in two cases, a 60 MW and a 100 MW turbine, while a circumferential crack developed in the third case, a 60 MW turbine, with the average depth of 1.6 mm [1.8].
- 1970, Nagasaki (Japan). A catastrophic accident occurred during the running test of a LP turbine. The machine burst, causing more than 60 casualties. After the investigation, the origin of the accident was identified in cracks originated by stress concentration at grinded holes on the turbine [1.9].
- 1970-1971, Japan. Four cases are reported respectively in two 125 MW and in two 375 MW steam turbines. These machines operated with frequent start-ups and shutdowns and the investigation reported that cracks were caused by the cyclic thermal stresses [1.9].
- 1972-1974, Ferrybridge power plant (UK). Three cracks in LP turbines were discovered in different units of that plant (in chronological order: unit 2, 1 and 3), all with a rated power of 500 MW [1.10].
- 1970-1980, USA. During this period, 28 cases of failures due to cracks are reported in [1.11].
- 1971-1981, UK. Another four cracked rotors of different type were discovered in this period. The first was a transverse crack in the generator of a 500 MW steam turbine, with the depth of 40% of the diameter. The second was a circumferential crack of about 70 mm of length in the IP turbine of a 60 MW steam turbine. The third was again a transverse crack in the generator of a 660 MW steam turbine but the depth reached 45% of the diameter. The last was a circumferential crack in the coupling between the HP and IP turbines of a 350 MW steam turbine [1.12].

- 1976, Cumberland power plant (USA). As presented before in figure 1.5, the increasing of vibration level was caused by the transverse crack. Also a 2X excitation was observed during the coast-down of the machine [1.13]. The crack was discovered close to the rigid coupling between the IP and the LP3, after 18000 h of operation. The failure analysis of the shaft showed that the crack originated from the shoulder in the coupling face (figure 1.10) and grew along a conical surface over 360°. During this phase there were not significant dynamical effects. Then the crack started to propagate towards the shaft skin and appeared on the surface with a circumferential extension of 230°, causing the vibration increasing.

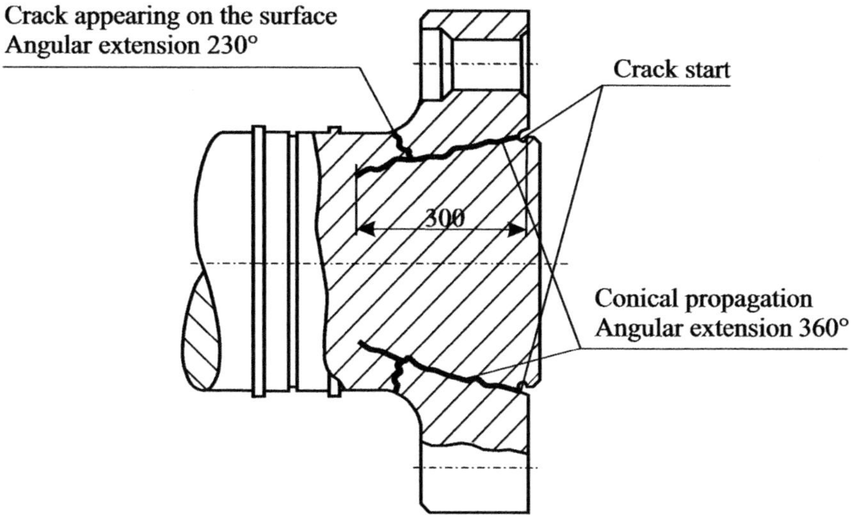

Fig. 1.10. Shape and extension of the crack on coupling between IP and LP3 of Cumberland power plant.

- 1977, Torvaldaliga power plant (Italy). A transverse crack was discovered in the high pressure – intermediate pressure (HP-IP) steam turbine of a 240 MW unit. This case is discussed in section 2.3.1.
- 1978, Wuergassen nuclear power plant (Germany). The vibration trend is presented in figure 1.6 and the case is discussed in [1.13].
- 1982, Finland. Transverse cracks were discovered in the generators of two turbo generators of 735 MW [1.14]. This case is discussed in section 2.2.
- 1983, Lavrion power plant (Greece). Thanks to the developing of modern condition monitoring systems, crack symptoms allowed the crack identification in a 300 MW steam turbine [1.15]. The shape of the crack is shown in figure 1.11 after the operation of 80000 h.

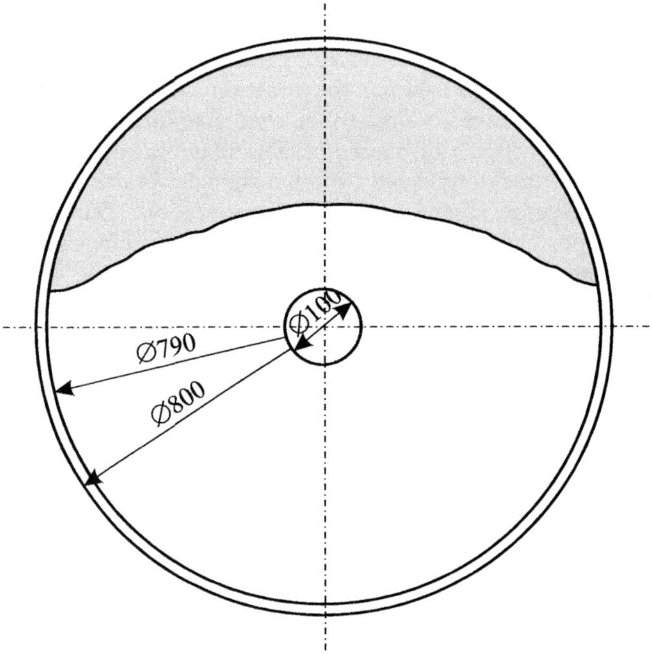

Fig. 1.11. Shape of the crack of Lavrion power plant.

- 1986, Crystal River nuclear power plant (USA). The shaft of a RCP broke due to a fatigue crack started from a groove and the impeller separated completely from the shaft [1.16].
- 1989, Crystal River nuclear power plant (USA). A twin failure to that of 1986 occurred [1.17].
- 1989, Grand Gulf nuclear power plant (USA). The failure of a RCP shaft is reported also in this case [1.18].
- 1990-2000, USA. Circumferential cracking of RCP and RRP shafts reported at several facilities including Sequoyah, Palo Verde [1.19] and St. Lucie [1.20].
- 1990, Japan. A transverse crack with the circumferential extension of 120° was found on the shaft of a wind tunnel fan for car testing [1.21], see figure 3.16 of section 3.2.2.1.
- 1990, Darlington nuclear power plant (Canada). A transverse crack that had the circumferential extension of about 25% was found on the generator of the 936 MW turbo generator [1.22]. Also in this case, sudden vibration increasing allowed the crack presence to be suspected, see figure 1.7. The crack started from the threads of a drilled hole aimed to connect the bars of the two pole-pair generator [1.23]. The crack started by fretting fatigue, then increased dangerously its depth also under the surface, see figure 1.12.

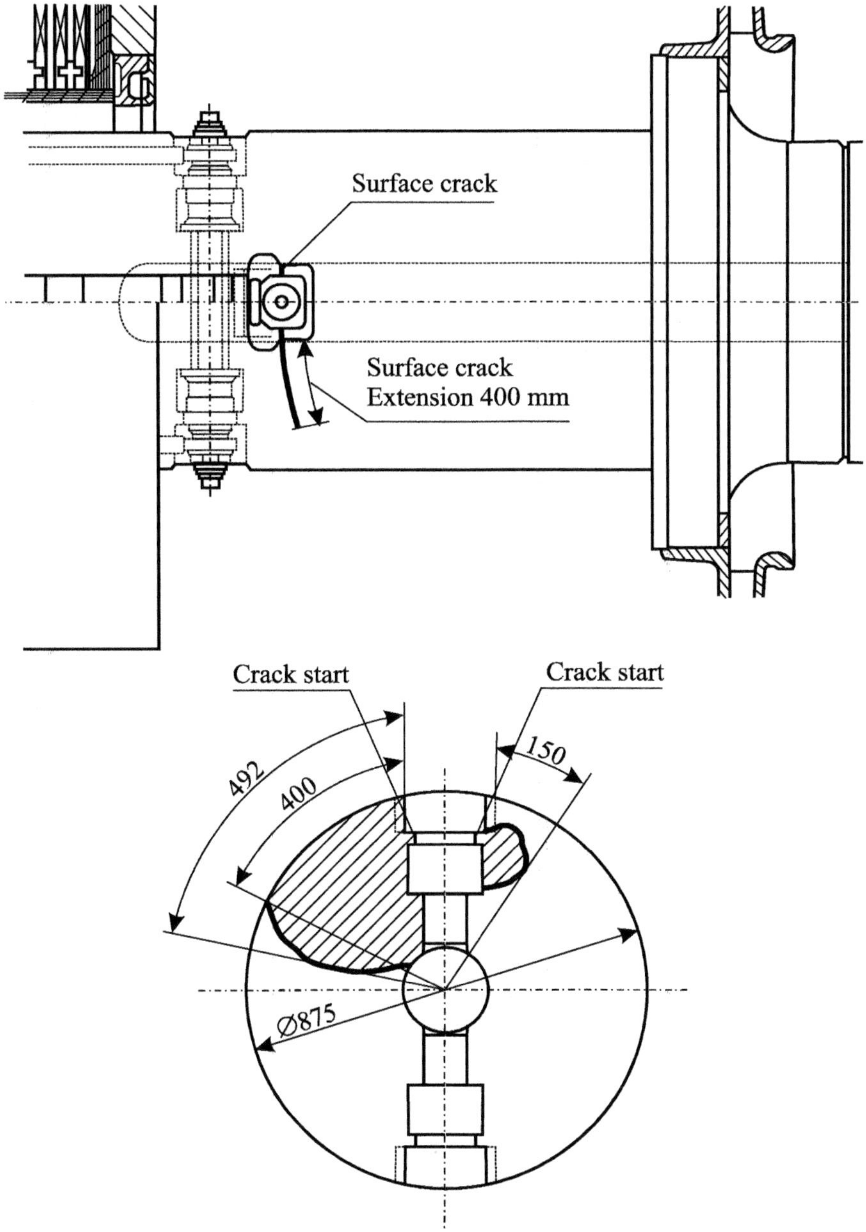

Fig. 1.12. Position and extension of the crack on the generator of Darlington nuclear power plant.

- 1991, Turbigo power plant (Italy). Figure 1.4 shows the crack surface in the generator of a 320 MW steam turbo generator, discovered after about 133000 h of operation [1.24]. This case is analyzed in detail in sections 2.3.2 and 2.4.1.

References

[1.1] Allianz Berichte (1987) Nr. 24, Nov. 1987, ISSN 0569-0692.

[1.2] Mayes IW, Davies WGR (1980) A method of calculating the vibrational behaviour of coupled rotating shafts containing a transverse crack. Institution of Mechanical Engineers Conference Publication, 2nd International Conference Vibration in Rotating Machinery, Cambridge, UK, paper C254/80, 17-27.

[1.3] DeForest DH, Grobel LP, Schabtach C, Seguin BR (1957) Investigation of the Generator Rotor Burst at the Pittsburg Station of the Pacific Gas and Electric Company. ASME Paper No. 57-PWR-12, 3-14.

[1.4] Ishida Y (2008) Cracked Rotors : Industrial Machine Case Histories and Nonlinear Effects Shown by Simple Jeffcott Rotor. Mechanical Systems and Signal Processing, 22(4), 805-817.

[1.5] Rankin AW, Seguin BR (1956) Report of the Investigation of the Turbine Wheel Fracture at Tanners Creek. Transactions of the ASME, 78(10), 1527.

[1.6] Schabtach C, Fogleman EL, Rankin AW, Winne DH (1956) Report of the Investigation of Two Generator Rotor Fractures. Transactions of the ASME, 78(10), 1567.

[1.7] Emmert HD, (1956) Investigation of Large Steam-Turbine Spindle Failure. Transactions of the ASME,78-10(1956-10), 1547-1565.

[1.8] Whyte RR (1975) Engineering Progress Through Trouble. The Institution of Mechanical Engineers, Chap.10.

[1.9] Yoshida M (1976) Steam Turbine Rotor Accidents and its Countermeasure. Turbomachinery, 4-11, 728.

[1.10] Jack AR, Paterson AN (1976) Cracking in 500MW LP Rotor Shafts, 1st Mech. Eng. Conference, The Influence of the Environment on Fatigue.

[1.11] Muszynska A (1982) Shaft Crack Detection. 7th Machinery Dynamics Seminar, Edmonton, Canada.

[1.12] Laws CW (1986) A Brief History of Cracked Rotor Saves – Turbine Generators in the U.K. Rated 60MW to 660MW (1971-1981). Bently Rotor Dynamics Co Seminar on Shaft Crack Detection, Atlanta GA, USA.

[1.13] Ziebarth H, Baumgartner RJ (1981) Early Detection of Cross-Sectional Rotor Cracks by Turbine Shaft Vibration Monitoring Techniques. ASME Paper 81-JPGC-PWR-26.

[1.14] Nilsson LRK (1982) On the Vibration Behaviour of a Cracked Rotor. IFToMM International Conference on Rotordynamic Problems in Power Plants, Rome, Italy, 515-524.

[1.15] Papadopoulos CA, Dimarogonas W (1988) Stability of Cracked Rotors in the Coupled Vibration Mode. ASME Journal of Vibrations, Acoustics, Stress and Reliability in Design, 110, 356-359.

[1.16] United States Nuclear Regulatory Commission (1986) Information Notice 86-19: Reactor Coolant Pump Shaft Failure at Crystal River.

[1.17] United States Nuclear Regulatory Commission (1989) Information Notice 89-15: Second Reactor Coolant Pump Shaft Failure at Crystal River.

[1.18] The Tampa Tribune (1989), January 20, 14-B.

[1.19] Maxwell JH, Rosario DA (2001) Using modelling to predict vibration from a shaft crack. Condition Monitoring and Diagnostic Engineering Management (COMADEM 2001), Elsevier, 243-250.

[1.20] United States Nuclear Regulatory Commission (2005) Information Notice 2005-08: Monitoring Vibration to Detect Circumferential Cracking of Reactor Coolant Pump and Reactor Recirculation Pump Shafts.

[1.21] Inagaki T, Hirabayashi M (1990) Transverse Vibrations of Cracked Rotors (Examples of Crack Detection and Vibration Analysis). Transactions of JSME, 56-523, 582.

[1.22] Sanderson AFP (1992) The Vibration Behaviour of a Large Steam Turbine Generator During Crack Propagation Through the Generator Rotor. IMechE International Conference on Vibrations in Rotating Machinery, Bath, UK, paper C432/102, 263-273.

[1.23] Schöllhorn K, Ebi G, Steigleder K (1993) Frettinganrisse in einem 936-MW Turbogeneratorrotor. VGB Kraftwerkstechnik, 73(4), 340-344.

[1.24] Bicego V, Lucon E, Rinaldi C, Crudeli R (1999) Failure analysis of a generator rotor with a deep crack detected during operation: Fractographic and fracture mechanics approach. Nuclear Engineering and Design, 188(2), 173-183.

Chapter 2
Typical Dynamic Behaviour of Cracked Shafts

2.1 Introduction

Case histories of chapter 1 have shown that the growth of a crack in a rotating shaft causes a change in its dynamic behaviour, providing that a condition monitoring system is installed and vibrations are analyzed. This chapter is focused on the typical symptoms that can be noticed in industrial rotating machinery and specifically three aspects are considered: the vibrations at rated speed, the so-called *thermal sensitivity* that affects cracked rotors and the vibrations during speed transients.

2.2 Analysis of the Vibrations at Rated Speed

Figures 1.5-1.8 have shown that the presence of a crack caused a remarkable change in the normal vibration level of the machines affected by the crack. This fact can appear obvious and the idea, about vibration trend increasing as a consequence of a crack, can form. Anyhow this idea is not exact, because it is more correct to analyze not simply the vibration levels, but rather their harmonic components, when diagnosing rotating machinery.

In this case, the sentence that *"in general all the vibration harmonic components are increasing as a consequence of a crack developing"* is not always confirmed by the experience, as shown hereafter, and a more precise definition of the symptoms at rated speed is *"the change of the amplitude and phase of the vibration harmonic components"*.

A very interesting example of dynamic behaviour at rated speed of a cracked generator rotor was reported in [2.1], related to a turbo generator of 735 MW. Rotor sketch is shown in figure 2.1, whilst the time histories of the 1X, 2X harmonic components and active power are shown in figure 2.2 as measured by the condition monitoring system before machine shutdown.

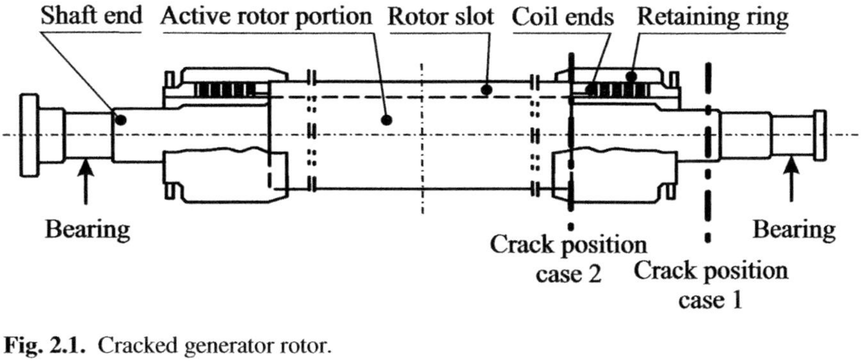

Fig. 2.1. Cracked generator rotor.

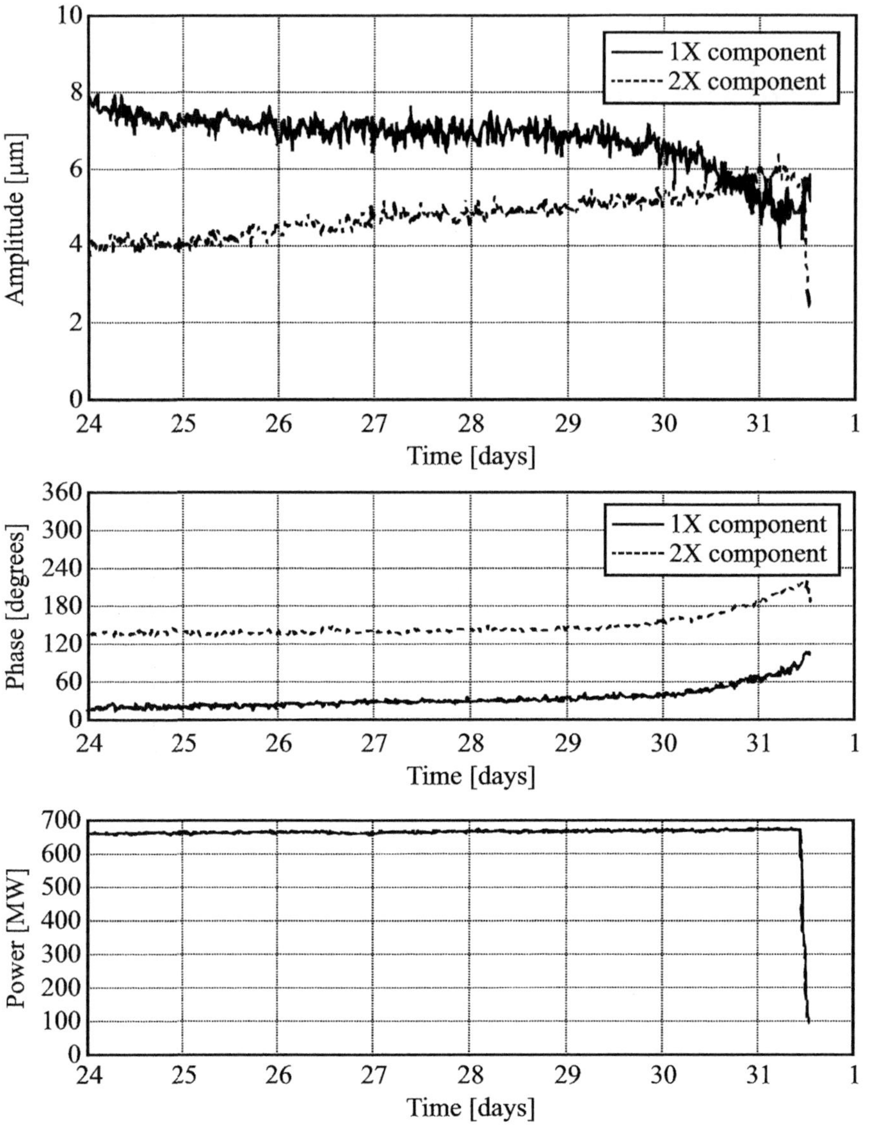

Fig. 2.2. Time histories of the 1X, 2X harmonic components and active power before shutdown.

The shaft was removed from service, inspected and cut in order to analyse the crack depth and shape. The crack surface, represented in figure 2.3, shows that the crack had propagated to a consistent depth, affecting more than 50% of the shaft section. The generator has operated for 4140 h. The position of the crack was that one indicated by case 1 in figure 2.1. In a twin machine, the crack had developed

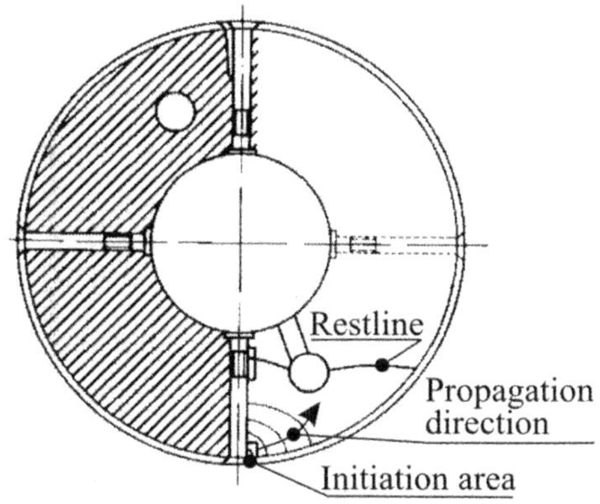

Fig. 2.3. Cross section of rotor in correspondence of the crack position – case 1.

in another location, indicated as case 2 in figure 2.1, and was discovered after 6800 h of operation.

From the analysis of trends of the harmonic components shown in figure 2.2 some interesting comments can be made:

- the reduction of the synchronous (1X) vibration amplitude is not at all a symptom of a healthy machine, but indicates in this case that something dangerous is happening. 1X vibrations are generated by residual unbalances or permanent bows; in this particular case the 1X vibration component generated by the propagating crack was in phase opposition with respect to the vibration component due to unbalance and bow, producing a reduction in the total 1X component;
- the 2X component is increasing continuously;
- the phase of both 1X and 2X vibration component is changing;
- the crack can propagate very rapidly in its final stage, as it was already evident in the cases presented in figures 1.5-1.8.

The plant manager took the right decision in stopping the machine (as can be seen from the sudden drop of the active power), despite the fact that the vibration level was much less then the alarm level, because the crack had already grown up to a dangerous depth.

Some important lessons can be drawn from this example:

- overall vibration level monitoring is not enough for an effective surveillance of the machine, but vibration components and related phases must be monitored;
- 2X component trend analysis is the best tool for discovering a crack.

Other cases in which the analysis of the vibrations at rated speed allowed crack detection are those of reactor cooling pumps.

Vertical shafts of RCPs of nuclear plants, a typical lay-out of which is shown in figure 2.4 (taken from [2.2]), are often affected by micro cracks generated by thermal fatigue in correspondence of a seal where the hot water is prevented from flowing upwards by cold water injection.

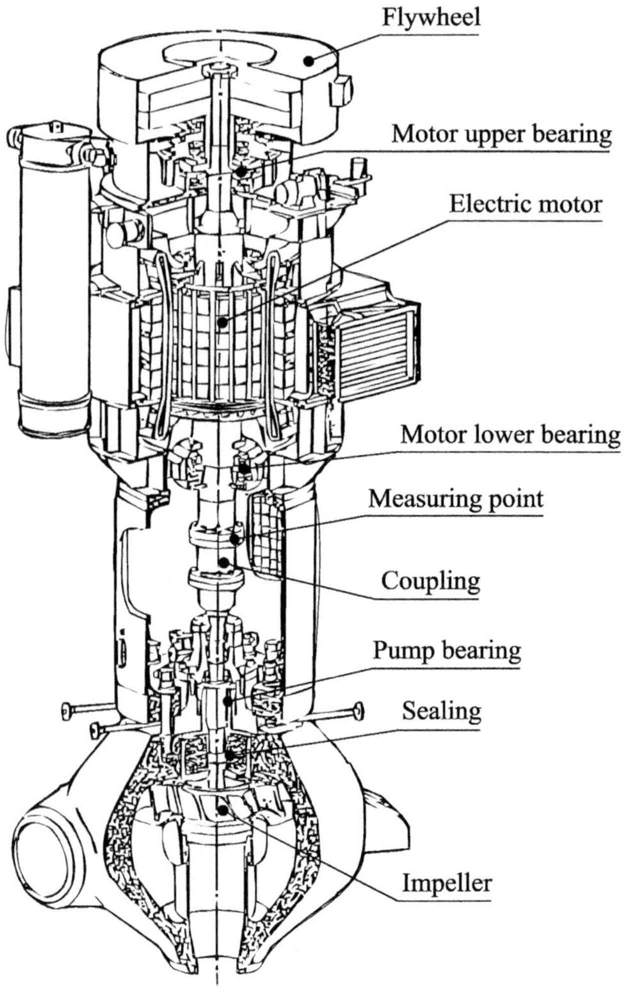

Fig. 2.4. Reactor cooling pumps (typical).

The presence of a protective sleeve is not sufficient to avoid completely the generation of a large quantity of micro cracks with random distribution. One of these cracks may then develop to a transverse crack and the vibration behaviour, similar to that one shown in figure 2.5, may appear. Some of these pumps have undergone to catastrophic failures, some others have been stopped and removed in time [2.3][2.4][2.5][2.6]. In this kind of machines with vertical shafts, it is difficult to forecast stresses and propagation velocities, because the axial and bending stresses are generated by axial and radial stationary hydraulic loads, which are in general roughly estimated and depend strongly on the operating conditions of the pump. The problems of this kind of machines are monitored by the Electrical Power Research Institute [2.7] during periodical meetings of utilities and manufacturers.

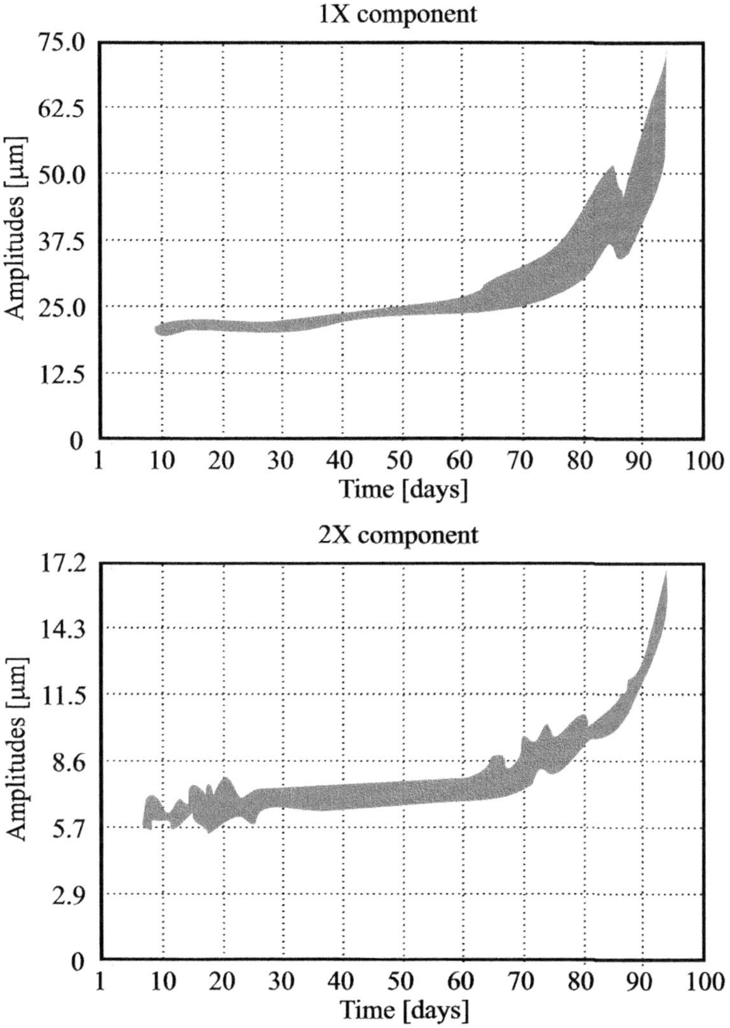

Fig. 2.5. Typical vibration behaviour in a cracked RCP.

2.3 Thermal Sensitivity of Cracked Rotors

The terms *thermal sensitivity* indicate a peculiar behaviour typical of cracked rotors: the vibrations change as a consequence of thermal transients applied to the rotor. These last can be caused for instance by changes in the steam temperature or in its distribution in turbines or by changes in the thermal state in generators caused by the temperature change of the cooling fluid or by the variation of the excitation current.

2.3.1 Case A – Steam Turbine

A typical example is represented by the dynamic behaviour of a high pressure – intermediate pressure (HP-IP) steam turbine of a 240 MW unit installed in the Torvaldaliga (Italy) power plant, the sketch of which is shown in figure 2.6, that had been submitted to a thermal shock during an accident, as described in [2.8].

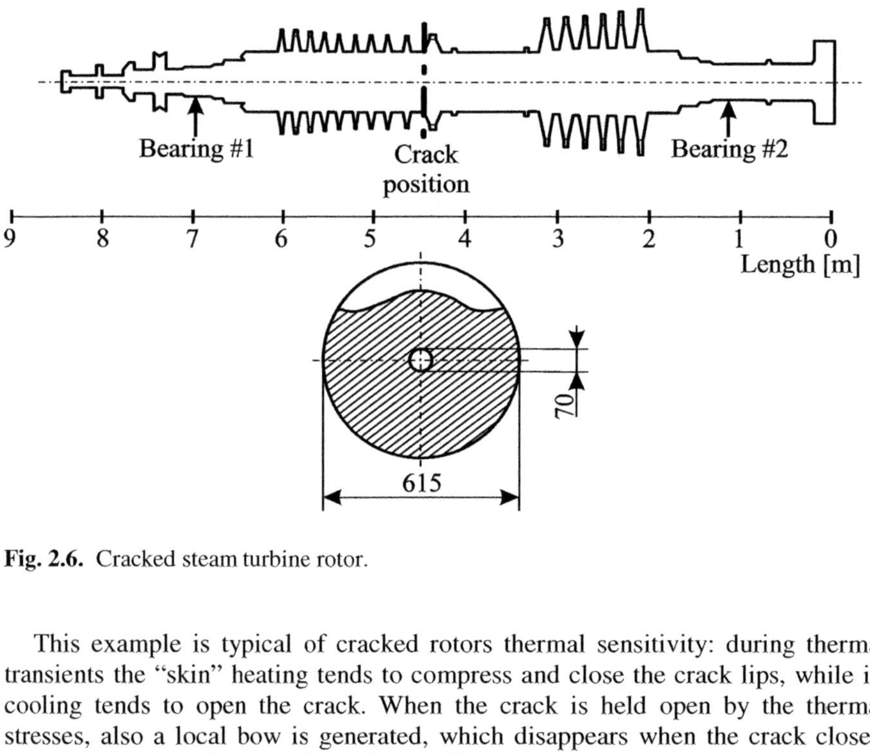

Fig. 2.6. Cracked steam turbine rotor.

This example is typical of cracked rotors thermal sensitivity: during thermal transients the "skin" heating tends to compress and close the crack lips, while its cooling tends to open the crack. When the crack is held open by the thermal stresses, also a local bow is generated, which disappears when the crack closes. The increasing and decreasing of overall vibration (mainly due to 1X and 2X components) is due respectively to cooling and heating thermal transients (see figure 2.7).

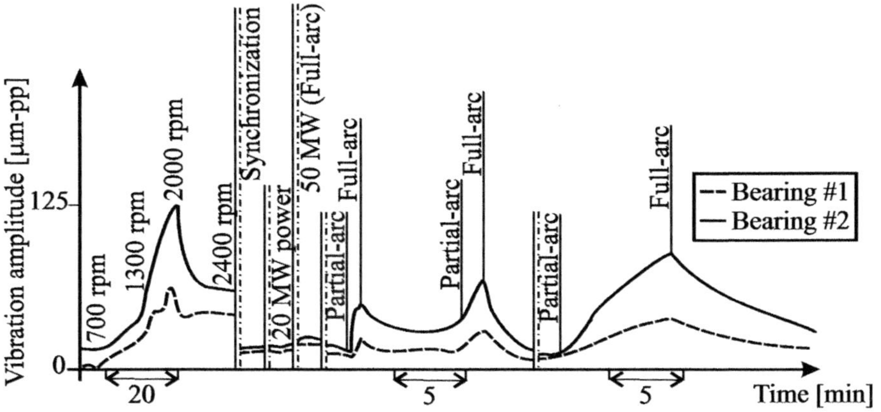

Fig. 2.7. Overall vibration amplitudes in bearings #1 and #2 during thermal transients.

When the operating condition is switched from full arc admission to partial arc, the 1st stage steam temperature lowers and the rotor is submitted to a cooling transient. The bow and the vibration amplitude increase quickly. The opposite happens when the operating condition switches from partial arc to full arc (see figure 2.7).

This particular thermal sensitivity is one of the most clear and reliable symptoms of a transverse crack in a rotor.

The same rotor, after removal, was then submitted to several tests in the factory: vibrations were measured during thermal transients at the different rotating speeds. Heating transients and cooling transients have been compared and are shown separately for 1X and 2X components in figure 2.8.

During the cooling transient, a local bow is generated and it can be recognized by means of rather high 1X vibration amplitude at low rotating speed. When passing the critical speed the superposition of the local bow to the original unbalance condition resulted in a "negative" peak of the vibration, instead of the classical resonance peak: the vibrations generated by the unbalance were compensated partially by the vibrations excited by the bow and the crack.

The heating transient instead closes partially the crack; this reduces the bow and the excitation of the 1X component. The vibration amplitudes at low speed become much smaller and, when passing the critical speed, the classical resonance peak due to the unbalance excitation appears, because the bow disappeared.

Regarding the 2X component due to the presence of the crack, during the cooling transient higher vibrations are measured due to the open crack at half of the critical speed; during the heating transient the crack partially closes and the vibrations become lower.

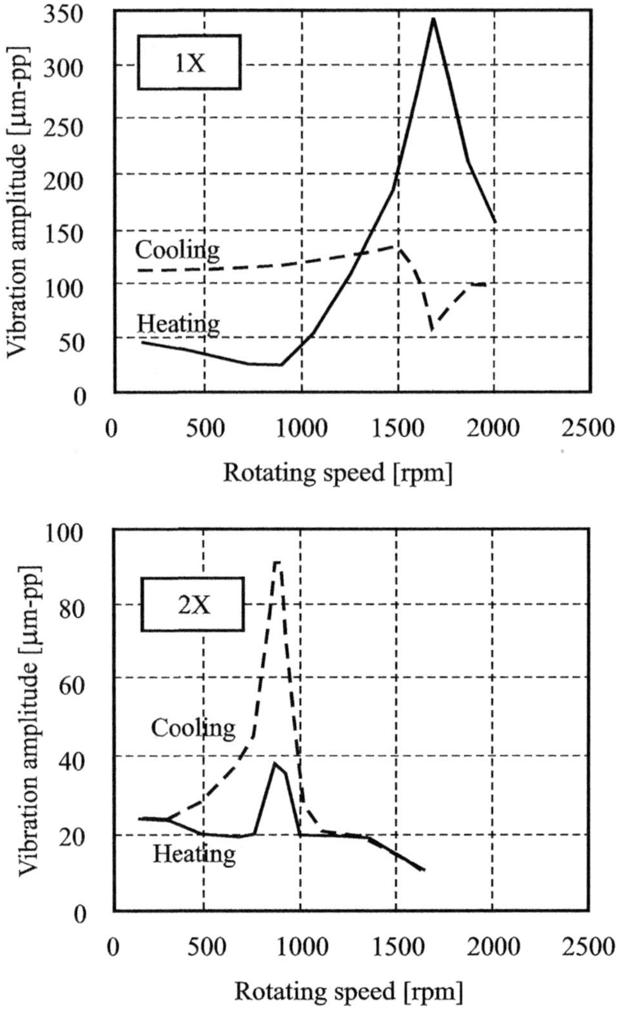

Fig. 2.8. 1X and 2X vibration components measured close to crack position during speed transients in different thermal conditions.

2.3.2 *Case B – Generator*

Cracks can also develop in the core of the rotor, under shrink fitted rings or similar parts: fretting corrosion might have been a concurring cause. In this case the above described thermal sensitivity could disappear completely or could be reduced.

On the contrary, a typical example in which thermal sensitivity was very high is the generator of a 320 MW turbo generator, shown in figure 2.9 and installed in the Turbigo (Italy) power plant, affected by a suspected crack, as described in [2.9]. Generator vibrations are measured by means of couples of proximity probes installed ±45° apart from the vertical in correspondence of the bearings.

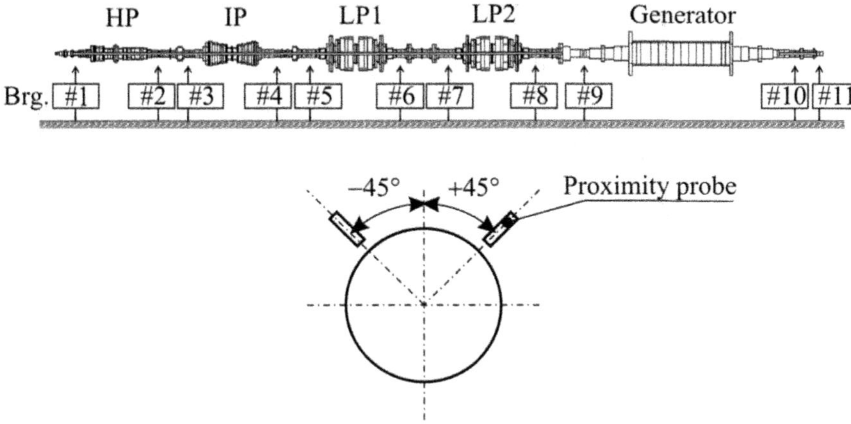

Fig. 2.9. Sketch of the 320 MW turbo generator shaft line, affected by a suspected crack.

The presence of the crack was confirmed by the generator thermal sensitivity that was observed and studied during thermal transient tests performed in normal operating conditions by changing the windings cooling fluid (H_2) temperature, as shown in figure 2.10, or by varying the excitation current and consequently the temperature distribution due to Joule effect, figure 2.11.

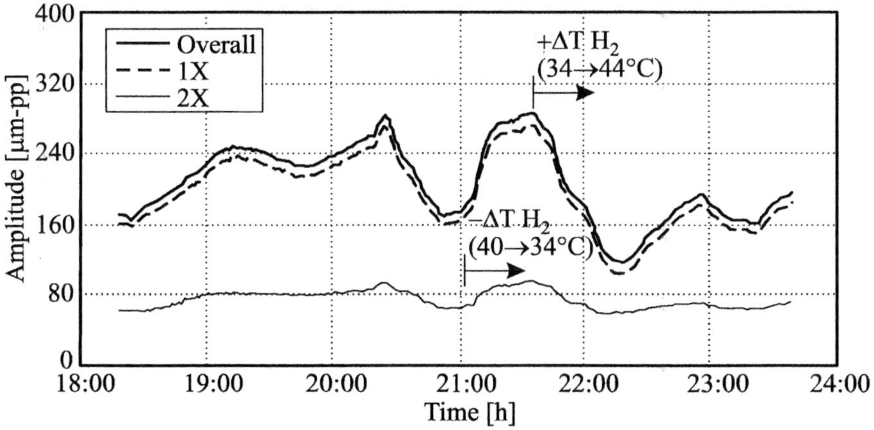

Fig. 2.10. Vibration time history during a windings cooling fluid temperature variation in a cracked generator.

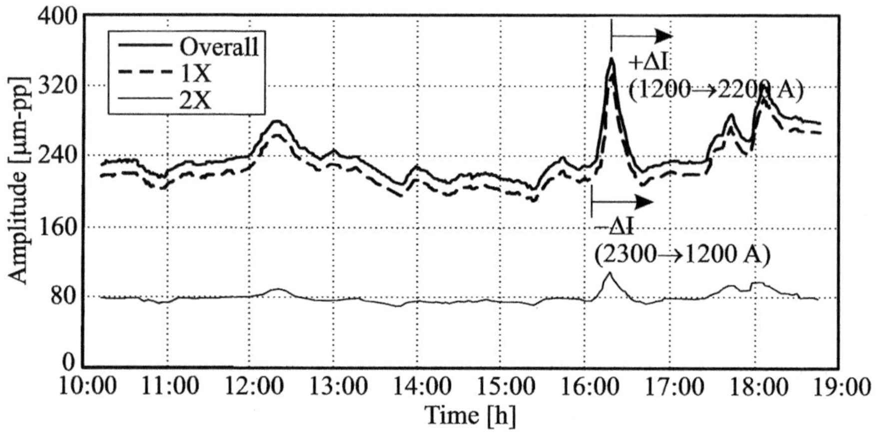

Fig. 2.11. Vibration time history during a variation of the excitation current in a cracked generator.

During the cooling transient the crack is forced to open by the thermal stresses, a local bow is also generated in correspondence of the crack, both 1X vibrations due to crack and bow, and 2X vibrations due to the crack are excited; they sum up to the original values due to unbalance, which acts on 1X vibrations, and to the residual axial stiffness asymmetry, which acts on 2X vibrations and that are unavoidable in one pole-pair generator.

During the following heating transient the thermal stresses, which are generated, force the crack to close: the local bow decreases or disappears, the 1X excitation due to crack and bow diminishes, as well as the 2X component.

2.4 Dynamic Behaviour during Speed Transients

The analysis of dynamic behaviour during speed transients provides typical symptoms of the presence of a crack. The case histories introduced hereafter allows them to be described in detail.

2.4.1 Case A – Generator

A remarkable example of the dynamic behaviour of cracked rotors during a speed transient can be shown analysing again the vibrations measured by the condition monitoring system of the Turbigo 320 MW turbo generator shown in figure 2.9, which was considered in the previous section.

Due to the suspected crack, the unit was shut down and terrible vibration levels were measured in correspondence of the generator first critical speed during the run-down transient. This is shown in figure 2.12 and figure 2.13 where the total vibration and the 1X as well as the 2X components are shown in the two main bearings of the generator, according to +45° measuring direction.

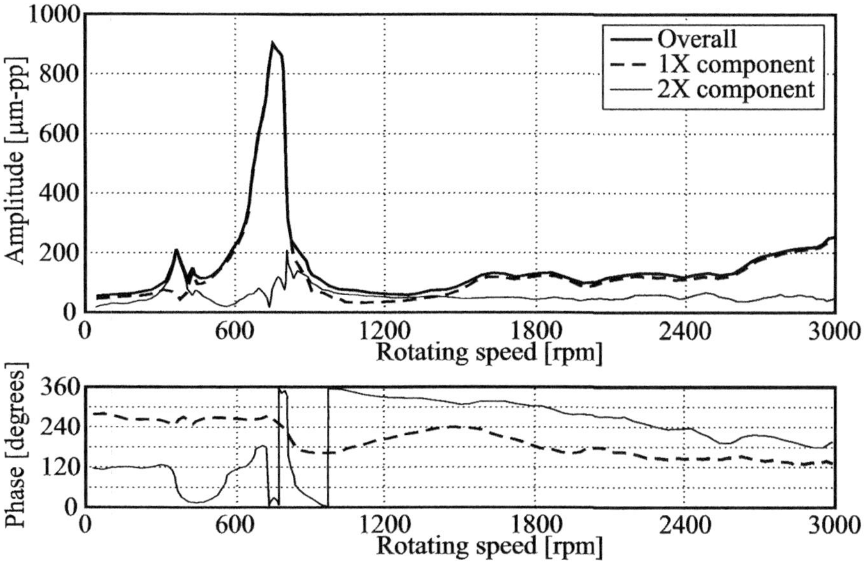

Fig. 2.12. Vibrations measured in bearing #9 (driven end of the generator) during the run-down transient.

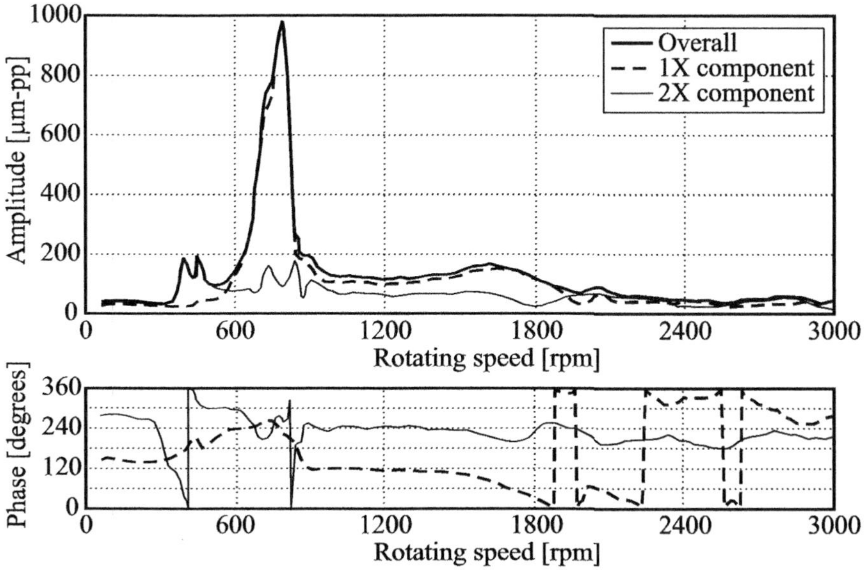

Fig. 2.13. Vibrations measured in bearing #10 (non-driven end of the generator) during the run-down transient.

In correspondence of the 1X critical speed at about 700 rpm (this value was probably lower than the original one due to the presence of the crack) more than 900 μm were measured in the bearings; this means that the clearances in correspondence of the seals have been overrun, a severe full annular rub occurred during the passing of the critical speed.

The rotor probably experienced a cooling transient during the run-down (electrical load was removed but the cooling fluid was still circulating) that forced the crack to open completely, generating a remarkable local bow; this can explain the very high values of 1X harmonic components. Due to the completely open crack also the 2X component reached almost 200 μm in correspondence of the rotating speed of about 350 rpm, i.e. half of the 1X critical speed. The 3X harmonic component could not be measured and this confirms the hypothesis of the completely open crack, because this component is due to the breathing mechanism of the crack that will be described in detail in chapter 5.

The generator was then removed from plant and inspected in the factory. Standard non-destructive techniques like dye penetrant tests and ultrasonic inspections (see section 3.2 for a detailed description of these techniques) on the outer surface of the generator rotor did not reveal any defect.

The results of the vibration test performed on the rotor, when it was suspended by means of ropes, were rather unclear. Static deflection measurements, which are described in the section 3.1.2.2, gave instead an indication of the presence and the location of the crack. Vibration measurements on the shaft rotating at low and medium speed in the factory test cell showed also typical symptoms, but since a reference situation of the shaft without crack was not available, these results could not be used for crack position and depth identification, using the method described in section 8.4.

Finally ultrasonic inspections, carried out from inside of the rotor body by introducing a probe in the longitudinal hole of the generator shaft, revealed the crack and its extension (see figure 2.14). The surface of the crack was not perfectly plane, especially close to the "skin".

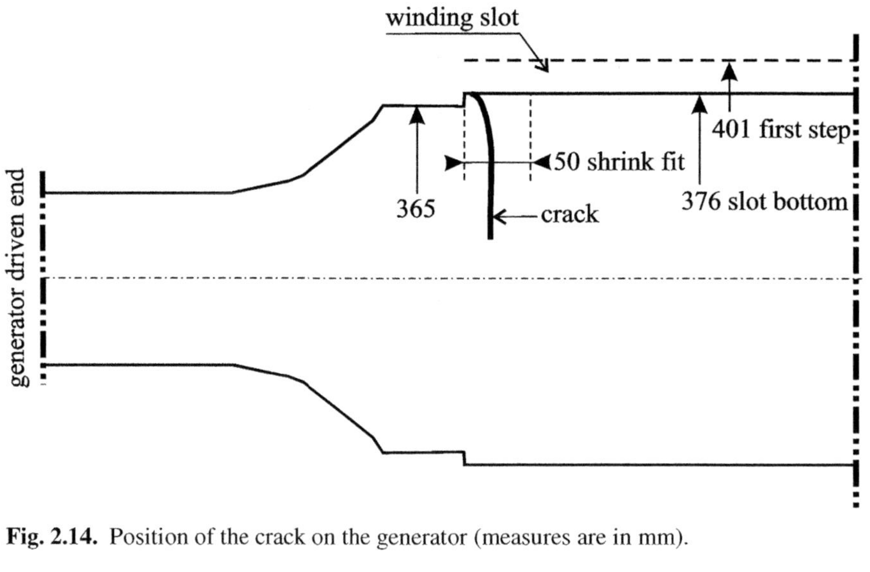

Fig. 2.14. Position of the crack on the generator (measures are in mm).

The thermal sensitivity experienced by the rotor in the power plant could not be reproduced in the factory test facility, despite the fact that it was possible to heat or cool down the surrounding air, because the crack had developed under the "caps" and therefore was rather insensitive to the surrounding air temperature. At the end, the actual shape of the crack was found definitively after dismantling the rotor and had propagated up to roughly a 50% of the cross section area, as shown in figure 2.15

Fig. 2.15. Crack depth with respect to rotor section.

where the cracked portion of the rotor cut in slices is shown. The starting point was in correspondence of a defect in a slot for the windings (figure 2.16).

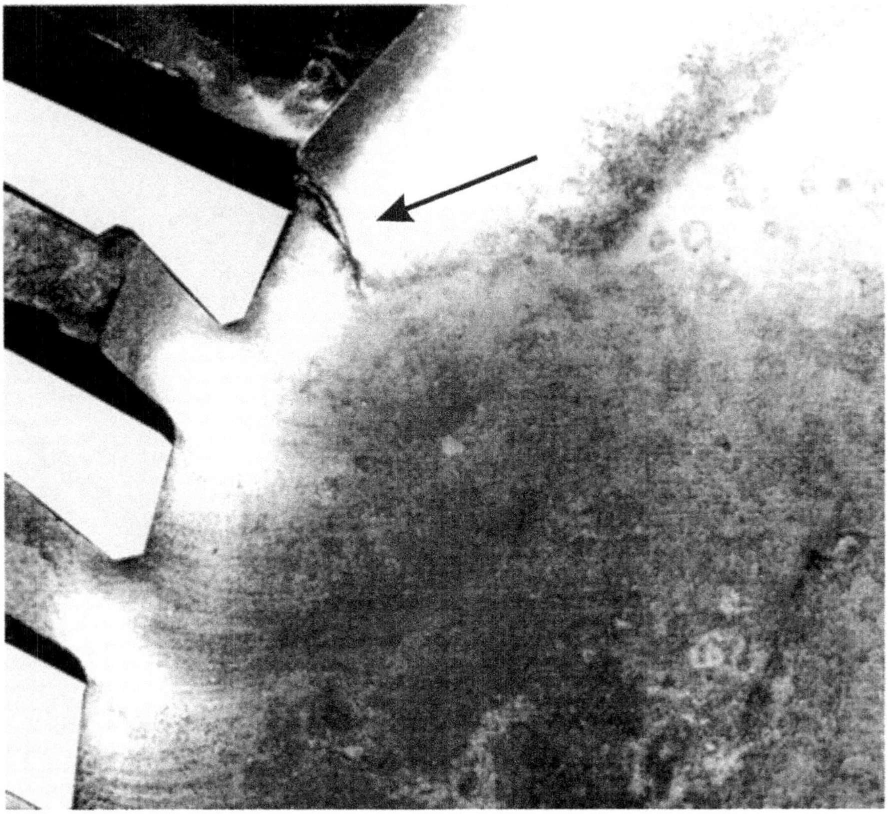

Fig. 2.16. Starting point of the crack.

In this case, as well as in the previous ones, the propagation of the crack is due to the alternating stresses induced on the rotating shaft by the static bending load due to the rotor weight and the bearing alignment conditions.

By analysing the operation history of the turbo generator, it came out that the crack started propagating from the defect shown in figure 2.16 after several years of safe operation, in concomitance of a change of bearing alignment conditions, which was performed to reduce the vibration levels of one bearing.

The change of the alignment conditions changed also the stress distribution in the rotor body and this was sufficient to let the crack starting its propagation. After the change of the alignment, the turbo generator was operated for more than 4 years, during which the crack was sometimes growing, sometimes resting, as it can be seen from the beach marks in figure 2.15.

Cracks have been found also under shrink fitted rings in steam turbines or under shrink fitted coupling flanges. In this last case, the position makes the detection of the crack increasingly difficult, because the bending moment is generally vanishing small and its effect on the vibrations is also rather small.

2.4.2 Case B – Generator

Another very interesting case of dynamic behaviour of a cracked rotating machine during a run-down is that of the Darlington generator [2.10], already considered in section 1.2. Figure 1.7 showed the vibration trend at the operating speed in the bearing on the driven-end side of the generator and the quick increase of the vibration level caused the machine to trip and the crack to be discovered in the following (see figure 1.7). The condition monitoring system of the machine allows to measure absolute and relative vibrations of the shaft in its seven bearings, by means of two composed sensors (proximity probe and seismic probe), which are ±45° apart from the vertical (see figure 2.17).

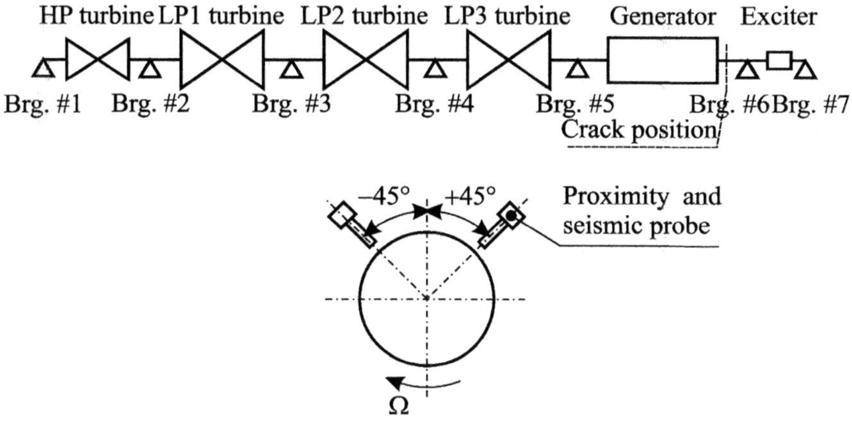

Fig. 2.17. Measuring planes of Darlington turbo generator.

Reference behaviour of the machine during a speed transient is normally stored after any major overhaul as the *reference* condition. This allowed the comparison of 1X and 2X components of the vibrations during the run-down after the crack disclose and those measured during a run-down before rotor cracking. The comparisons shown in figure 2.18 and figure 2.19 are of big interest to understand the dynamic behaviour of cracked shafts.

In the reference condition, at the operating speed of 1800 rpm, the machine steady state vibrations are rather low, with typically less than 40 μm for the relative vibration, with a predominant 1X component and a minor contribution by 2X component. Generator has two pole-pairs and has a negligible stiffness asymmetry. On the contrary one pole-pair generators have instead intrinsic stiffness asymmetry that could cause 2X vibration component also during normal operating. During the reference run-down the following characteristics are highlighted on the generator dynamic behaviour:

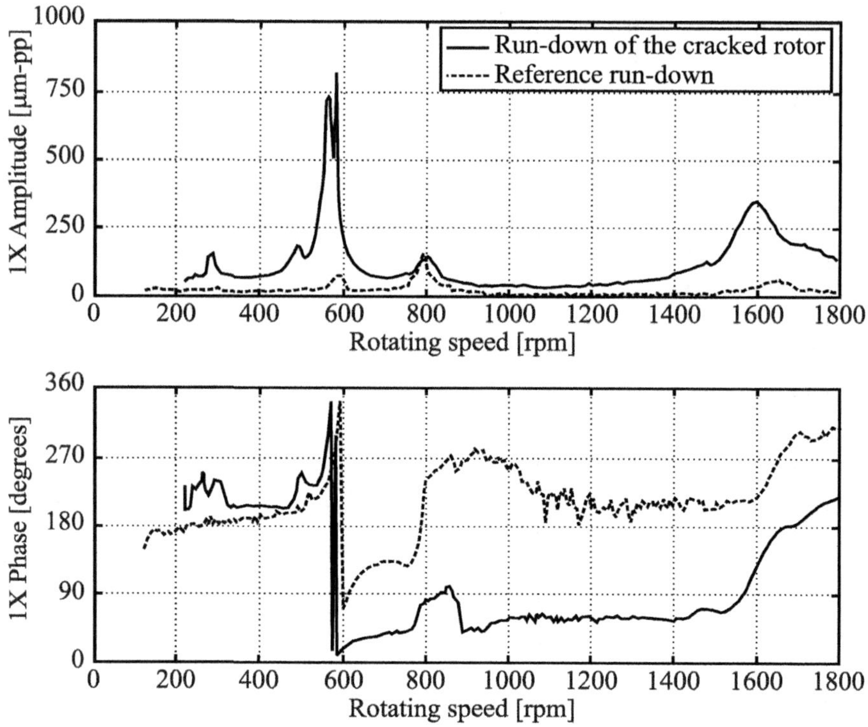

Fig. 2.18. Comparison of 1X vibrations after and before crack disclosure in bearing #5, direction –45° apart from vertical.

- the supporting structure resonance at 290 rpm, which has a strong vertical component measured by the seismic probes is not affecting the shaft relative vibration;
- the first critical speed of the generator is at about 590 rpm;
- there is a remarkable influence on generator vibrations caused by the dynamic amplification corresponding to the first critical speed of the LP turbines at 780 rpm (consider that the vertical scale in figure 2.18 and figure 2.19 compresses the diagram of the reference condition).
- The second critical speed of the generator is at about 1630 rpm.

Taking into consideration the run-down after the crack disclosure, some evident differences are immediately observed in both 1X and 2X components:

- 1X component amplitude is considerably increased in correspondence of the first and the second generator critical speed (see figure 2.18). This can be explained by the developing of the bow caused by the crack;
- the actual value of the critical speeds is slightly reduced, of about 2.5-3%, as a consequence of the local effect on the shaft stiffness of the crack;

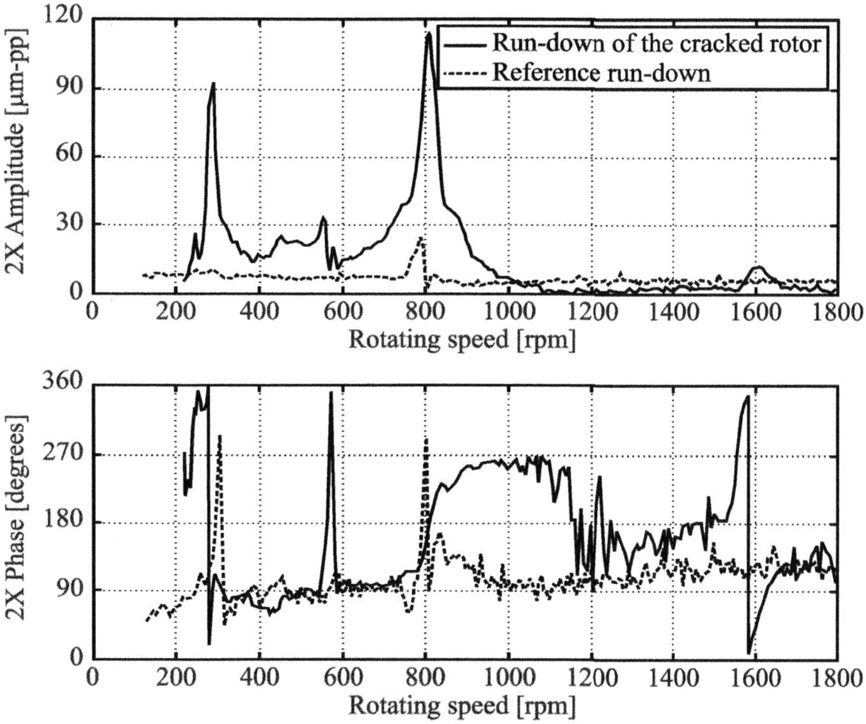

Fig. 2.19. Comparison of 2X vibrations after and before crack disclosure in bearing #5, direction –45° apart from vertical.

- first critical speed is split in two: this may be due to oil-film and bearing housing stiffness asymmetry, the effects of which are highlighted by the high vibration amplitude in the bearing.
- remarkable 2X peaks appear in correspondence of ½ of both critical speeds (at about 295 and 815 rpm respectively, see figure 2.19). These are caused by the stiffness asymmetry introduced by the crack in the generator with two pole-pairs and are important symptoms of the crack presence.
- there is a not negligible response of the 2X component at the second critical speed (at about 1600 rpm) and more evident at the first critical speed (at about 580 rpm). This is due to non-linear effects caused by the oil-film and is rather common: it was also present, but not so evident, also in the previous case (see figure 2.12 and figure 2.13).

2.4.3 Case C – Test Rig

Other symptoms besides 1X and 2X components can be excited by the crack during speed transients of rotating shafts. The change in the mean stiffness should lower the natural frequencies, therefore the peaks in vibration amplitude at critical speeds appear at lower rotating speeds as shown in figure 2.18. This effect is difficult to be measured when the crack is not very deep, because the shift in natural frequency is small and the resonance peak at critical speed is rather wide in well damped big sized industrial machines.

Another symptom is the excitation of 3X components caused by a breathing crack. The excitation is rather small and can be recognized only when resonance amplifies the response.

Moreover, condition monitoring systems of industrial machines sometimes limit the availability of data to only 1X and 2X harmonic components. In order to show the effect also on higher components, another case is presented, which is relative to the test-rig EURoPE (Ensemble Utilisant un Rotor Pour Essais) of EDF – Électricité de France, described in detail in section 8.5.2. In this case, the laboratory environment allowed accurate measurements to be performed and also the effect on the 3X resonance at ⅓ of the critical speed could be observed.

Figure 2.20 shows the first three harmonic components measured in correspondence of the crack in the shaft during a run-down. The excitation of the three harmonic components is clearly emphasized and also a not negligible response of the 2X component at the first critical speed, caused by non-linear effects, can be observed.

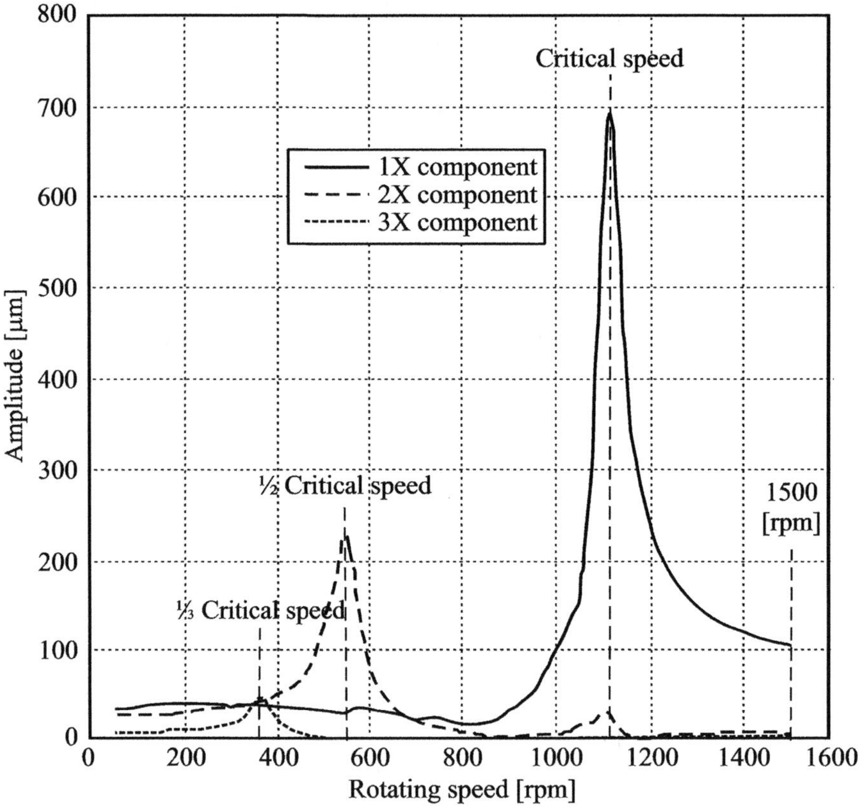

Fig. 2.20. Harmonic components of the vibration measured during a run-down in a big size test-rig.

2.4.4 Final Comments

The presented cases highlight the typical dynamic behaviour of cracked shaft during speed transients, which can be recorded by condition monitoring systems that measure usually the lateral vibrations of the machines. Also changes in torsion and axial natural frequencies, caused by the local effect of the crack on the shaft stiffness, could be measured with suitable instruments, which are unfortunately generally not installed on industrial machines.

Finally it is known that also sub-harmonic components as well as instability regions could also appear, but in well damped industrial machines both effects are generally completely cancelled by damping.

References

[2.1] Nilsson LRK (1982) On the Vibration Behaviour of a Cracked Rotor. IFToMM International Conference on Rotordynamic Problems in Power Plants, Rome, Italy, 515-524.

[2.2] Bachschmid N, Pennacchi P, Tanzi E, Verrier P, Hasnaoui F, Aabadi K (2004) Crack Detectability in Vertical Axis Cooling Pumps During Operation. International Journal of Rotating Machinery, 10(2), 121-133.

[2.3] United States Nuclear Regulatory Commission (1986) Information Notice 86-19: Reactor Coolant Pump Shaft Failure at Crystal River.

[2.4] United States Nuclear Regulatory Commission (1989) Information Notice 89-15: Second Reactor Coolant Pump Shaft Failure at Crystal River.

[2.5] The Tampa Tribune (1989), January 20, 14-B.

[2.6] United States Nuclear Regulatory Commission (2005) Information Notice 2005-08: Monitoring Vibration to Detect Circumferential Cracking of Reactor Coolant Pump and Reactor Recirculation Pump Shafts.

[2.7] Electrical Power Research Institute (1989,1990,1991,1992,1993,1994) EPRI Workshops on Main Cooling Pumps.

[2.8] Passleva G, Pira G (1982) Cracked shaft vibration sensitivity to steam temperature variations. IFToMM International Conference on Rotordynamic Problems in Power Plants, Rome, Italy, (poster presentation).

[2.9] Lapini GL, Zippo M, Bachschmid N, Collina A, Vallini A (1993) Experimental tests and model based calculations for the diagnosis of a crack in a 320 MW generator. Diagnostics of rotating machines in power plants, CISM/IFToMM Symposium Proceedings, Udine, Italy, 87-97.

[2.10] Sanderson AFP (1992) The Vibration Behaviour of a Large Steam Turbine Generator During Crack Propagation Through the Generator Rotor. IMechE International Conference on Vibrations in Rotating Machinery, Bath, UK, paper C432/102, 263-273.

[2.11] Audebert S, Stoisser C (2008) A comprehensive theoretical, numerical and experimental approach in view of crack detection in power plant rotating machinery. Mechanical Systems and Signal Processing - Special Issue Crack effects in Rotordynamics, 22(4), 818-844.

Chapter 3
Rotor Testing for Crack Detection

3.1 Dynamic and Static Tests for the Detection of Cracks in Rotors

Once the crack presence is suspected in a rotor and the rotor is removed from service, it can be tested in different ways to ascertain the presence, the position and the depth of a crack.

3.1.1 Dynamic Tests

The simplest dynamic test, also called *bump test*, consists in measuring the natural frequencies of the rotor in correspondence of different angular positions when the rotor is supported on roller bearings or suspended by ropes.

Natural frequencies should be independent on the angular position in axial-symmetric shafts without defects. If instead the natural frequencies depend on the position, this validates the existence of a crack that causes the stiffness asymmetry.

Lower frequencies correspond to angular positions in which the crack is open, higher frequencies to those in which the crack is closed. Since the variation of natural frequency is small, care should be taken to have in all the different positions the same boundary conditions as e.g. same supporting roller stiffness and same roller-shaft contact surface. By combining modal shapes measured by a set of accelerometers applied to the shaft and natural frequencies, also the position of the crack and its depth can be estimated, with the aid of the model of the shaft.

3.1.1.1 Case A: Impact Tests on a Test-Rig

The first example is relative to the test-rig MODIAROT of Dept. of Mechanical Engineering of Politecnico di Milano. A crack had been produced starting from a notch by means of alternate bending excitation and reached the depth of about 25%-30% of the diameter (see figure 3.1), as it was deduced from circumferential crack length as revealed by dye penetrant. The position of the crack is practically in correspondence of the antinode of the first lateral mode.

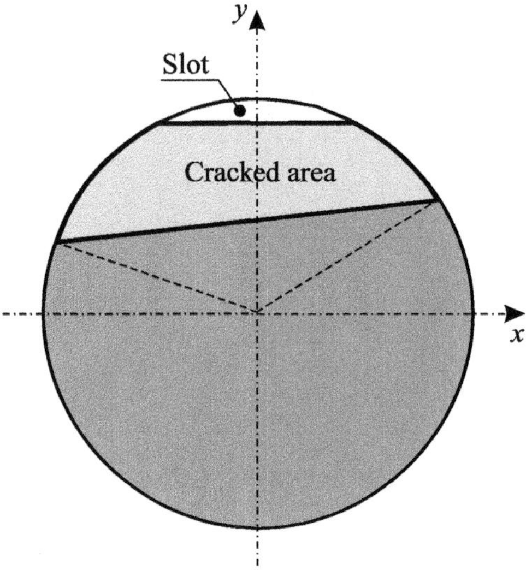

Fig. 3.1. Extension of the crack in the test-rig shaft.

The shaft has been set in a pinned-pinned configuration (see figure 3.2 and figure 3.3) and a couple of accelerometers were located in the antinodes of the first and second mode deformed shapes. Note that the antinode of the first mode corresponds to the node of the second mode.

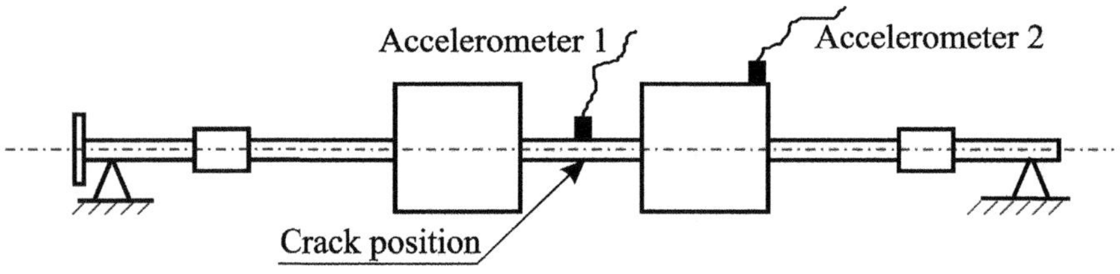

Fig. 3.2. Shaft configuration on the two supports and location of the accelerometers.

The natural frequencies were excited by means of impact tests and have been measured in two configurations, corresponding to the complete open crack (when crack is located in the lower part, the weight opens the crack) and to the complete closed crack (when it is located in the upper part). Their values are reported in Table 3.1 together with eigenfrequency calculated values assuming a stiff supporting structure and using the crack model FLEX described in section 5.4.3.

Fig. 3.3. Shaft support reproducing a pinned configuration.

Table 3.1. Natural frequencies of the cracked test-rig shaft (values in Hz).

Crack	Experimental		Calculated (25%)		Calculated (30%)	
	1st	2nd	1st	2nd	1st	2nd
Closed	12.72	59.00	12.30	59.04	12.30	59.04
Open	12.42	59.00	12.14	59.04	12.06	59.04

The experimental values, which were measured with a resolution of ±0.025 Hz, show a variation in the 1st natural frequency of roughly 2.3% between open and closed crack configuration. The calculated values have instead a variation of 1.30% for 25% depth and 1.93% for 30% depth. No variation at all has been found for the 2nd mode since the crack, located in correspondence of the node of the mode shape, does not affect its bending stiffness (or modal stiffness).

3.1.1.2 Case B: Sinusoidal Excitation Tests on a HP-IP Turbine

The HP-IP turbine considered in section 2.3.1 was also dynamically tested in the factory. The close-up of the crack position is shown in figure 3.4, whilst crack profile and depth is shown in figure 2.6.

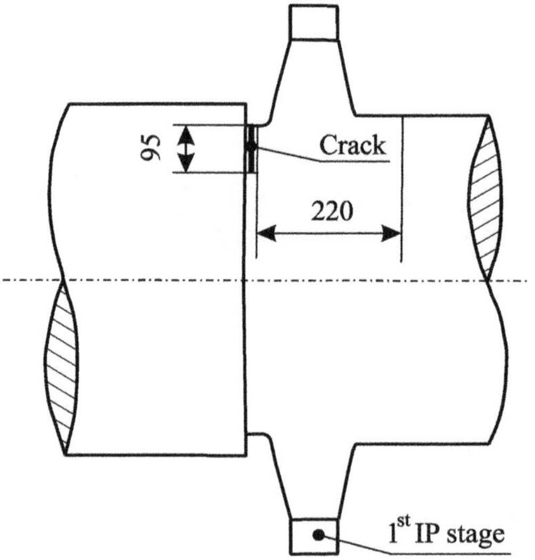

Fig. 3.4. Crack position on the HP-IP turbine (measures in mm).

The results plotted in figure 3.5 were obtained as resonance frequencies during sinusoidal excitation tests, in the different angular positions of the suspended steam turbine shaft in horizontal position. For a comparison, the same test results were available for the new (un-cracked) shaft and have been also plotted in figure 3.5.

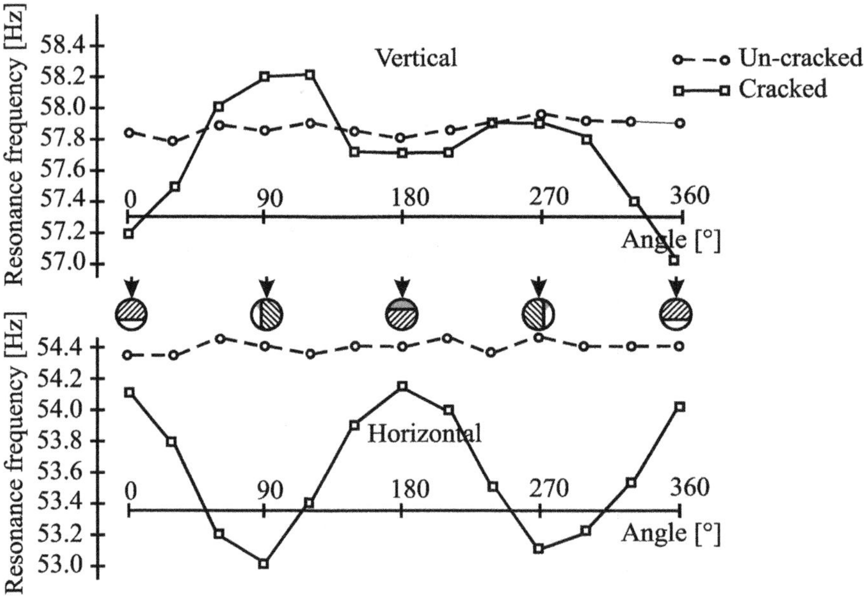

Fig. 3.5. Natural frequencies vs. angular position of the shaft.

The change of the natural frequencies in horizontal and vertical direction is anyway rather small, reaching a maximum variation of about 2%.

It is more difficult to use this technique for generators with one pole-pair, because in this case the generator has an "original" axial stiffness asymmetry, as shown in figure 3.6. Therefore they have different natural frequencies in different angular positions, with a 2X variation; this can mask or make hard to extract the contribution of the crack-induced asymmetry.

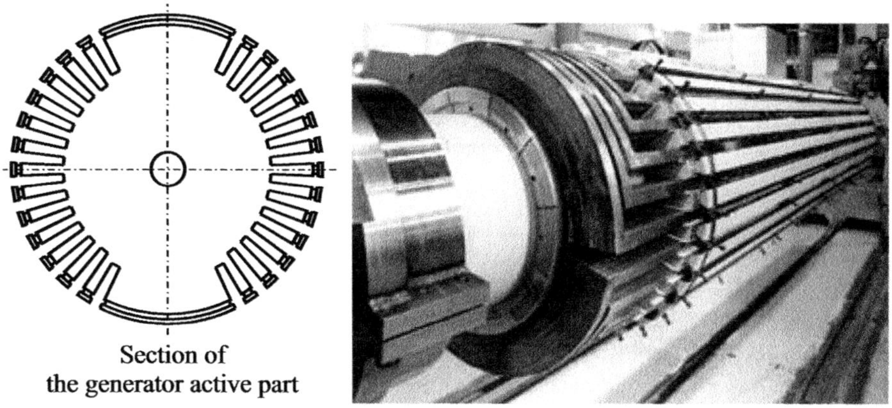

Fig. 3.6. One pole-pair generator.

3.1.2 Static Tests

Cracks generate a local reduction of the shaft stiffness. Natural frequencies measured by means of dynamic tests are proportional to the square root of the shaft stiffness. If the complete shaft stiffness varies of 1% between a minimum (open crack) and a maximum (closed crack), the frequency changes of only 0.5%.

On the contrary, static deflections are proportional to the inverse of the stiffness: a change of 1% of the shaft stiffness causes 1% of change of the deflection.

Therefore higher accuracy in crack detection can be expected from deflection measurements rather than from natural frequencies measurements.

When a crack is developing, a small discontinuity in the slope of the deflection may appear in correspondence of the crack position, which is more evident when the crack is open. When deflections are measured along the shaft loaded by its weight during a complete revolution of the shaft, the Fourier expansion allows the definition of 1X, 2X and 3X deflection shapes. All are due to the local flexibility of the crack only and have therefore a bi-linear shape with a pronounced curvature in correspondence of the crack.

3.1.2.1 Case A: Static Deflection of a Test-Rig

In this first example, the static deflection measurements are shown by means of the results obtained with the test-rig cracked shaft of the test-rig already considered in section 3.1.1.1. The shaft is mounted on a lathe for commodity, in order to perform the measurements during a slow rotation of the mandrel carrying the cracked shaft (see figure 3.7 and figure 3.8).

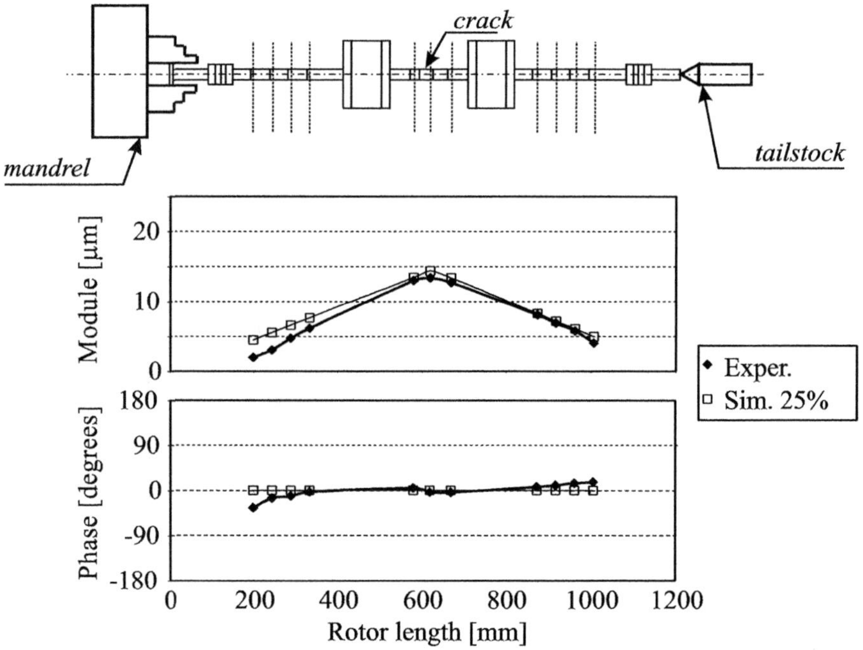

Fig. 3.7. Horizontal 2X deflections along the cracked test-rig shaft.

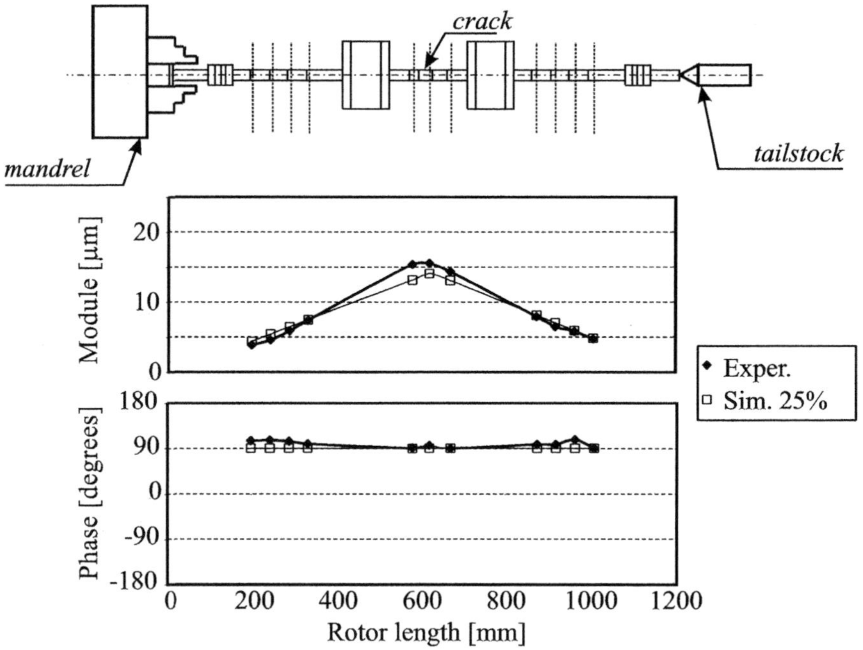

Fig. 3.8. Vertical 2X deflections along the cracked test-rig shaft.

Accurate measurement required correction of geometrical surface errors (shaft waviness or shaft ovalization), therefore 4 sensors were used on a frame, which was moved along the shaft from time to time, as shown in figure 3.9.

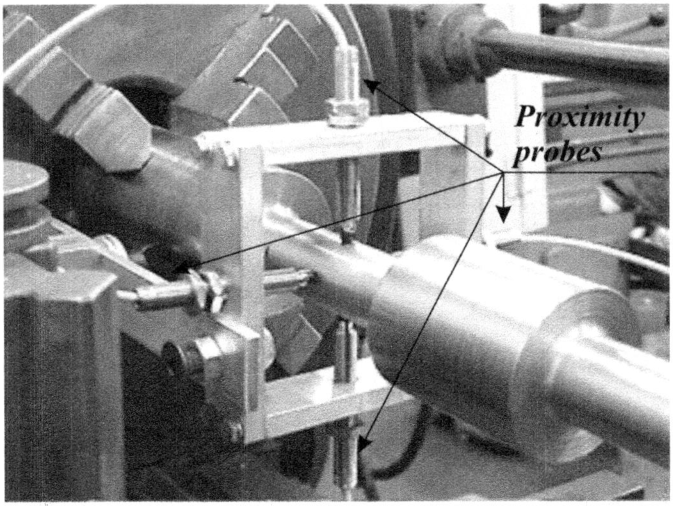

Fig. 3.9. Deflection measurements of the cracked shaft during slow rotation on a lathe.

The measured displacements have been compared with those calculated theoretically for a crack depth of 25%, using the FLEX model described in sections 5.4.3 and 5.6. The vertical and horizontal 2X deflections should be equal, according to the results of the model.

The 2X measured deflections both in the horizontal and in the vertical direction are rather linear, except for a small part in correspondence of the crack, and fit very well with the calculated deflection.

With regard to the other harmonic components of the deflection, the 1X component is masked by a high static bow, which was generated during the crack propagation, while the 3X component is so small that its presence is masked by geometric surface errors and by the measuring errors.

3.1.2.2 Case B: Static Deflection of a Generator

The cracked generator considered in section 2.3.2 underwent also to static tests in the factory. The rotor was supported by means of roller bearings. The measurements of the static deflections in different angular positions and in different positions along the shaft allow to identify the 1X bow, due to an open crack. The 1X component of the deflection shape is shown in the upper part of figure 3.10.

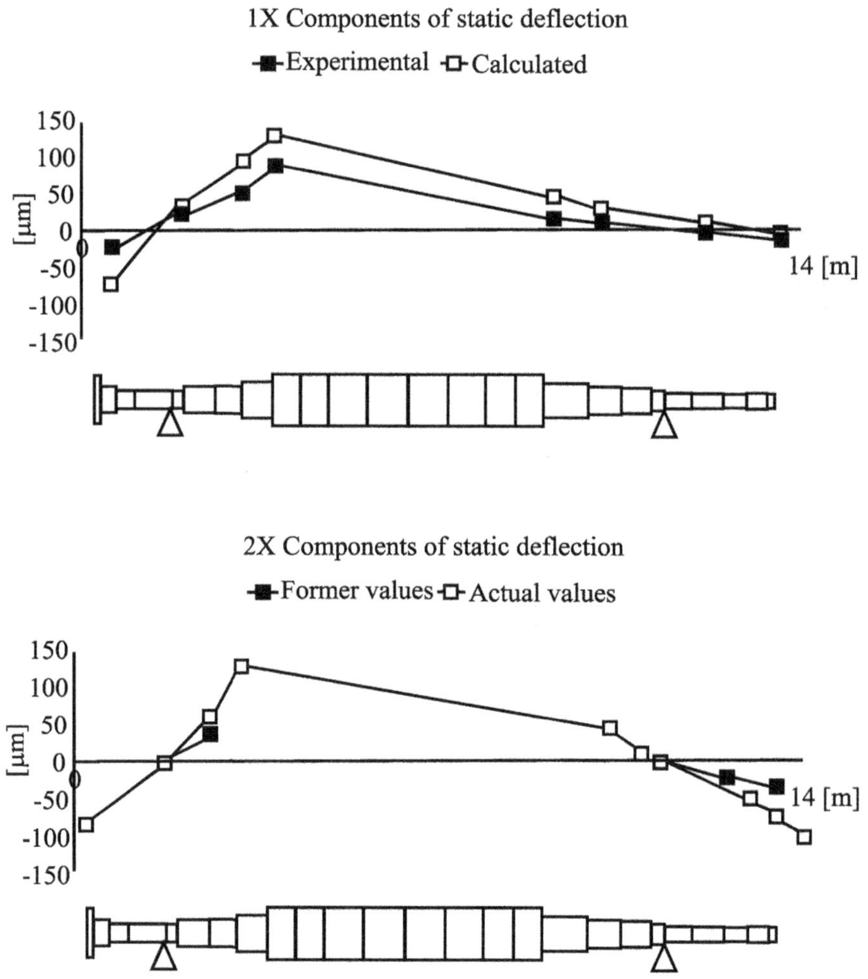

Fig. 3.10. Static deflections of a generator cracked rotor: 1X (upper part) and 2X (lower part) components.

The 2X component of the deflections generated by the crack, is shown in the lower part of figure 3.10. This component, extracted from static deflection measurements, allows inferring presence and position of crack. Former values indicate original measurements taken on the new rotor (without crack). Both components (1X and 2X) have a maximum in correspondence of the crack (compare figure 2.14) and decrease linearly towards the supporting rollers.

This allows the position of the crack to be identified. With the aid of the identification procedure described in section 8.4 it is also possible to identify its depth.

3.2 Non-destructive Testing for Power Generation Rotors

Contributed by Michele Carboni and Michele Sangirardi, Department of Mechanical Engineering, Politecnico di Milano, Milan, Italy.

Life cycle engineering and life management of power plants, and consequently of turbo generator rotors, involves a number of critical factors (figure 3.11), one of which is non-destructive evaluation (NDE) of quality and integrity [3.1].

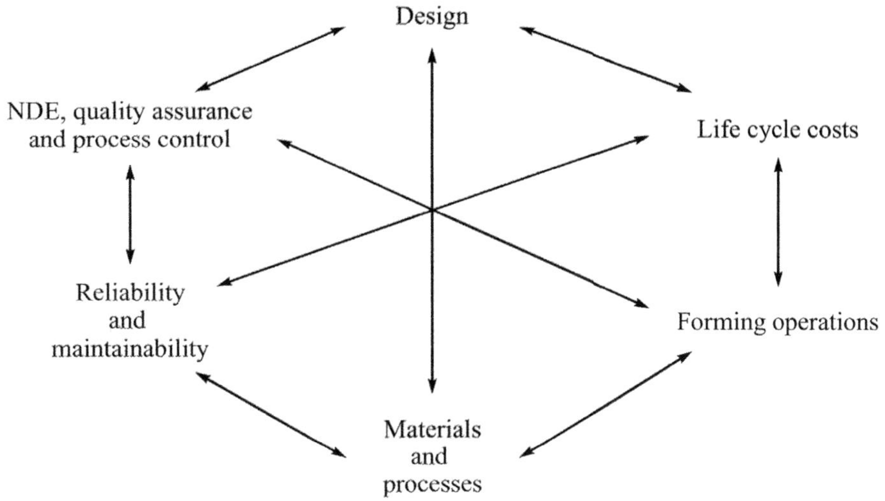

Fig. 3.11. The main factors in life cycle engineering.

In recent years, the role of NDE has become even more important considering that the strong competition in the energy industry and the compromise between the demand of electricity (overcapacity) and the new environmental regulations have made more economically advantageous to keep older power plants running beyond their design life (considering also technological and economic revamping or conversion to different power generation methodologies and combined cycles) in respect to build new structures. In particular, the traditional demand for useful life was fixed to 200000 hours (about 25 years) plus a set number of start/stop cycles, while the requests of the today market can be longer than 400000 hours plus a proper number of start/stop cycles [3.2].

To this aim, early detection of damage is critical to maintain safety operation of plants and, consequently, NDE is usually carried out [3.3] both: i) at the building stage (in order to define a "starting condition" for the beginning of service of possible new plants) and ii) during service (in terms of "inspection intervals" carefully chosen for a suitable maintenance of new or extended plants). In any case, the NDE process involves (figure 3.12) measuring and interpreting the response of the examined object to a proper non destructive stimulus. This stimulus-piece interaction leads to an "indication" to be interpreted by certified inspectors who must distinguish between "relevant", "nonrelevant" and "false" indications.

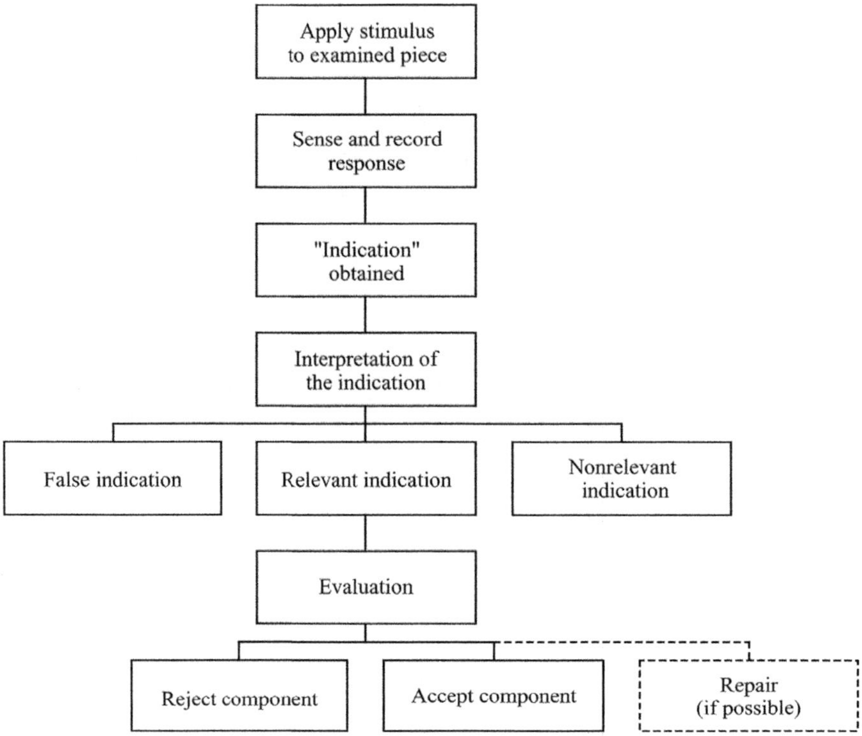

Fig. 3.12. Overview of the NDE (modified from [3.4]).

Considering manufacturing, NDE is applied in order to find all those volumetric and surface defects prospectively detrimental in service and due to the technological process (typically foundry, forging and machining) and subsequent assembly operations. For example, it is worth noting [3.5] that quite often a bore is introduced in the shaft in order to remove the concentration of segregations present on the longitudinal axis and formed during the foundry process, but a significant probability that a defect is not removed still remains. The non-destructive testing (NDT) methodologies [3.6][3.7] traditionally applied during the production stage are [3.5]: visual testing (VT), magnetic particles (MPI), dye penetrant (PT) and ultrasonic (UT); in most recent years, MPI has been substituted by arrays of eddy currents (ET) probes in automatic production lines.

During service, the relevant application of NDE consists in the detection of cracks whose nucleation can have different causes. The most dangerous (and stressed) section of a rotor is the bore [3.5][3.8][3.9] where cracks can initiate during both steady state regime due to creep fatigue and cyclic plasticity (low cycle fatigue) and start/stop cycles due to thermal fatigue and centrifugal forces. Possible defects not removed during the production stage or characterized by a size below a given level of acceptance can also initiate cracks due to high cycle fatigue. Other critical sections [3.5][3.10] are blades (stress corrosion cracking, erosion, unexpected impacts), blade root fixings and grooves (creep) and stress concentrations (creep, thermal fatigue and stress corrosion cracking). The goal is then to localize and size (in terms of length and depth) such cracks in order to apply dedicated design or life prediction techniques (such as the "Damage Tolerance" approach [3.11], see section 3.2.5) for the determination of residual life and proper inspection intervals. From this point of view, the choice of the proper NDT technique is critical. The NDT methodologies traditionally applied during service are the same already seen for manufacturing: VT, MPI (or ET), PT and UT.

Since fatigue cracks are generated by stresses, in-service inspections are more interesting and trivial than inspections during the manufacturing process. In-service inspections are generally carried out "off-line" (i.e. stopping the plant and, sometimes, disassembling the interested parts) by means of traditional NDT due to several reasons like the presence of working fluids or the fact that the surfaces inside the casing are not accessible.

Recently [3.8], "on-line" (real time) measurements are being introduced borrowing fundamental concepts from the medical, automotive and aeronautical fields. Such concepts involve condition monitoring and prognostics based, for power plants, onto methodologies like thermography (very promising), ultrasonic guided waves (difficult application and strongly influenced by high structural noise), acoustic emission (difficult application and strongly influenced by high structural noise) and vibration monitoring (very promising, see section 8.4). On-line measurements are one of the most important challenges for the future application of NDT, especially considering that the periods of downtime, required to apply off-line inspections, are evaluated to be in the range between 7% and 15% of the total operational costs of a power plant [3.1].

Alternatively (and more feasible at the moment), costs can be diminished trying to act on the length of inspection intervals. In particular, considering the Italian case [3.2], the traditional demand for the length of the inspection interval was 40000 hours (5 years) of service involving a downtime of 12 weeks, while the request of the today market is more than 70000 hours with a maximum downtime of 6-7 weeks. It is also worth noting that a single NDT technique can hardly satisfy adequately all the requirements, but these can be fulfilled by superimposing the results provided by the coordinated applications of different methods (as actually it happens during inspections). Such an approach requires long periods of downtime since different NDT techniques can be seldom applied simultaneously.

NDT and NDE are heavily standardised due to the criticality of their application. The most important references for application methodologies and acceptability criteria are those proposed by ISO [3.12], ASNT [3.7], ASTM [3.13], ASME [3.14] and ASM [3.15]. Very important and mandatory is also the qualification and certification of personnel applying NDT during inspections, so relevant standards are available also considering this topic [3.16][3.17][3.18][3.19].

3.2.1 Basic Criteria for the Selection of Inspection Techniques

A reasoned choice of the inspection methods to be applied in the examination of a given rotor should be based onto different factors. The first one is related to the observation that every NDT method has its peculiarities, advantages and limits (Table 3.2 shows this info for the traditional NDT techniques usually involved in inspection of rotors).

Table 3.2. Peculiarities of traditional NDT techniques for inspection of rotors.

	Ultrasonics (UT)	**Eddy currents (ET)**	**Liquid penetrants (PT)**	**Magnetic particles (MPI)**
Cost of instrumentation	Average-High	Average-Low	Low	Average
Cost of consumables	Very low	Low	Low	Low
Time to get results	Immediate	Immediate	Low	Low
Influence of geometry	Important	Important	Not so important	Not so important
Accessibility	Important	Important	Important	Important
Detectable defects	Internal	Surface breaking and sub-surface	Surface breaking	Surface breaking and sub-surface
Sensibility	High	High	Low	Low
Automated	Yes	Yes	No	No
Influence of material	High	High	Low	Only ferromagnetic
Operator's skill	High	Average	Low	Low
Portability	High	Average-High	High	Average-High
Peculiar advantages	Deep penetration into the material, single side access, provides distances information, minimum part preparation	Probe contact is not required, minimum part preparation	Large areas inspected rapidly, indication given directly on the part	Large areas inspected rapidly, indication given directly on the part
Peculiar limits	Surface roughness, need for couplant, defects parallel to sound beam are not detected, reference standards are needed	Limited depth of penetration, defects parallel to coil are not detected	Needs care preparation of the surface, non-porous materials, post cleaning and chemicals handling precautions	Proper, alignment of magnetic field and defect is critical, paint can affect sensitivity, demagnetization is usually needed

The second factor is instead related to the specimen condition to be inspected in terms of:

1) knowledge of the stress distribution during service and in case of different expected loads. In general, the use of the inspection methods does not require very accurate stress analysis, or rather it is enough to know the areas of stress concentration;
2) knowledge about rotor manufacturing technology, in order to be able to assess the influence of residual stresses;
3) knowledge of assembly and fitting operations of the different parts of the machine: supports, shrink fittings, couplings, etc.

It is obvious that, due to completely different geometries and accessibility, some NDT techniques can be more suitable and effective for some parts of the rotor, while other techniques for other parts. Moreover, as already introduced above, the synergistic application of different techniques can permit to inspect also very difficult components and improve significantly the relevance of results, but downtimes get usually longer.

The last factor is experience, when available. The detailed knowledge of previous failures of similar components can be very useful in designing inspections, exactly as the availability of skilled operators working in the field since many years.

3.2.2 Methods Based on Visual Checks

The simplest investigation techniques are based on visual checks, which are mainly devoted to locate surface or, in some cases, sub-surface cracks and to evaluate their surface extension (but not their depths). For all the techniques presented in the following, some uncertainty on the surface extension can be found at the surface crack tips.

3.2.2.1 Liquid (or dye) penetrant method (PT)

Liquid (or dye) penetrant testing [3.20], [3.21] is a non-destructive method used to detect surface breaking defects in any non-porous material. The steps are the following (figure 3.13):

- the surface is cleaned and prepared for inspection;
- liquid penetrant is applied to the surface and is drawn into cracks and pores by capillary action;

- after a preset dwell time, the liquid penetrant surplus is wiped off and the surface is dried;
- a developer (usually a chalk powder suspension) is sprayed on the surface in order to extract the liquid penetrant from possible defects;
- after a proper time for the development of indications, visual inspection is carried out in order to reveal defects;
- final cleaning of the surface.

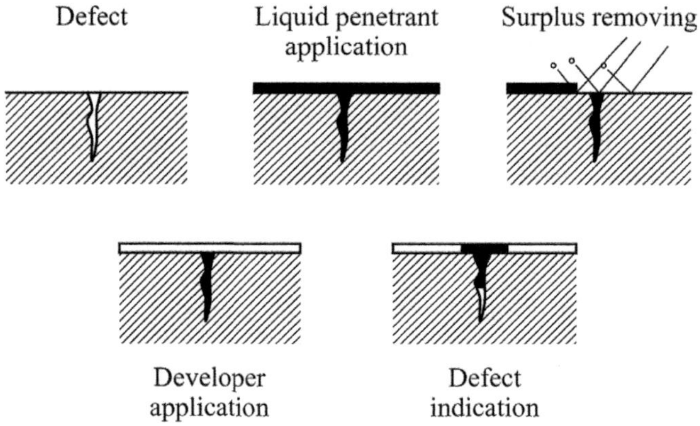

Fig. 3.13. Application steps of liquid penetrant NDT technique.

If liquid penetrants are of the coloured type, they are visual checked in white light and the indications are revealed by the strong contrast between the colour of the penetrant (typically red) and the white colour of the chalk. If they are fluorescent, they are checked in black light (realised by means of a Wood's lamp) which evidences the yellow-green ultraviolet (UV) shade of the penetrant. The fluorescent type of penetrant is more sensible than the coloured one, but it also requires higher investments (curtains in order to shadow the inspected parts and the lamp) in order to be carried out. Figure 3.14a shows the inspection by means of coloured penetrant of a shaft with a surface crack, while figure 3.14b shows a crack, related to blade root groove, revealed by UV active dye.

(a)

(b)

Fig. 3.14. Crack extension highlighted by: a) red dye penetrant and white chalk powder developer on the shaft; b) fluorescent penetrant on blade root groove.

Dye penetrant applies to all situations without presenting limitations related to the material. Anyway, it is suggested to reduce the roughness of the surface to be inspected (an excessive roughness would produce dye spreading) by finishing. This is a weak constraint, as shown in figure 3.15a that presents a rather rusted shaft of a horizontal Pelton turbine and in figure 3.15b that reveals the crack. The only actual limit is the accessibility of the surfaces, that can be difficult like in the case of figure 3.16.

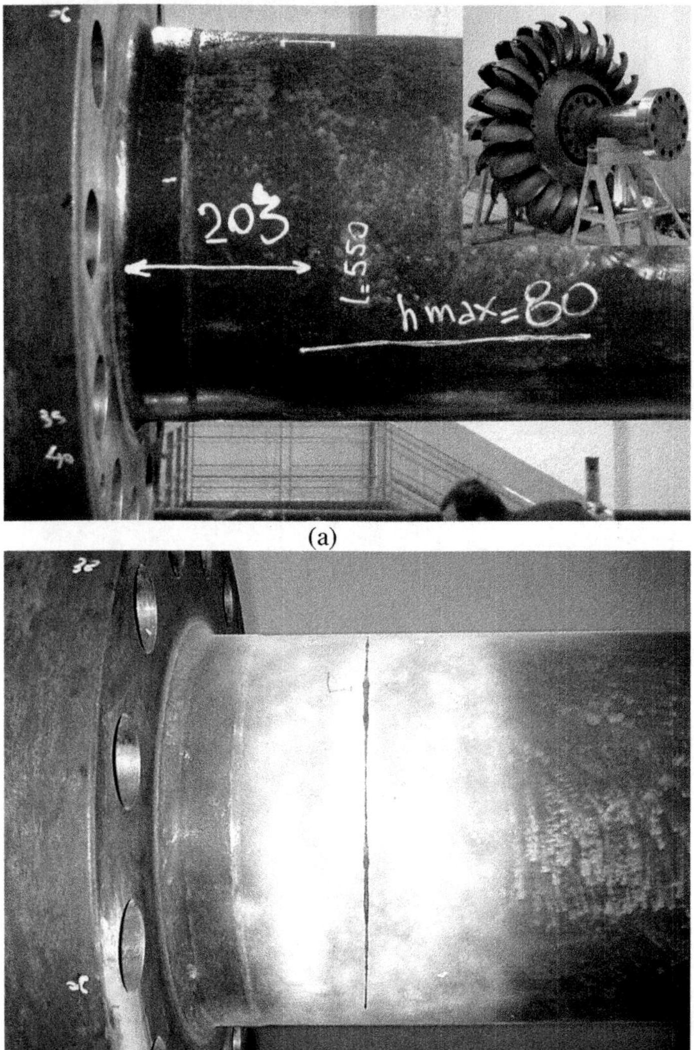

(a)

(b)

Fig. 3.15. Horizontal shaft of a Pelton turbine tested during maintenance: a) starting surface conditions; b) crack revealed.

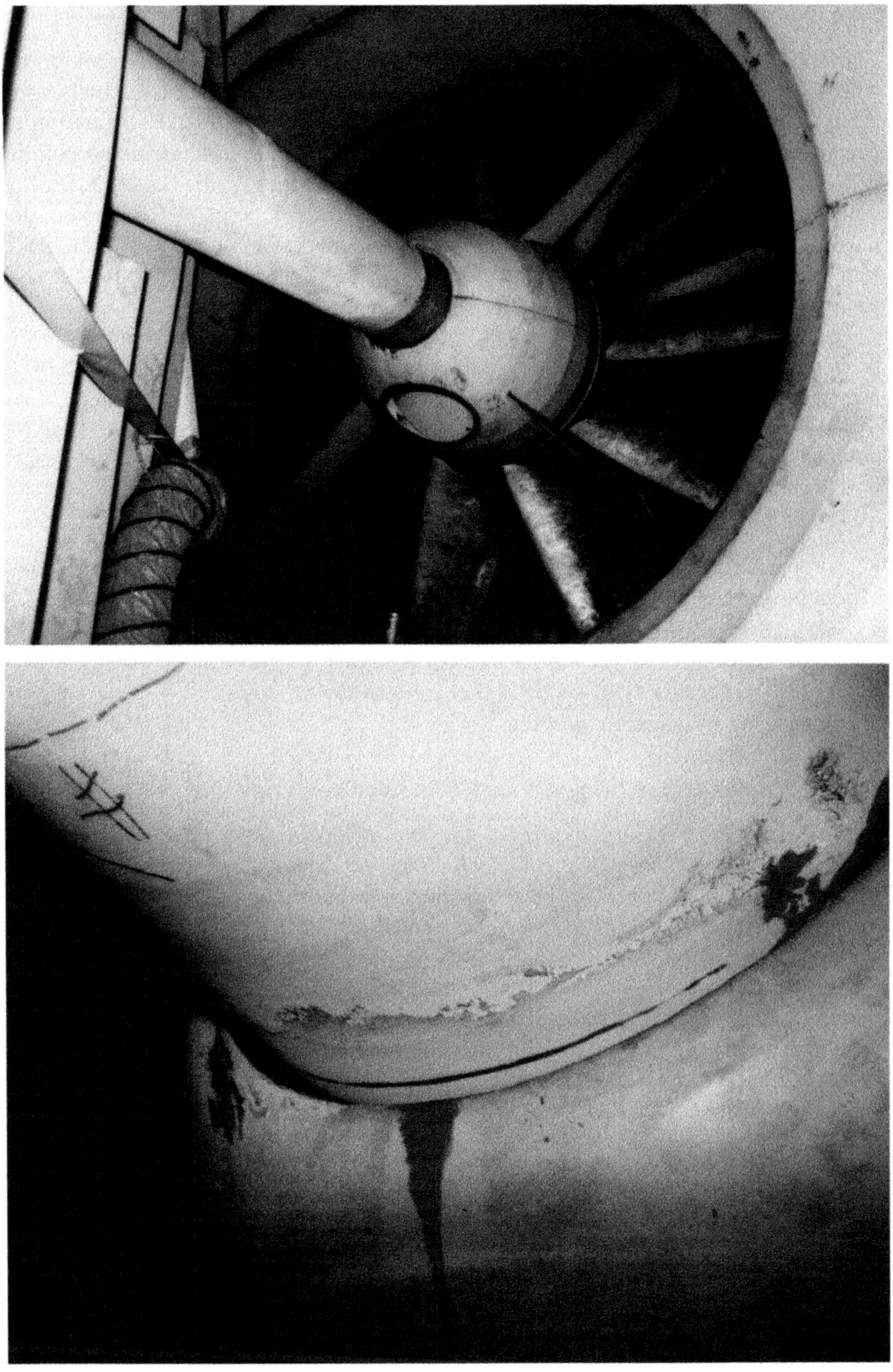

Fig. 3.16. Crack detected in a power shaft of a wind tunnel fan (courtesy of Prof. Yukio Ishida, Nagoya University, Japan).

3.2.2.2 Magnetic Particle Test (MPI)

This NDT technique is based [3.22], [3.23] on the concept that when the continuity of a magnet is broken (due to cracks, non-ferromagnetic inclusions, etc.), a north and south pole will form at each edge of the discontinuity. Consequently, a so-called "flux leakage field" due to this double localized polarity is generated. If magnetic particles are sprinkled, they will be attracted, by magnetic field gradients, to and cluster not only at the poles at the ends of the magnet, but also at the poles at the edges of the crack. This cluster of particles is much easier to see than the actual discontinuity and this is the basis for magnetic particles inspection (figure 3.17a).

It is then obvious that, for a successful application of MPI, the material has to be ferromagnetic. It is useful to remember that some stainless steels have a magnetic permeability too low to allow the use of this technique.

It is also worth noting that MPI permits to reveal not only surface breaking defects but also sub-surface ones, at least until the flux leakage, and not the defect, is surface breaking (figure 3.17b).

The steps for the inspection are the following:

- preparation of the surface to be inspected (this step is much less critical than in the case of PT);
- magnetisation of the component to be inspected (both AC and DC can be applied, where the sensibility of AC is higher for surface breaking defects, while the sensitivity of DC is higher for sub-surface ones);
- sprinkling of magnetic particles;
- visual inspection;
- demagnetization of the component (if needed).

Different methods are available for the magnetization of the component: permanent magnets, magnetic yokes, electrical prods, coils and conductive cables. The most important aspect to be remembered is that magnetic inspection is more sensitive to discontinuities that are orthogonal to the magnetic flux direction. In the case of rotors, this means that the inspection, to be fully effective, should be carried out considering both longitudinal and circular magnetization. Examples of circular and longitudinal magnetization of a shaft using a permanent magnet are shown in figure 3.18. Just like liquid penetrants, also magnetic particles can be coloured or fluorescent. An example of a crack revealed at the shrink-fit seat of a shaft is shown in figure 3.19.

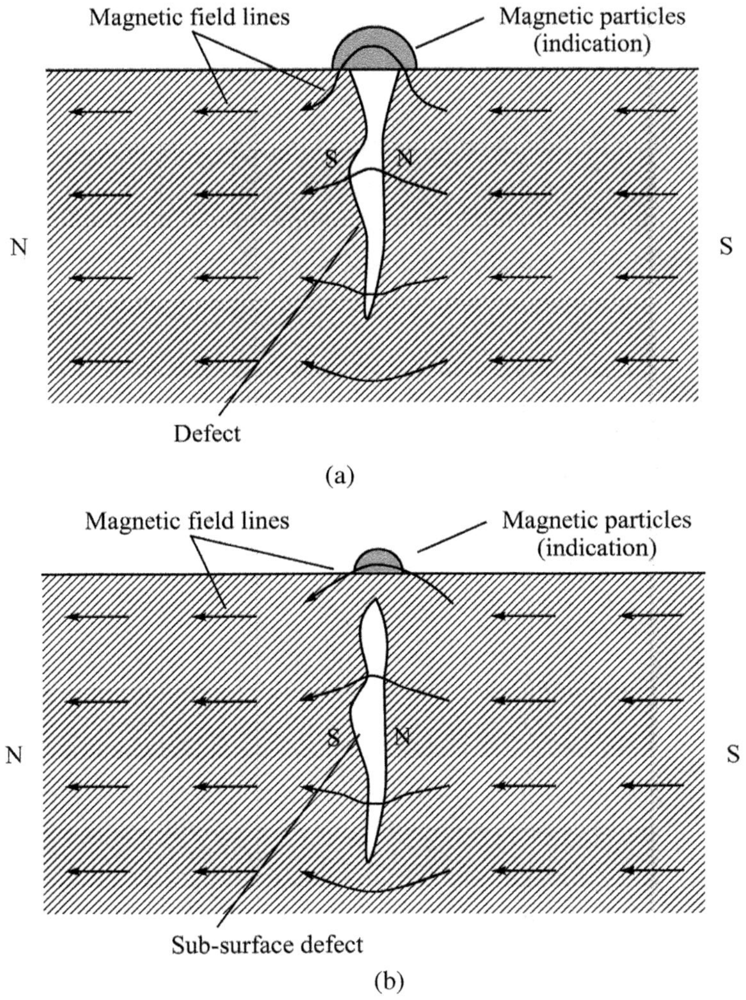

Fig. 3.17. Principle of magnetic particles inspection: a) surface breaking defect; b) sub-surface defect (surface breaking flux leakage).

(a)

(b)

Fig. 3.18. Magnetization of a shaft by a permanent magnet: a) transverse; b) longitudinal.

Fig. 3.19. Crack revealed at the shrink-fit seat of a shaft by application fluorescent MPI.

3.2.2.3 Direct Vision Method (VT)

The visual testing method [3.24] uses, together with direct visual checks, optical microscopes for external surfaces and endoscopes for the surface of the bore (borescopes). The lighting can be provided by an optical fibre source. The best results are obtained when the light is directed almost normally to the crack.

Figure 3.20 shows an example of visual inspection, by an optical microscope, of a shaft in which a crack was started, in a given section, by a drilled micro-hole.

Fig. 3.20. Optical microscope used for visual inspection of a shaft surface (top). The crack starts from a drilled micro-hole (bottom).

3.2.2.4 Comparative Considerations

Although VT, PT and MPI methods appear equivalent from various points of view, it is appropriate to highlight some important differences in their application.

VT is more suitable for inspection of small areas, whilst PT and MPI are suitable also for large structures or surfaces.

Comparing PT to MPI in regards of the time needed, magnetic particle test is faster but may require two different inspections (longitudinal and circular) and the complete demagnetization of the rotor (this operation is quite delicate and not straightforward). A comparison between the sensibility of PT and MPI is shown in Figure 3.21 considering a crack, generated by artificial micro-notches, in a shaft. As already pointed out, fluorescent inspections are more sensible than coloured ones, but require higher investments.

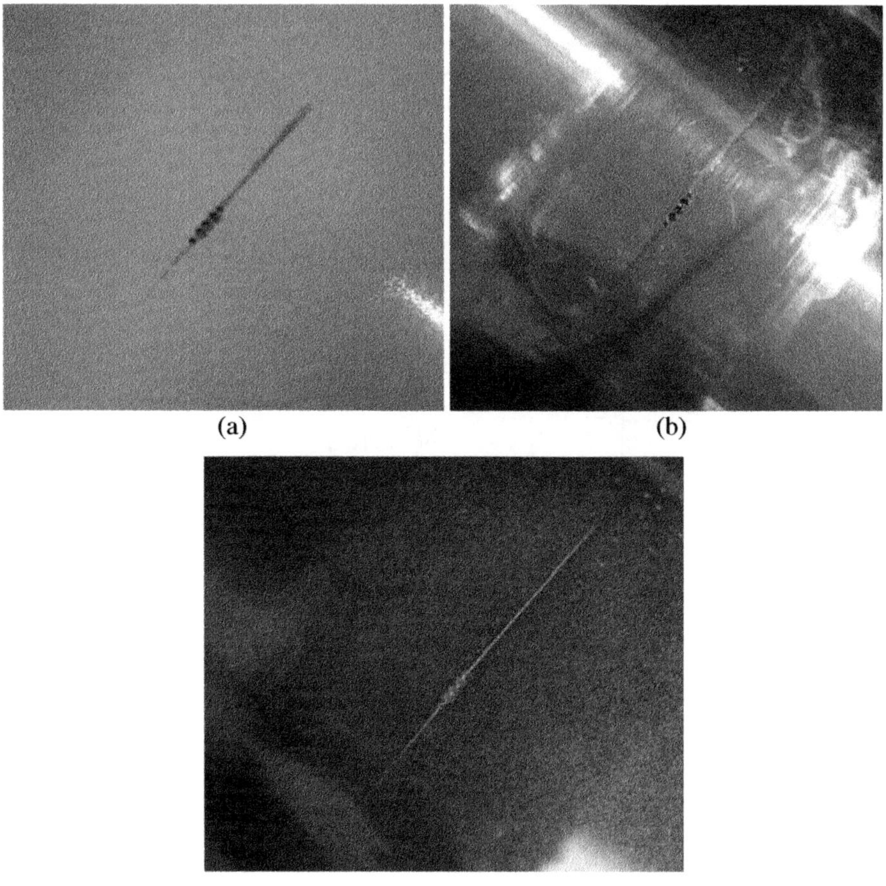

(c)

Fig. 3.21. Comparative analysis of the sensibility of different NDT methods: a) colored PT; b) colored MPI; c) fluorescent MPI.

For PT and MPI, a sensitivity check is currently made by means of normalized blocks with known cracks, sometimes also a prototype rotor containing artificial discontinuities can be used.

3.2.3 Ultrasonic Inspections (UT)

Ultrasonic inspections [3.25][3.26][3.27] are based onto the measurement of the time of flight of high frequency sound waves introduced into the component at various surface locations. In particular, these beams of sound waves (that can be longitudinal, shear, surface, etc.) are transmitted from a probe and detected by the same (pulse-echo technique, figure 3.22a) or other probes (through-transmission technique, figure 3.22b). The interpretation of the reflection, refraction and diffraction echoes at particular boundaries leads to information about the discontinuities in the test piece. Since sound travels into many materials, the UT inspection method is the most widely used NDT technique and the only one, in particular, able to detect deep volumetric discontinuities and in conditions of difficult accessibility. The use of these techniques does not have limitations related to materials, except for those particularly porous or characterised by an inherently discontinuous structure (austenitic stainless steels are a classical example).

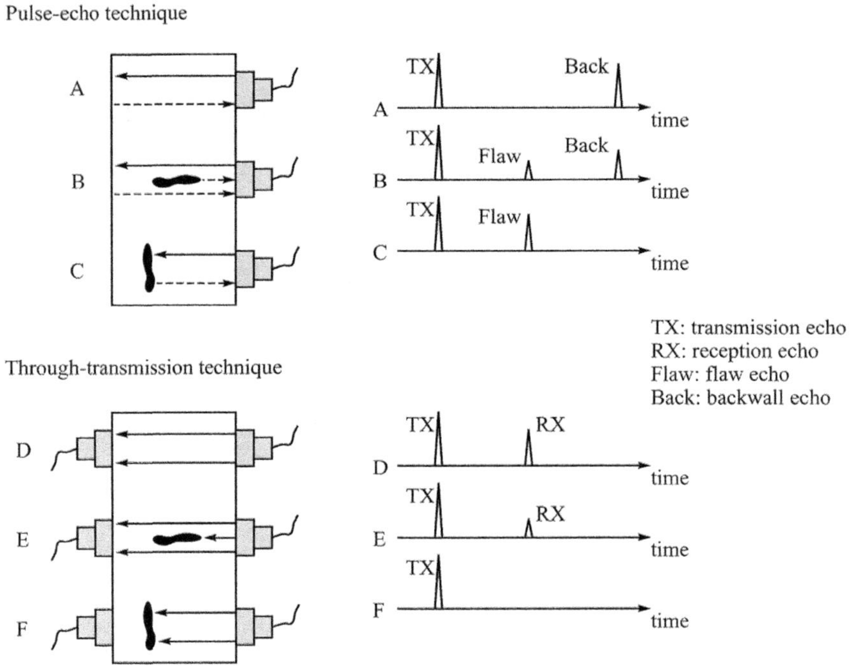

Fig. 3.22. Typical UT techniques for defect detection.

Piezoelectric crystals are currently adopted in probes (figure 3.23). They are fed by suitable voltage pulses of short duration repeated at proper time pitch so to excite packages, or trains, of mechanical elasto-dynamic waves in the tested part. Such ultrasonic wave trains are typically characterised by frequencies in the range between 200 kHz and 25 MHz and, in the most traditional application, are applied by a hand-held single or double crystals probe placed on the specimen surface. It

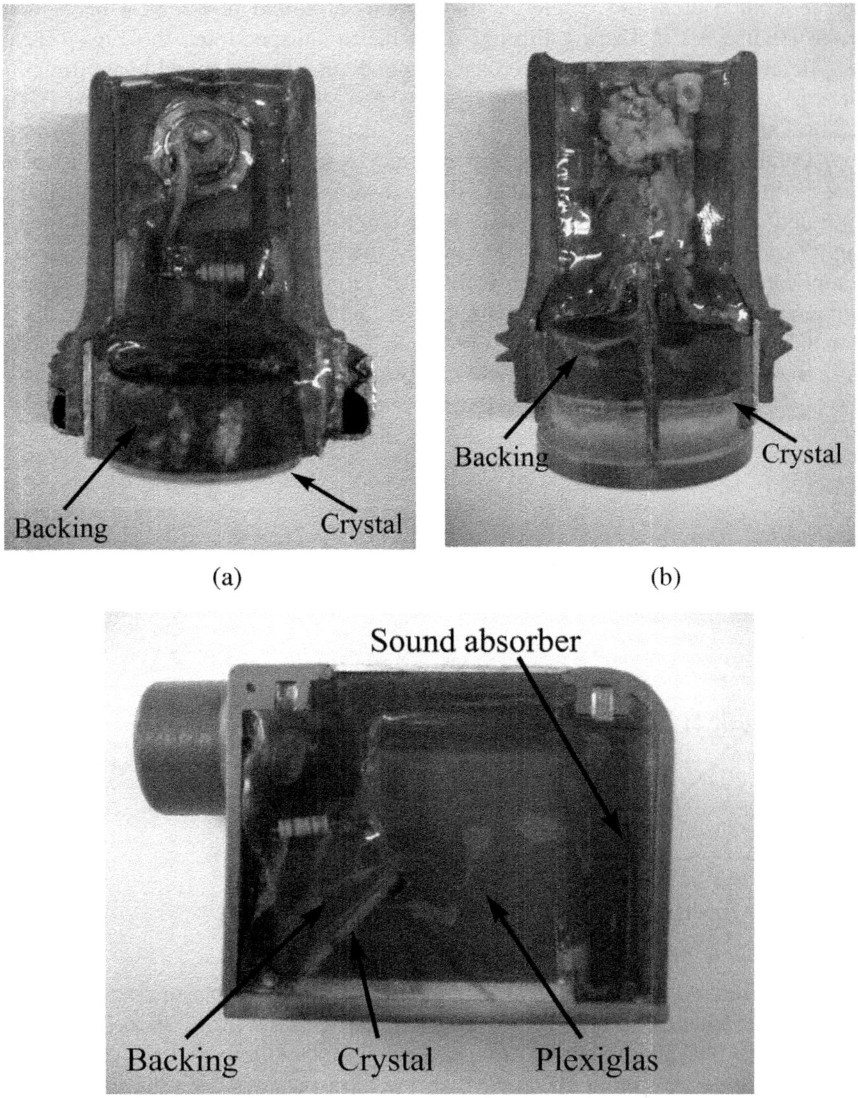

(c)

Fig. 3.23. Sectioned single crystal probes for UT inspections: a) single; b) double; c) inclined.

is worth noting that, especially during the manufacturing stage, automatic multi-probe UT systems and water immersion inspections are often adopted.

An oscilloscope is used to show the time of flight taken by an ultrasonic pulse to travel from the transmitter to a reflector (a flaw or a crack, the backwall or any other free boundary of the inspected body) and then to a receiver. During pulse-echo inspections, the amplitude of the received pulse is related to the dimension of the portion of the defect area perpendicular to the sound beam direction: the bigger is such area, the greater is the amount of sound power sent back to the probe (figure 3.22). During through-transmission inspections, the bigger is the total defect area projected onto a plane perpendicular to the sound beam direction, the smaller is the amount of sound power reaching the receiving probe (figure 3.22). The relationship between flaw size, distance and reflectivity is complex and, usually, considerable skills are required to interpret the displayed response echoes. Such echoes represent, then, the received sound pressure as a function of time and are called echograms or A-Scan (figure 3.24). Time is called "time of flight" and is usually measured in μs. Echogram interpretation often is complicated by: i) echoes coming from shaft geometry variations, like changes of diameter, machined surfaces for connections with other parts and so on; ii) the actual behaviour of the complex ultrasonic field, which is somewhat different from the behaviour of the single field components (surface, longitudinal and shear waves).

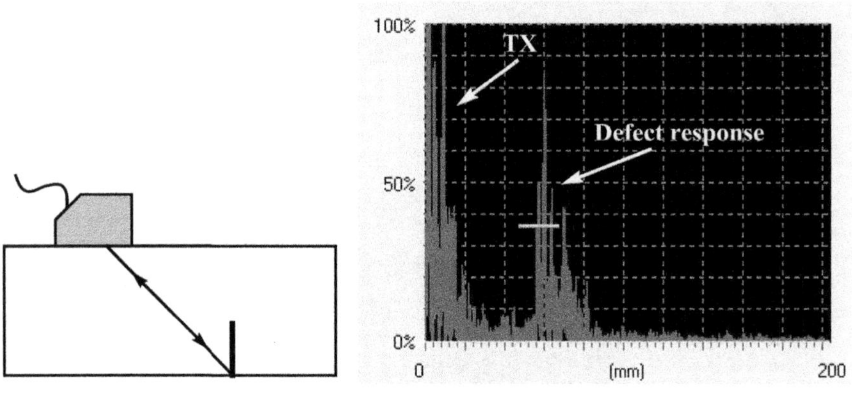

Fig. 3.24. Example of the typical A-Scan (echogram) of a defect.

For an effective inspection, some fundamental operating parameters need to be discussed and kept in mind. The central frequency of the probe should be chosen in relation to the minimum size of the expected discontinuity to be highlighted: in particular, this size should be larger than the ultrasonic wavelength, calculated as follows [3.27]:

$$\lambda = \frac{v}{f} \tag{3.1}$$

where λ is the wavelength, v is the (longitudinal, shear or surface) sound speed in the considered material, f is the central frequency of the probe. Since frequency is proportional to the inverse of wavelength, the smaller is the size of the defect, the higher is the needed frequency. However, it should be remembered that the increasing of the frequency implies the increasing of the wave attenuation into the material, which produces a shorter propagation path of ultrasonic energy.

The size of the crystal defines the depth into the material of the most sensitive section of the sound beam. Such a distance corresponds to the transition from the acoustic near field to the far field and can be calculated as [3.27]:

$$N = \frac{D^2 - \frac{\lambda^2}{4}}{\lambda} \tag{3.2}$$

where N is the distance of the most sensitive section of the sound beam and D is a characteristic dimension of the crystal. Moreover, this size defines the lateral divergence of the wave train [3.27]:

$$\sin \alpha_{10} = 0.835 \frac{\lambda}{D} \tag{3.3}$$

where α_{10} is the half-angle at which the sound pressure is decreased to 10%. Practically, the crystal size determines the limit beyond which the detection of the flaw is possible as well as the spatial resolution of the test. The last is important to determine with high precision the borders of an extended defect.

The filter must be chosen carefully, during both the calibration and the inspection, considering that this has effect on the useful part of the echogram, on the parts related to the structure of the material and on the surface roughness. The aim of the filter selection is to enhance the useful signal component reducing what can be considered as background noise (usually called "grass").

Finally, curvatures and surface conditions are similarly important for the proper application of the method, because probes usually present flat coupling surfaces and bad contact conditions greatly reduce the ultrasonic energy introduced in the component. Considering surface finish, the surface must then be covered with a thin layer of oil or grease to ensure acoustic continuity that a layer of air would undermine definitively. In the case of geometrical curvatures, proper wedges can be applied to probes in order to maximise the acoustical coupling (figure 3.25).

Fig. 3.25. Application of an ultrasonic probe on a geometrical curvature by a plastic wedge.

It is useful to remember that UT is a relative method: the estimation of the defect size is possible only by the comparison of the actual measured response with the one generated by sample reflectors (flaws, cracks, notches, etc.) characterised by known geometry and placed at the same depth. Figure 3.26 shows a sample block, for shaft inspection, characterised by known surface notches (holes, saw-cuts) obtained by traditional machining. Recent results [3.28] have shown that responses from artificial notches characterised by completely different shapes and obtained by traditional machining or EDM are not discernable from real fatigue cracks if their reflecting area is taken as their characteristic parameter (so far, the traditional parameter taken has always been crack depth or length).

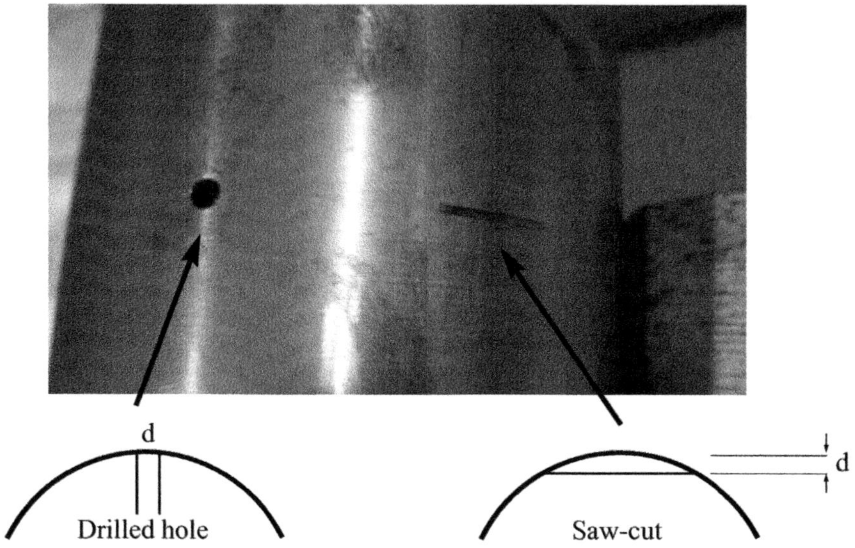

Fig. 3.26. Sample block for shaft inspection.

3.2.3.1 Traditional Approaches to UT Inspections

In relation to ultrasonic inspections of rotors, special modes of wave propagation are frequently used. The first mode is represented by the surface waves (known as Rayleigh waves), which can locate surface breaking defects. In these applications, the surface acts as a sort of wave guide, but the probe must not be too far from the defect, because this type of propagation is characterised by high attenuation, caused by the surface roughness.

The other modes correspond to the techniques mostly used for detecting cracks in rotors. The first one applies waves emitted by probes, characterised by normal sound incidence in respect to the surface (figure 3.23a and figure 3.23b), generating a train of longitudinal waves, i.e. vibrations parallel to the direction of propagation. These probes can be used by scanning along a grid put on a shaft end in order to obtain an axial sound propagation able to detect transverse reflectors,

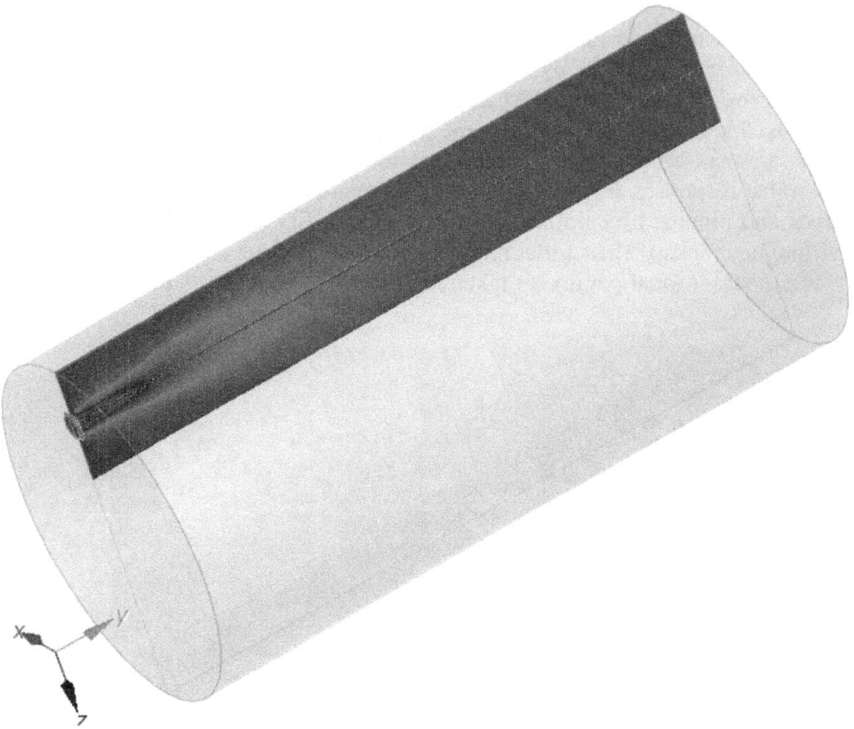

Fig. 3.27. Crack scanning of a shaft using longitudinal waves.

as shown in figure 3.27. Figure 3.28 shows the results of the longitudinal scanning of the same shaft in presence and in absence of a transverse surface crack. The A-Scans are given in radio frequency representation, i.e. the signal was not rectified.

The second technique applies waves emitted by angle probes inclined in respect to the contact surface (figure 3.23c). If the refraction angle of the longitudinal component is less than the first critical angle, two different wave trains are present in the component: longitudinal waves, with a higher angle of propagation, and shear waves, characterised by vibrations perpendicular to the propagation direction and by a lower angle of propagation. If the refraction angle is higher than the first critical angle, the wave train is formed only by shear waves. Two particular variations of this technique are: i) the "inner diameter" (ID) inspection: ultrasonic trains are sent directly towards the defect as shown in figure 3.29a; ii) the "outer diameter" (OD) inspection: ultrasonic trains are sent to the surface opposite to the probe and then to the defect (in a so called "2nd leg" inspection), as shown in figure 3.29b. The complete inspection requires to cover with the probe a strip of circumferential external surface of the shaft, at a distance from the reflector crack that depends on the angles of propagation of the wave trains, on the depth of the defects and on the size of the probe.

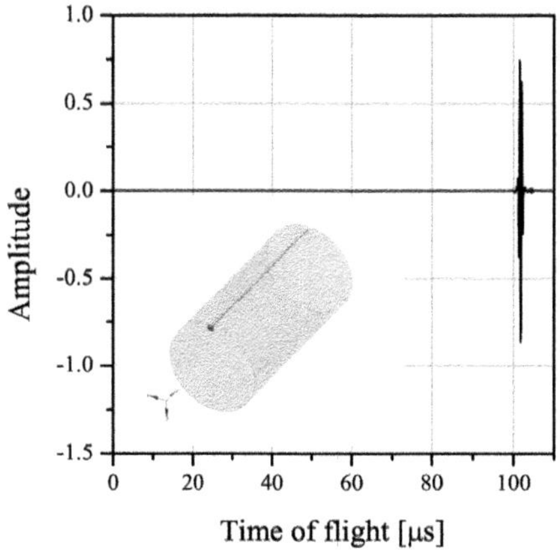

(a)

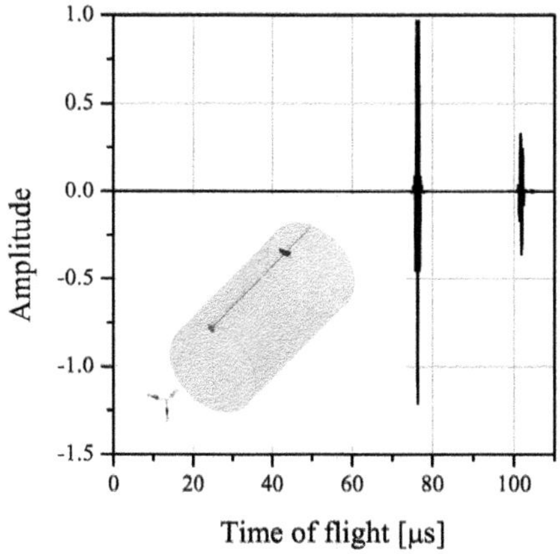

(b)

Fig. 3.28. A-Scans (given in radio frequency) of an inspected shaft: a) without crack; b) with crack.

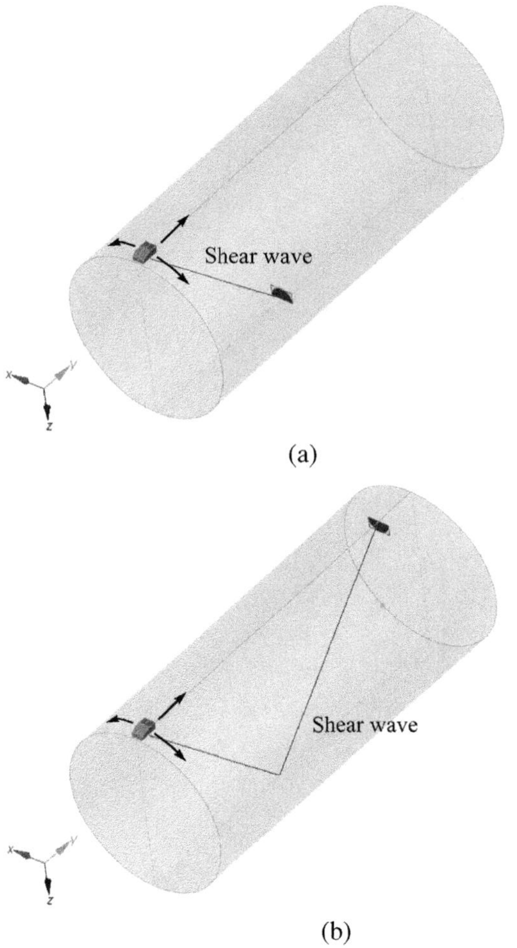

Fig. 3.29. Angle probe inspection: a) ID (*inner diameter*) direct method for crack detection; b) OD (*outer diameter*) indirect method for crack detection.

In order to speed up UT inspections, special multi-probe systems can be applied to rotors. The most interesting application is the wide spread boreprobe [3.2], [3.5], [3.29]: a probe head arrangement containing various single crystal angle probes mounted on a centralised rod to be inserted into the bore with the aim to inspect the whole volume of the shaft (figure 3.30). Such approach is adopted (considering manual, and automatic inspections) in a lot of mechanical fields were hollow rotating shafts are involved (an interesting example are railway axles) because it is proven [3.29] that the "Probability of Detection" (POD) (see Section 3.2.4) of this system is significantly higher than the one obtained adopting single crystal traditional inspections.

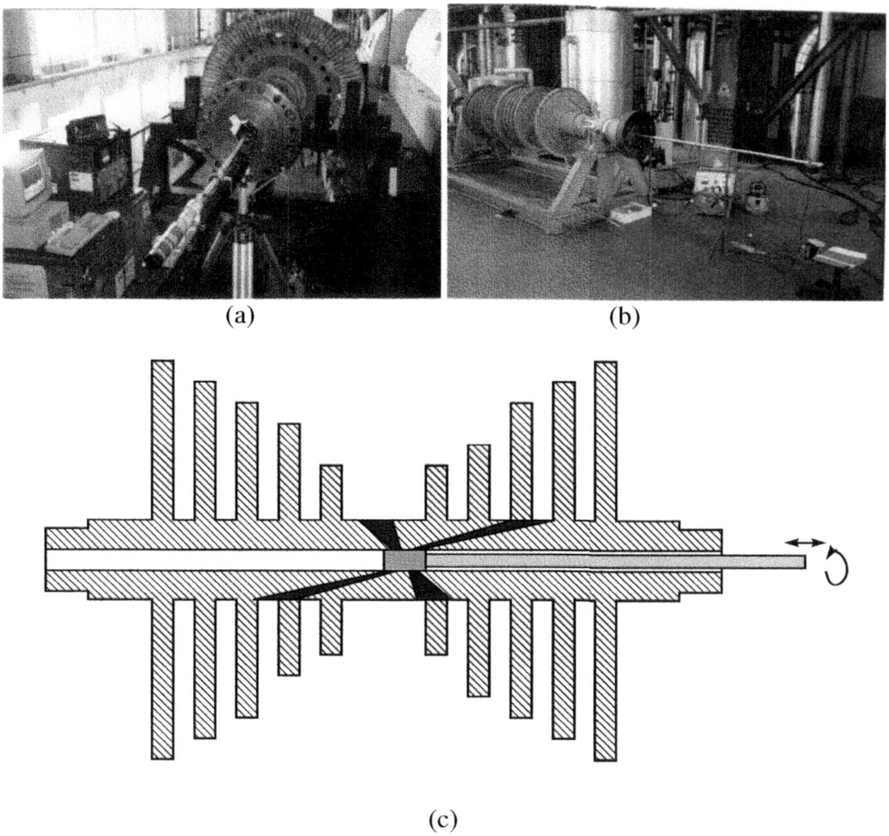

Fig. 3.30. Application of boreprobes to rotor inspection (pictures are courtesy of G. M. Crespi, [3.2]).

3.2.3.2 Application of Ultrasonic Phased Array Systems to Rotor Inspection

A multi-probe approach is also the idea at the base of phased array (PA) systems. This technology adopts single probes whose crystal is divided into smaller identical elements each excited by an independent pulser [3.30]. In particular, the excitation time of each crystal is computer-controlled and -delayed according to specific electronic focal laws able to provide the parameters that affect the forming of the ultrasonic beam and define specific focusing, steering and beam width without the application of mechanical wedges (figure 3.31). Arrays can be one-dimensional (a linear sequence of crystals) or two-dimensional (a matrix of crystals): while the former applies focal laws along a given direction permitting the instantaneous inspection of areas, the latter solution is able to apply focusing and steering along two independent directions permitting the fast inspection of volumes.

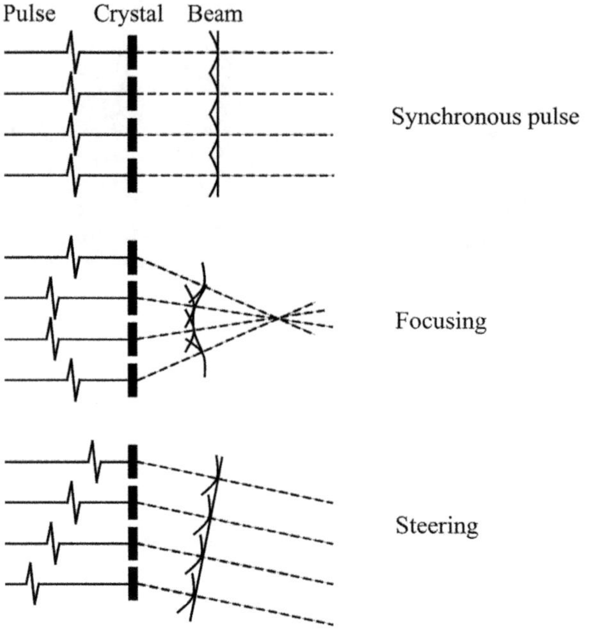

Fig. 3.31. Focusing and steering of phased array sound beams.

Another very important technical feature available adopting PA systems is the possibility to represent ultrasonic maps of the inspected regions. Since every crystal transmits and receives a different sound beam (i.e. a different A-Scan), all the response echoes can be combined to form two-dimensional scans like the sectorial scan (S-Scan) or the linear scan (L-Scan). As an example, the most famous S-Scans obtained from an ultrasonic phased array system are the well-known echographs taken during pregnancy.

Considering the application to rotors, all of these features make PA:

- much faster than traditional single crystal raster scanning (saving plant downtime and operator costs);
- more flexible, since a single PA probe can inspect many different component with different inspection patterns, using proper electronic set-up files (saving costs for a large number of probes and control units);
- able to inspect geometrically complex components. In particular, PA has been successfully applied to the inspection of blade attachments [3.10], [3.31] (where a very interesting example is shown, see figure 3.32), while boreprobes using PA instead of multi-probe systems are being developed [3.32];
- more powerful and effective in crack sizing, since a map representation of the inspected area permits to evidence both corner echoes and diffractions at defect's tips. In particular, figure 3.32a shows the analysis of a blade attachment without any defect or notch, while figure 3.32b the same

attachment with artificial notches. As it can be seen, in the latter case, some new indication spots are present at the notch tip, so that measuring the distance between these echoes and the corresponding boundary an estimation of the notch depth is possible. To carry out an analysis like this by means of traditional single crystal is not straightforward;

- smaller in size in respect to a number of single crystal probes equal to the number of elements;
- the detectability of mis-oriented defects increased;
- the mechanical reliability higher since less moving parts are present;

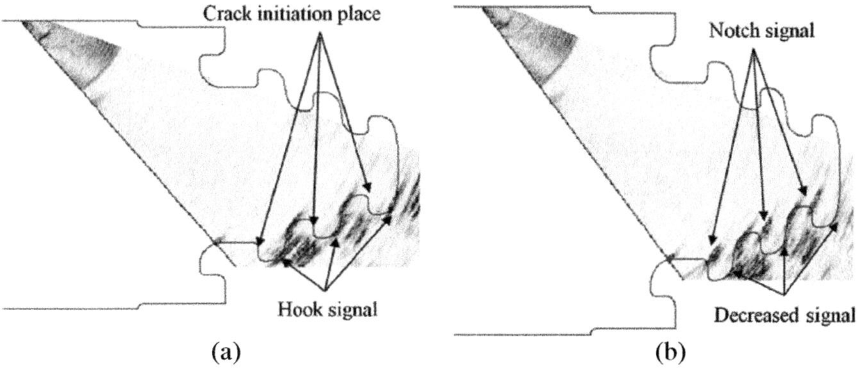

Fig. 3.32. Phased array signal from a blade attachment [3.31]: a) without notch; b) with notch.

Some disadvantages are also to be kept in mind during the applications of PA systems in respect to single crystal probes:

- the initial investment is significantly higher (usually one order of magnitude);
- they are not yet included in standards, so the final official inspection has to be carried out based onto a traditional A-Scan (or at least considering the single A-Scans of the PA elements without taking advantage of the ultrasonic map);
- technicians must be instructed, since operating a PA system is more complicated;
- sometimes in the literature [3.31], a defect depth equal to 2 mm is indicated as the lowest inspectable value (but some unpublished data obtained by the authors have shown that this lowest value can be between 0.5 and 1 mm).

3.2.3.3 TOFD Technique for Rotor Inspection

This method has gained popularity especially considering the inspection of welds (for it is being incorporated into relevant codes and standards) and differs in that diffracted, and not reflected, sound waves are used to locate and size defects [3.11]. In particular, longitudinal and surface waves emitted by an angle probe and detected by a separate receiver allow to define the "time of flight diffraction" (TOFD) [3.33]: the probes are coupled and positioned in longitudinal direction on the shaft at equal distance from crack, see figure 3.33 and figure 3.34. The two probes can be gradually rotated around the shaft to investigate the circumferential extension of the discontinuity. In this case, an ultrasonic map of the inspected region can be defined putting together all the A-Scans recorded in time sequence by the two probes during their movements (this map is called B-Scan) instead of the A-Scans obtained at the same time from different crystals as in PA systems. The output processing should highlight the position of the echo from the top surface (coming from the surface wave), the echo from the bottom surface (coming from the longitudinal wave reflected by the back-wall) and especially those of the longitudinal wave diffracted by the crack boundaries. Figure 3.35 and figure 3.36 show two examples for an internal and for a surface breaking crack, respectively. It is intuitive, considering the more general case of an internal crack, that the surface and back-wall echoes are always present during the inspection, while the diffracted echoes only if the defect is present.

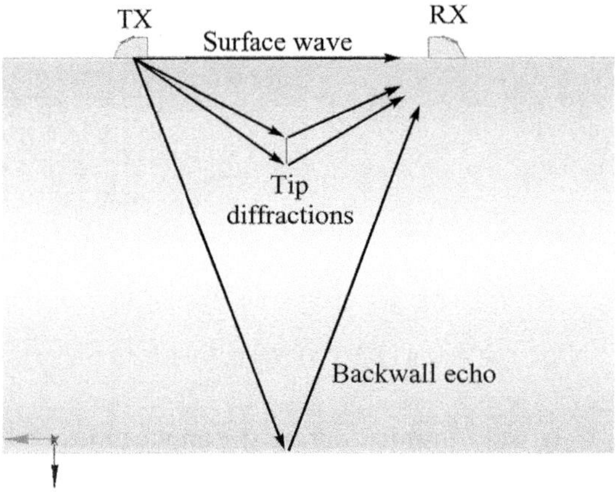

Fig. 3.33. Detecting internal cracks using TOFD technique.

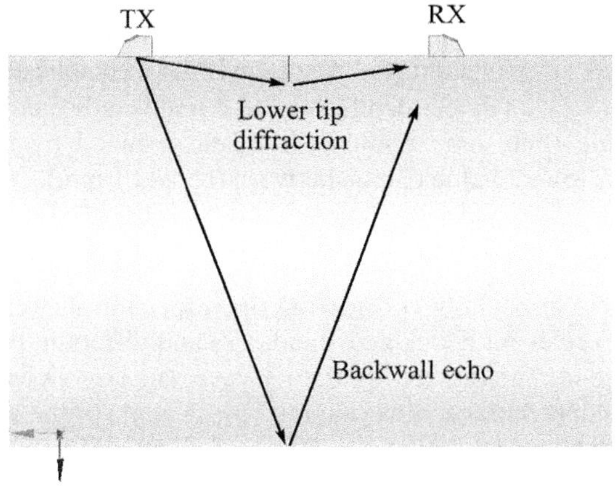

Fig. 3.34. Detecting surface breaking cracks using TOFD technique.

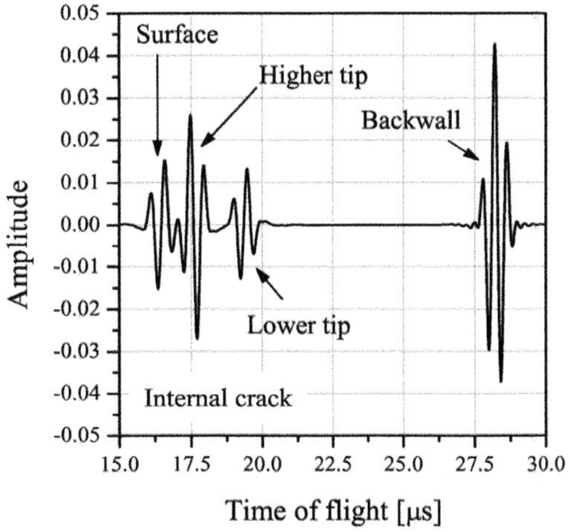

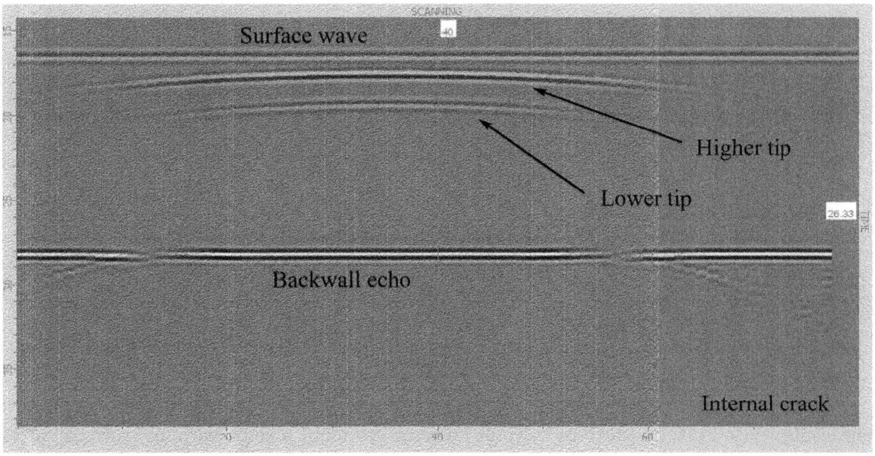

Fig. 3.35. A-scan (top) and B-scan (bottom) echograms of a shaft with an internal crack.

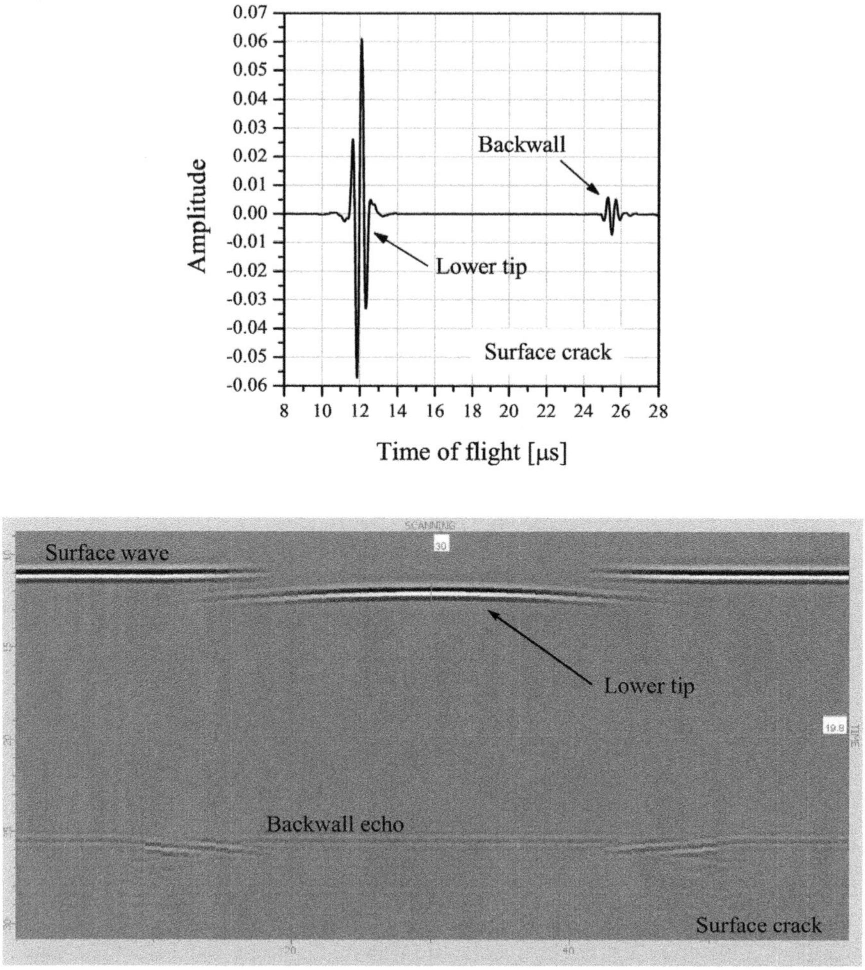

Fig. 3.36. A-scan (top) and B-scan (bottom) echograms of a shaft with a surface breaking crack.

The ultrasonic wave train time of flight from transmitter to receiver can then be post-elaborated and transformed into paths by means of the knowledge of the propagation speed, which is $c_l = 5920$ m/s for longitudinal waves and $c_t = 3230$ m/s for shear waves in carbon steel. The images shown in figure 3.35 and figure 3.36 show then the possibility to precisely locate and size the defect by calculating distances between diffractions or between a diffraction and the backwall or surface echoes. It is important to add that the TOFD method does not depend on crack orientation, but requires full accessibility from both sides of the expected defect.

The results of A-scan and B-scan echograms at a frequency of 5 MHz, obtained [3.34] by CETIM (Centre Technique des Industries Mécaniques) on cracked shaft specimens especially prepared for the EDF test-rig described in detail in section 8.5.2, are shown in figure 3.37a and figure 3.37b for closed and open crack configurations, respectively. Shape, depth and circumferential extension of the crack seem to be defined very well in open crack configuration, while, in closed crack configuration, the results are more confused: diffraction occurs in different positions inside the cracked area, while unaffected propagation of waves takes place in points where a good contact is present. It can be deduced that, in general, highly skilled operators are needed for interpreting the accurate ultrasonic measurements, and computerized analysis greatly helps in this task. Echograms of the same shaft without cracks are required to assess the echoes generated by different surfaces, sometimes also the calibration of the procedure by the aid of a notched sample block is required.

3.2.3.4 Numerical Tools for Designing UT Inspections

A different approach (today very interesting due to the development progress of calculation tools and instruments) to NDT is the numerical one, useful to design inspection strategies before carrying out real measurements on-site. This usefulness comes from the opportunity to test probe configurations, inspection approaches to complex geometries and to interpret incomprehensible responses.

Only some NDT techniques can be numerically simulated, i.e. those, like UT, ET and RT, based onto physical and mathematical laws implementable in algorithms. A good survey of numerical tools for simulating different NDT techniques is reported in [3.35], while in the following this topic will be faced, as an example, considering only the most wide spread and developed case of UT inspections.

A part from general numerical tools, like finite or boundary elements sometimes used for simulating also ultrasonic waves propagating in materials, a lot of free or commercial dedicated software packages are available. Each of them starts from solving the elasto-dynamic equations of equilibrium, but is different for the numerical methods involved, for the calculating strategies and for pre-processing and the post-processing tools provided to the user (CAD environment, material database, etc.).

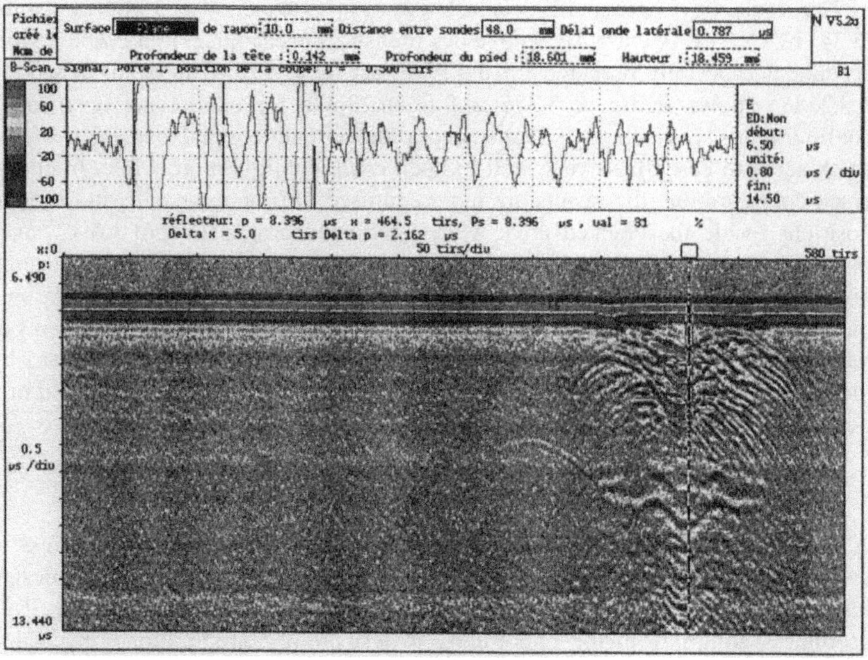

(a)

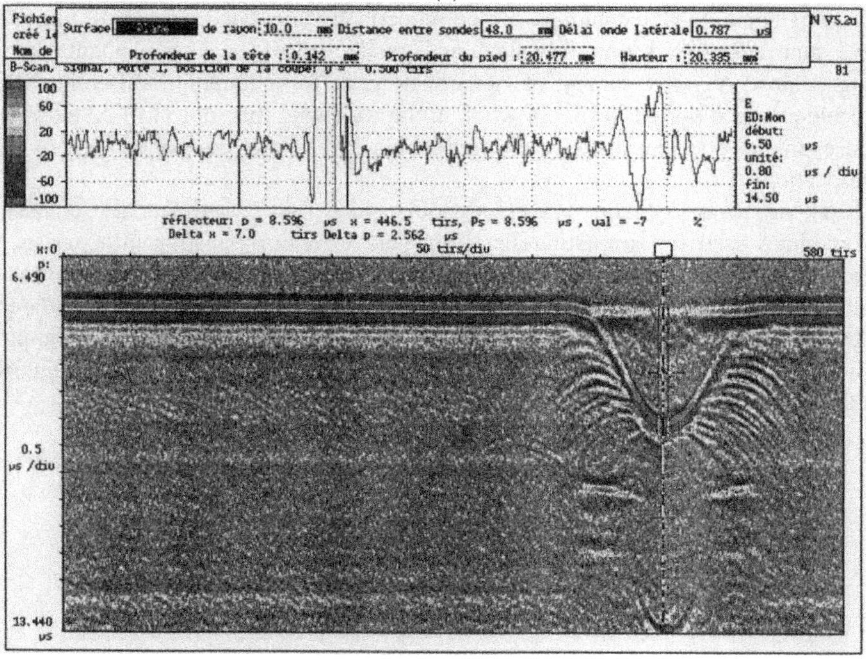

(b)

Fig. 3.37. TOFD echograms of a cracked shaft specimen: a) in closed crack configuration; b) in open crack configuration.

One of the first dedicated approaches is the so-called Elasto-dynamic Finite Integration Technique (EFIT) [3.36], [3.37], [3.38], not strictly a tool, but a procedure for a 2D integration of the elasto-dynamic equations of equilibrium based onto subsequent integration on finite volumes of materials. Such procedure is still under development and has influenced all the tools come later [3.39]. Figure 3.38 shows the simulation of a simple inspection both in terms of the behaviour of the sound beam (figure 3.38a) and of the obtained A-Scan (figure 3.38b). Modelling proper probes and materials, it is also possible to simulate all the special UT techniques described in the present Chapter (PA, TOFD). As an example, the A-Scans and B-Scans shown in figure 3.28, figure 3.35 and figure 3.36 were obtained by numerical simulations carried out by means of one of the most powerful commercial software.

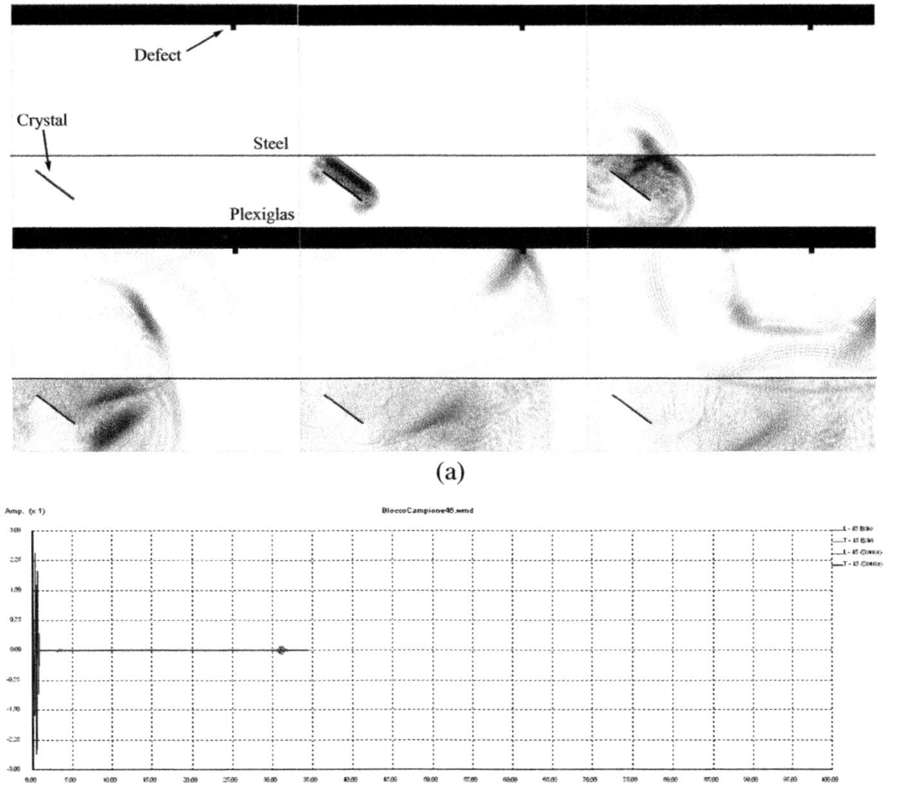

(a)

(b)

Fig. 3.38. Simulation of a UT inspection by means of an EFIT tool: a) beam development and behaviour; b) obtained A-Scan.

3.2.4 Reliability of NDT Inspections

In order to ensure structural integrity, inspections must yield accurate and reliable results. From this point of view, many factors influence such results and the consequent correct accept/reject decision [3.11]:

- choice of the best NDT method for the given inspection (see Section 3.2.1);
- correct application of the chosen technique (standards requirements, if available, should always be fully satisfied);
- environmental effects (accessibility, temperature, working conditions, etc.);
- variability in materials, crack shape, performance of equipments, component dimensions, etc.;
- human factor.

Due to such factors and considering that NDE systems and operators are usually driven to their limits in order to reveal the smallest possible crack, repeated inspections by the same or different operators, equipments, etc. can lead to inconsistent results. Consequently, inspection outcomes are not achievable in deterministic terms, but must be considered in terms of Probability of Detection (POD). The most important concept related to this statistical aspect is that, when evaluating NDE capabilities (reliability), the critical point is *not* to determine the smallest defect detectable by a method, but *the largest defect that can be missed and grow to failure during the desired life of the component* [3.11].

Repeated inspections on cracked and uncracked components taken from the same family can then be used to define the following useful quantities (figure 3.39):

- probability of detection (POD), i.e. the sensitivity of detection, when cracks are present and detected, so it represents the "correct rejection" of cracked specimens;
- probability of recognition (POR) when cracks are absent and, consequently, not detected, so it represents the "correct acceptance" of uncracked specimens;
- false negative response (FN) when cracks are present but not detected, so it represents the "incorrect acceptance" of cracked specimens and introduces Type II errors to which safety problems are related;
- false-call probability (FCP) when cracks are absent but detected, so it represents the "false rejection" of uncracked specimens and introduces Type I errors to which economic problems are related.

It is worth noting that increasing the sensitivity of the measuring system could lead to a significant increase of FCP and to the rejection of all the examined specimens so losing the meaning of inspections, while, ideally, a high POD is

		Defect condition	
		Present	Absent
Inspection result	Positive	POD "Correct rejection" True positive	FCP "False rejection" False positive Type I error Economic consequence
	Negative	FN "Incorrect acceptance" False negative Type II error Safety consequence	POR "Correct acceptance" True negative

Fig. 3.39. Possible scenarios for inspection results.

looked at together with a low FCP. The system efficiency can then be represented, relating all the probability quantities just described (figure 3.40), by the so-called "relative operating characteristic" (ROC) which plots POD vs. FCP [3.11], [3.15].

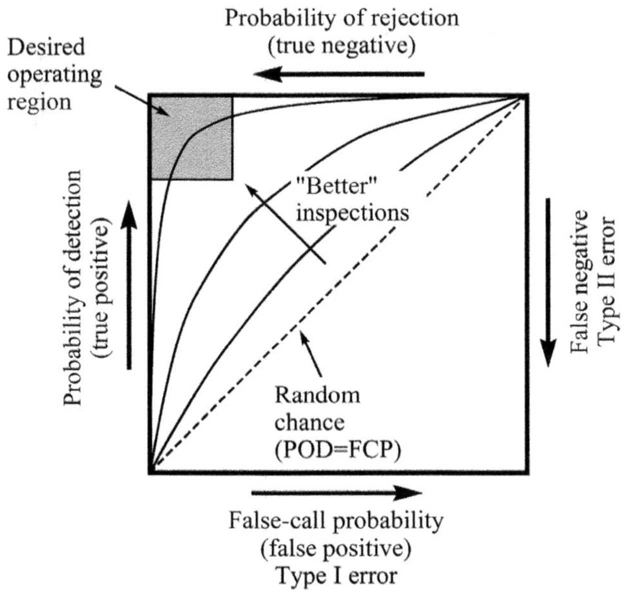

Fig. 3.40. ROC curves in terms of POD vs. FCP [3.11].

An alternative, and more common, way to quantify NDE performance is to consider POD vs. crack size "*a*" (length, depth or, recently, reflecting area [3.28]). This approach has some limitations since POD(*a*) curves are actually not dependent only on crack size [3.40], but also on other physical and operational parameters (material, geometry, operator, equipment,...). A complete description about how to set-up proper experiments and the consequent statistical elaborations in order to derive the POD curves is reported in [3.15] and [3.40].

Two different approaches are used in POD(*a*) curve determination: i) "hit/miss" data; ii) "signal response" data. The former is typically applied on data obtained by NDT methods such as MPI or PT where the inspector usually records only whether or not a crack is present without providing linear measurements of the detected cracks. By applying complex mathematical methods is then possible to derive the POD(*a*) curves.

In the case of the signal response approach (typically applied for UT or ET inspections), the operator records crack size information in terms of a response signal whose intensity is related, by means of a proper calibration, to the physical dimension of the inspected flaw. The signal response approach has several advantages in respect to the hit/miss one (fewer tests are needed, data can be processed to reflect various signal thresholds – "decision thresholds" – as well as saturation limits, it is possible to estimate results for other crack types difficult to introduce in sample blocks) and for this reason is more used and wide spread. Conceptually, POD(*a*) can be obtained from the relation between the signal response "*A*" and crack size (figure 3.41): if $g_a(A)$ is the probability density of the *A* values for a given fixed a value, then:

$$\mathrm{POD}(a) = \int_{A_{dec}}^{\infty} g_a(A)dA \qquad (3.4)$$

where A_{dec} is the decision threshold. Figure 3.41a illustrates this calculation, where the shaded area under the density functions represents the POD(*a*).

The ideal POD(*a*) curve obtained from calculations is shown schematically in figure 3.41b. All flaws larger than a_{NDE} would be detected all of the time (POD(*a*)=1), while all flaws smaller than a_{NDE} would not be detected at all (POD(*a*)=0). As already pointed out, no ideal and deterministic NDE system exists, and the POD(*a*) curve is represented by a continuous curve.

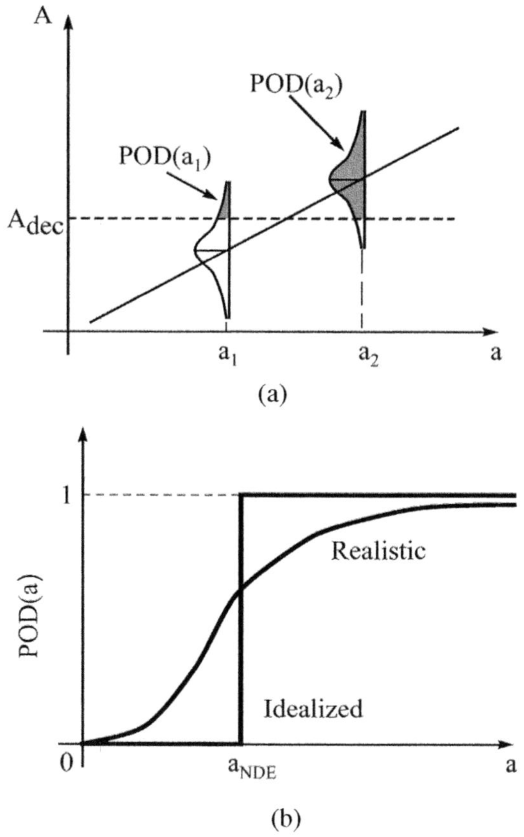

Fig. 3.41. POD(a) curves [3.15]: a) schematic of POD(a) calculation from A-a data; b) schematic of POD(a) curves.

An example [3.41] of real POD(a) curves for solid shafts made in carbon steel is shown in figure 3.42. Here, inspections were carried out by means of MPI, UT from the side close to the crack ("near end scan", NES) and UT from the side far from the crack ("far end scan", FES). A general observation coming from this figure is that the acceptability level of defects is strictly related to the performance of the adopted NDT method: as an example, the smallest crack size detectable by FES is more or less 2 mm, while by MPI is much less than 1 mm. Moreover, POD(a) in the case of MPI arrives at 100% quite quickly (at a crack size about 4-5 mm), while FES never arrives to this POD value. On the other hand the application of MPI is significantly more time consuming in respect of UT since the complete disassembling of the part must be carried out.

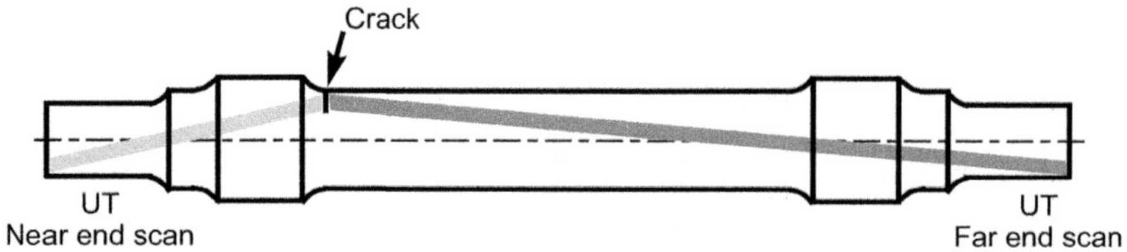

(a)

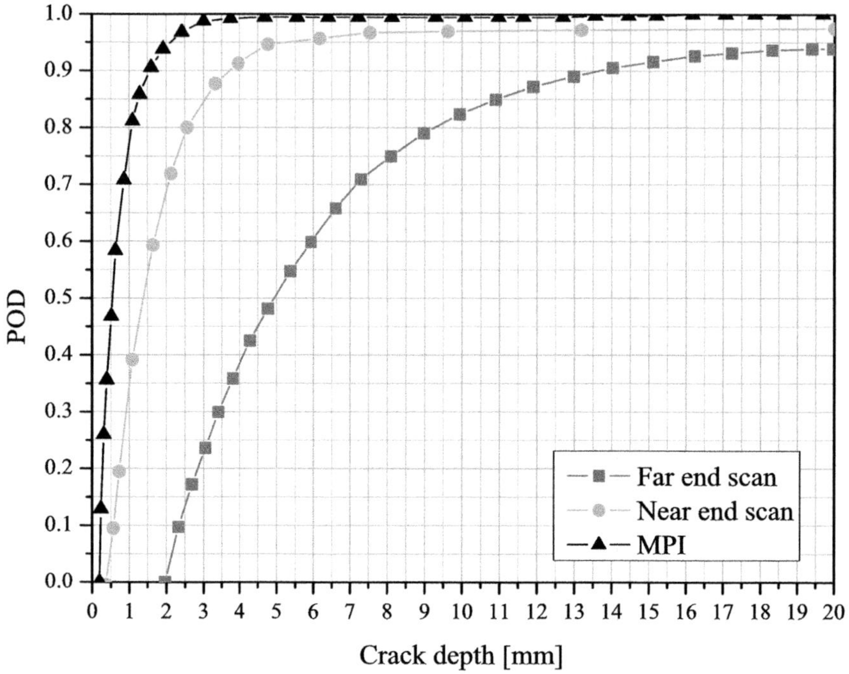

(b)

Fig. 3.42. Real POD(a) curves for solid shafts made in carbon steel.

Another very important topic, coming from the statistical nature of NDE results, concerns the confidence levels of POD(a) curves. Figure 3.43 shows schematically the mean POD(a) curve (50% confidence) together with the 95% confidence bound. An NDE crack size a_{NDE} may then be defined as the crack size having a given probability of detection with a given confidence level (in this example called the "90/95 crack size" since the probability of detection is 90% and the confidence bound is 95%). Alternative probability/confidence crack sizes can be defined for different applications. Keeping into account for confidence levels guarantees safety increasing the economic factor since at the same POD corresponds a larger defect size.

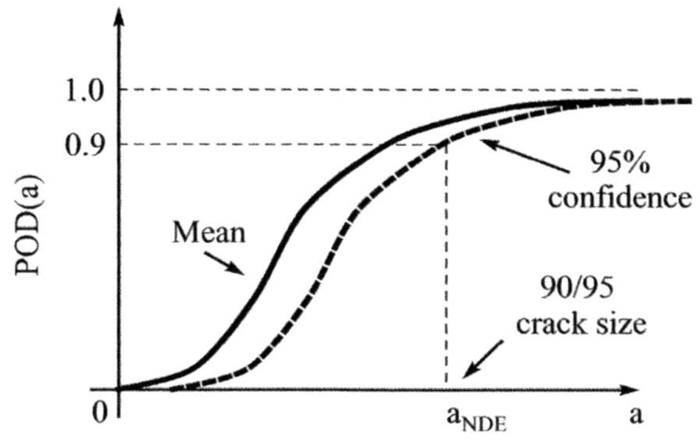

Fig. 3.43. Confidence levels for POD(a) curves [3.11].

3.2.4.1 Sensitivity to Human and Environmental Factors

Human and environmental factors are the largest variables in the effectiveness of NDE systems. This is the reason why a considerable effort is carried out in order to automate as much of the inspection procedure as possible.

Human factors (HF) include the inspector's proficiency (training, certification and continuous education), response to the work environment (outdoor inspections in inclement weather or temperatures, hot and humid buildings, etc.), mental attitude (concentration, boredom, alertness and focus, work breaks, rest, freedom of drugs, good health) and experience (prior knowledge), while environmental effects (EE) include surface conditions, liquid or solid couplants, geometries, temperatures, etc. It is worth noting that the analysis of the influence of human errors during technical operations is the main topic of the "Human Reliability Analysis" [3.42] [3.43], a research field dedicated to this problem.

The POD(a) curves which have been considered until now represent a "theoretical" or "physically achievable" asymptotic limit due to the intrinsic probability of failure of the given NDT methodology in detecting cracks or defects. A first simple approximation [3.44] suggests to assume HF and EE equal to a constant POD reduction factor H. This leads to:

$$\mathrm{POD}(a)_{total} = H \cdot \mathrm{POD}(a)_{NDT} \tag{3.5}$$

where H is the human and environmental factor (suggested to be equal to 0.95 [3.44]), while $\mathrm{POD}(a)_{NDT}$ is the intrinsic probability of detection of the given NDT methodology.

A more general equation for reducing the POD_{NDT} has been suggested again in [3.44]:

$$\mathrm{POD}(a)_{total} = \mathrm{POD}(a)_{NDT} - g(\mathrm{EE}) - h(\mathrm{HF}) \tag{3.6}$$

where g(EE) and h(HF) are opportune functions which take into account for the effects of EE and HF, respectively. Introducing the general functions g(EE) and *f*(HF) means that the trend of the human and environmental effects can be non-linear with defect dimensions (for example, human errors can be larger for small defects [3.45]). No data are available in the literature about the quantification of *g*(EE) and *f*(HF), but there is some ongoing research taking place in this field.

3.2.5 NDT as a Fundamental Aspect of Damage Tolerance Design

Some authors define [3.11] Damage Tolerance (DT) as "the ability of a structure or component to resist fracture from pre-existent cracks for a given period of time (i.e. the design life)", while others [3.46] as "the most opportune inspection interval given the POD of the adopted NDT technique or, alternatively, the needed NDT specifications given a programmed inspection interval". A part from definitions, DT and NDT are the key points in assuring the safe service of structures on long periods: starting from the hypothesis that pre-existent manufacturing or service-induced damage are always present, NDT represent the first step to structural integrity by identifying damaged components to be repaired or discarded; since, as already described, all NDT techniques have some limitations and, consequently, some components can be accepted even if containing cracks, DT design provides the second line of defense against premature failures.

The DT design approach is based onto the concept of "inspection interval" defined as the distance that can be safely run between two NDT inspections [3.47]. A scientific approach, for the determination of the length of the inspection interval, has been proposed only in recent years, but it is spreading very slowly among the industry: many quality managers are still fixed onto traditional and empirical indications. For example, the axles, characterized by a design life before retirement equal to about 1000000 km, mounted onto an Italian high speed train are typically inspected every 200000 km because experience showed that very few failures are thus guaranteed. Obviously, an empirical approach rarely is also an optimized one. From this point of view, a more scientific approach is shown in figure 3.44, where the inspection is fixed at half of the useful propagation life for simplicity. In particular, the inspection interval is a function of:

- the proper knowledge of service loads (including possible residual stresses);
- the crack propagation curve of the material (da/dN-ΔK);
- the minimum detectable crack (defect) size: proper levels of POD and confidence (for example, the 90/95 crack size for the adopted NDT method) should be chosen in order to assure safety for a given component;
- the suitable characterization of the stress intensity factor (SIF) range for complex geometries;
- the proper analytical or semi-analytical crack propagation algorithms in order to predict fatigue crack growth life of the component;
- the critical maximum SIF before rupture (K_{IC}).

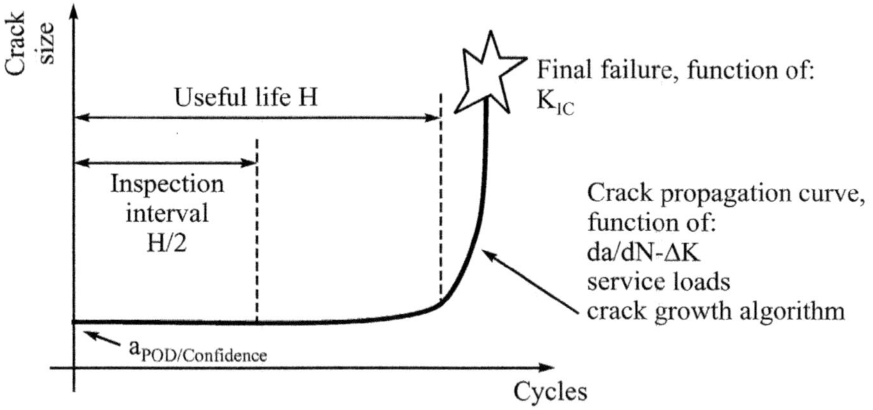

Fig. 3.44. Definition of inspection interval.

Acting properly on all of these parameters yields the reduction of the "total life cycle cost" of rotors, since an optimization of the length of inspection intervals is obtained. The most interesting parameter in the present discussion is, obviously, improving the NDT crack detection threshold in order to reduce the $a_{POD/Confidence}$ value, but also other actions (such as the development of materials with higher toughness and slower crack growth and creep crack growth, of optimized crack growth algorithms under variable loads, etc.) should be kept in mind [3.48][3.49][3.50]. It is interesting to add that repeated inspections and repair actions could, theoretically, keep the structure in service indefinitely [3.11]. Moreover, fatigue life is very sensitive to initial crack size: from this point of view, it worth noting that the choice of a few farer, detailed and difficult (i.e. able to detect small cracks) but very lengthy and time consuming inspections is not necessarily more reliable in respect to simpler NDE methodologies repeated often.

An inspection interval, leading to a trade-off between costs and safety, must then be always looked at, keeping in mind that the number of inspections carried out during the component life directly influences the probability of failure of the component itself. From this point of view, given the length of the inspection interval, a cumulative probability of detection of a defect can be calculated [3.51] based on a given number of inspections "i" derived from the POD(a) curve of the adopted NDT technique. Figure 3.45, in which three inspections are assumed, shows this approach and the concepts of "forward" and "backward" inspections. The former represent the "natural" approach and assumes to apply inspection intervals from detection to failure, while the latter assumes their application from failure going back to detection: such approach is sometimes applied during trials after serious accidents to define responsibilities. It is necessary to add that the latter is systematically not conservative in respect to the former and should always be considered also during design calculations. The cumulative POD of a defect can then be derived as:

$$\mathrm{PC}_{DET} = 1 - \left[\prod_i \mathrm{POND}(a)_i \right] = 1 - \left\{ \prod_i [1 - \mathrm{POD}(a)]_i \right\} \tag{3.7}$$

where PC_{DET} is the cumulative probability of detection and $\mathrm{POND}(a)_i$ ("Probability of Non Detection") represents the probability to fail the i-th detection. The probability of failure P_f of the component is then related to PC_{DET} by the following expression:

$$P_f = 1 - \mathrm{PC}_{DET} = \prod_i \mathrm{POND}(a)_i = \prod_i [1 - \mathrm{POD}(a)]_i \,. \tag{3.8}$$

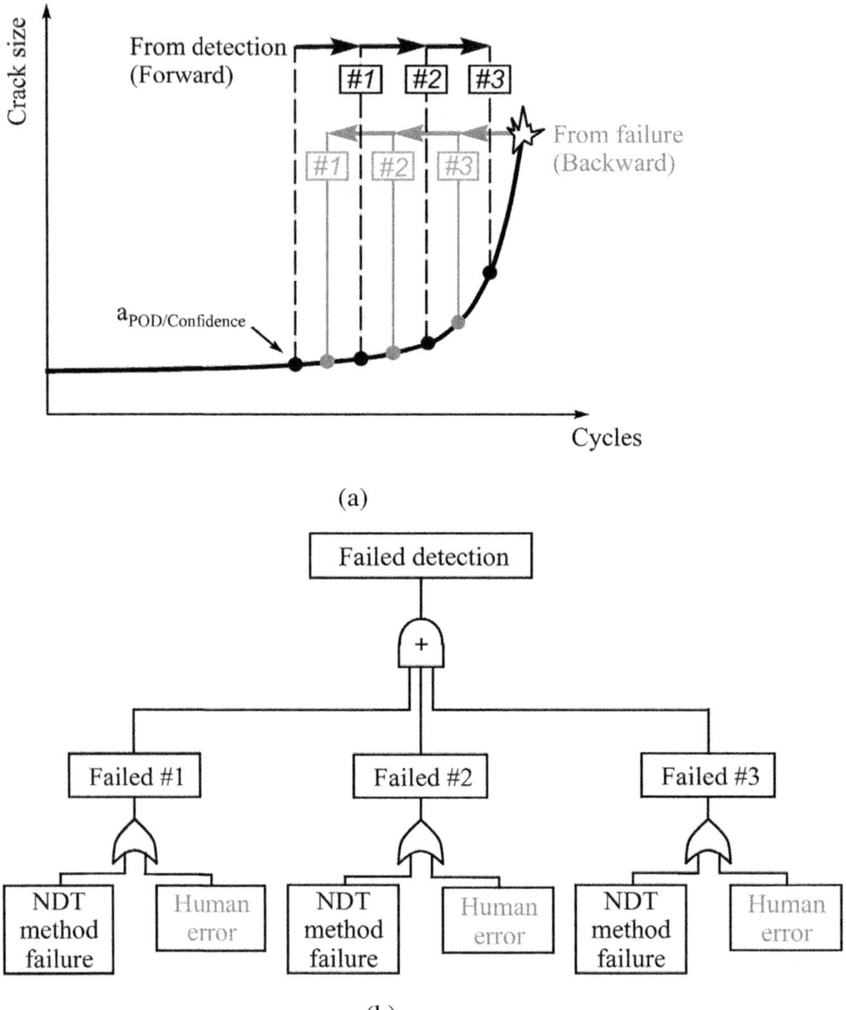

Fig. 3.45. Cumulative POD: a) definition of "forward" and "backward" inspections; b) fault tree.

The drawback of this approach is that the repetition of inspections on small cracks having very low POD (i.e. in the initial phases of propagation life) increases the PC_{DET} value, but introduces further influences (sizing errors, operator errors) typical of this stage focusing again the attention on the best trade-off between safety and costs. For this reason a different approach has been recently proposed: given an inspection interval, only the probability of detection P_{DET} of a defect corresponding to the last useful inspection before failure is considered (the so-called "last chance", figure 3.46). It results:

$$P_{DET} = \mathrm{POD}(a)_{\max(i)} \tag{3.9}$$

and

$$P_f = 1 - P_{DET} = 1 - \mathrm{POD}(a)_{\max(i)} \, . \tag{3.10}$$

The advantage of this approach (originally proposed for railway axles in [3.46]) consists in a simple relationship between the probability of detection and the inspection interval. Moreover, this methodology makes it possible to focus the attention on the dimension of the crack that must be observed in order to have a high probability of detection.

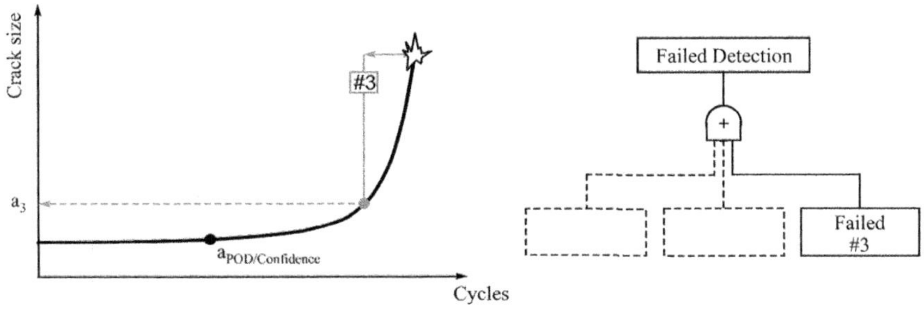

Fig. 3.46. "Last chance" inspection.

It is then finally clear and intuitive that the choice of the length of the inspection interval based onto the just presented methodologies is strictly related to the P_f level that design and maintenance personnel wants to undertake for the considered component: no general values can be proposed or provided, but the most possible exact and complete calculations and experiments should be carried out in order to define all the involved parameters.

References

[3.1] Auerkari P (1997) NDT for high-temperature installations – a review. Welding in The World, 39(6), 351-362.

[3.2] Crespi GM (2008) The determination of spent life and the prediction of residual life in the diagnostic of a steam turbine rotor. Workshop "Useful Life and Residual Life in transportation means", Milan (Italy). [In Italian]

[3.3] Carboni M, Reboa F, Sangirardi M (2008) NDT for rotating shafts. J. Italian NDT Association, 2, 18-26. [In Italian]

[3.4] ASTM E1316 (1995) Standard terminology for nondestructive examinations.

[3.5] Stubbs T (2004) The role of NDE in the life management of steam turbine rotors. Insight, 46(9), 529-532.

[3.6] McMaster RC (1963) Nondestructive testing handbook. The Ronald Press Company.

[3.7] ASNT (2007) Nondestructive testing handbook, 3rd Ed.

[3.8] Lyon R (2004) The development of condition monitoring tools for the power generation industry. Insight, 46(9), 516-519.

[3.9] Pignone E (2004) Enhancement in image quality in ultrasonic flaw detection process in rotor turbine using SAFT. International Workshop on Imaging Systems and Technique, Stresa, Italy.

[3.10] Dumas P, Clossen M, Opheys M, Rauschenbach H (2006) Nondestructive inspection of blade roots and blade attachment grooves – Ideal application of phased array inspection technique. Proc. 9th European Conference on NDT, Berlin, Germany.

[3.11] Grandt AF (2003) Fundamentals of structural integrity: damage tolerant design and non-destructive testing, Wiley.

[3.12] ISO/TS 18173 (2005) Non-destructive testing - General terms and definitions.

[3.13] ASTM (2008) Metal test methods and analytical procedures, ASTM Standards Section 3, Vol. 3.03.

[3.14] ASME (2007) Boiler and pressure vessel code, Sections V and VIII.

[3.15] ASM (1997) ASM handbook – Vol. 17: Non-destructive evaluation and quality control.

[3.16] ISO 9712 (2005) Non-destructive testing – Qualification and certification of personnel.

[3.17] ASNT SNT-TC-1A (2006) Recommended Practice, Personnel Qualification and Certification in Nondestructive Testing.

[3.18] ATA-105 (1997) Guidelines for Training and Qualifying Personnel in Nondestructive Testing Methods.

[3.19] AIA-NAS-410 (2003) National Aerospace Standard, NAS Certification and Qualification of Nondestructive Test Personnel.

[3.20] ISO 3452 (1984) Non-destructive testing - Penetrant inspection - General principles.

[3.21] ASTM E165-02 (2002) Standard Test Method for Liquid Penetrant Examination.

[3.22] ISO 9934-1 (2001) Non-destructive testing - Magnetic particle testing - Part 1: General principles.

[3.23] ASTM E709-01 (2001) Standard Guide for Magnetic Particle Examination.

[3.24] ISO 3058 (1998) Non-destructive testing - Aids to visual inspection - Selection of low-power magnifiers.

[3.25] ISO 5577 (2000), Non-destructive testing - Ultrasonic inspection – Vocabulary.

[3.26] ASTM E587-00 (2005) Standard Practice for Ultrasonic Angle-Beam Examination by the Contact Method.

[3.27] Krautkrämer J, Krautkrämer H (1990) Ultrasonic testing of materials, 4th Ed., Springer-Verlag.

[3.28] Carboni M (2008) An analysis of UT echoes coming from fatigue cracks and artificial defects on railway axles. 17th World Congress on NDT, Shangai, China.

[3.29] Carboni M, Rudlin J (2008) Analysis and determination of UT POD curves for railway axles. 3rd International Conference "Reliability, Safety and Diagnostics of transport structures and means 2008", Pardubice, Czech Republic.

[3.30] R/D Tech (2005) Phased array technical guidelines – Useful formulas, graphs and examples, R/D Tech Corp.

[3.31] Yang S, Yoon B, Kim Y (2009) Using phased array ultrasonic technique for the inspection of straddle mount-type low-pressure turbine disc. NDT&E Int., 42, 128-132.

[3.32] Boehm R, Spruch W (2006) Phased Array Rotation Scanner Probe System for Ultrasonic Testing of Sleeve Shafts. Proc. 9th European Conference on NDT, Berlin, Germany.

[3.33] Charlesworth JP, Temple JAG (2001) Engineering applications of ultrasonic time-of-flight diffraction. 2nd Ed., Research Studies Press Ltd.

[3.34] CETIM (2000) *Prefissuration en fatigue de 3 arbres pour essais vibratoires*, 4/054879/417.1A.

[3.35] Ginzel E (2007) NDT modelling: an overview. Proc. Conference on Modelling in Non-Destructive Testing, Pretoria, South Africa,.

[3.36] Marklein R (1997) Numerical Methods for the Modeling of Acoustic, Electromagnetic, Elastic and Piezoelectric Wave Propagation Problems in the Time Domain Based on the Finite Integration Technique, Aachen, Shaker Verlag. [in German]

[3.37] Marklein R, Mayer K, Ampha P, Langenberg KJ (2003) Computer Simulation of Ultrasonic Non-Destructive Testing of Concrete Using the Elastodynamic Finite Integration Technique (EFIT). Proc. International Symposium "Non-Destructive Testing in Civil Engineering", Berlin, German.

[3.38] Marklein R, Langenberg KJ, Mayer K (2003) EFIT Simulations for Ultrasonic NDE. NDT.net 8(3).

[3.39] Carboni M, Sangirardi M (2007) UT simulations for crack sizing. Proc. Biennal Conference of the Italian Ass. for NDT, Milano, Italy. [In Italian]

[3.40] Georgiou GA (2006) Probability of Detection (POD) curves: derivation, applications and limitations. Research Report 454, HSE Books, Health and Safety, Executive, UK.

[3.41] Benyon JA, Watson AS (2001) The use of Monte-Carlo analysis to increase axle inspection interval. Proceedings of the 13th Int. Wheelset Congress, Roma, Italy.

[3.42] Pyy P (2000) Human reliability analysis methods for probabilistic safety assessment. VTT Publications 422, VTT Technical Research Centre of Finland.

[3.43] NEA/CSNI/R(98)1 (1998) Critical operator actions: human reliability modelling and data issues. Nuclear Energy Agency OECD 62060, Paris.

[3.44] Wall M, Wedgwood FA, Burch S (1998) Modelling of NDT Reliability (POD) and applying corrections for human factors. Proceedings of the 7th European Conference on NDT, Copenhagen, Denmark.

[3.45] Murgatroyd RA, Seed H, Worrall G, Crutzen S (1994) Human Reliability in Inspection. Results of Action 7 of PISC III Programme.

[3.46] Zerbst U, Vormwald M, Andersch C, Mädler K, Pfuff M (2003) The development of a damage tolerance concept for railway components and its demonstration for a railway axle. Eng. Fract. Mech., 72, 209-239.

[3.47] Cantini S, Ghidini A, Beretta S, Carboni M (2004) "Safe life" inspection intervals of railway axles: comparison of crack growth properties of different steel grades. Proc of the 14th Int. Wheelset Congress, Orlando, Florida, USA.

[3.48] Beretta S, Carboni M (2005) Simulation of fatigue crack propagation in railway axles. J. ASTM Int., 2, 1-15.

[3.49] Beretta S, Carboni M, Cantini S, Ghidini A (2004) Application of fatigue crack growth algorithms to railway axles and comparison of two steel grades. J. Rail and Rapid Transit, 218, 317-326,.

[3.50] Beretta S, Carboni M (2006) Experiments and stochastic model for propagation lifetime of railway axles. Eng. Fract. Mech., 73, 2627-2641.

[3.51] Carboni M, Beretta S (2007) The effect of "Probability of Detection" upon the definition of inspection intervals for railway axles. J. Rail and Rapid Transit, 221, 409-417.

Chapter 4
Laboratory Tests on Cracked Shafts

4.1 Introduction

The results of two different laboratory test campaigns on cracked shafts are described because they illustrate two peculiar characteristics of cracked shafts: the *breathing mechanism* and the *thermal sensitivity*.

The description of the dynamic behaviour of horizontal heavy rotors presenting a transverse crack has been presented by many researchers (see e.g. the overviews in [4.1][4.2] and [4.3]), particularly as regards the breathing mechanism. This characteristic behaviour takes place during the rotor rotation: the crack moves from the upper position in which the static bending moment forces the crack to be "closed" to the opposite position in which the crack is forced to be "open". The gradually opening and closing of the crack is called the *breathing mechanism*. Therefore the stiffness changes periodically during one rotation, and correspondingly also the static deflection due to the weight (and to the bearing alignment conditions). In fact the stiffness of the rotor with the open crack is smaller with respect to the closed crack situation, in which the stiffness equals the value of the un-cracked shaft. The periodical deflection - and actually the periodical stiffness too - can both be expanded in a Fourier series as seen in section 3.1.2. It can be shown that the same forces which excite the static deflection components, excite also the vibrations, when the shaft is rotating at higher speeds. Therefore the study of the breathing mechanism is of great importance to understand the cracked shaft dynamic behaviour.

Besides the effect of the static bending moments, also thermal transients have an influence on the breathing mechanism. Often, in real machines, the sudden change in vibrational behaviour during thermal transients allowed crack in rotating shafts to be discovered, as shown in section 2.3. The experimental effects of the thermal sensitivity are described in detail by means of laboratory tests in section 4.3.

4.2 Laboratory Test about Breathing Mechanism on a Specimen

A cracked specimen prepared by CETIM for Électricité de France - EDF has been submitted to some static test in the laboratory of the Dept. of Mechanical Engineering of Politecnico di Milano. Due to the interest in defining the true breathing behaviour of cracked shafts it was decided to measure experimentally the strains in different points of the cracked specimen and in different load conditions (see figure 4.1). To this aim, a series of strain-gauges have been applied close to the crack and also directly across the crack lips (see figure 4.2).

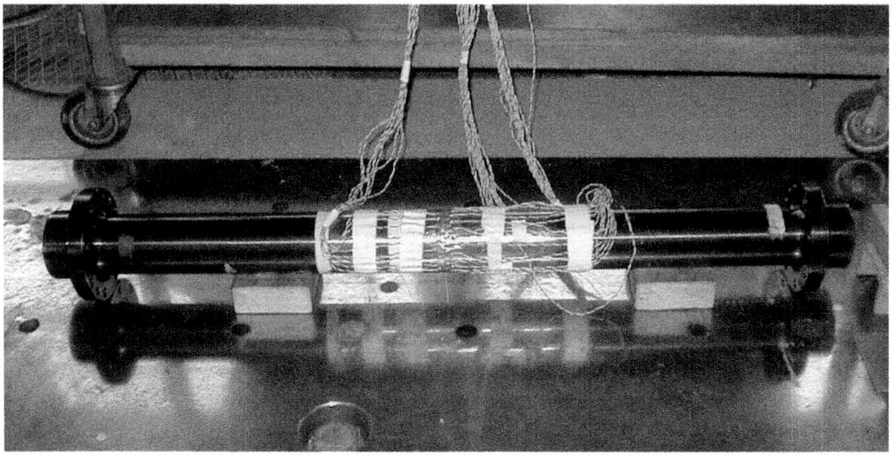

Fig. 4.1. Instrumented specimen.

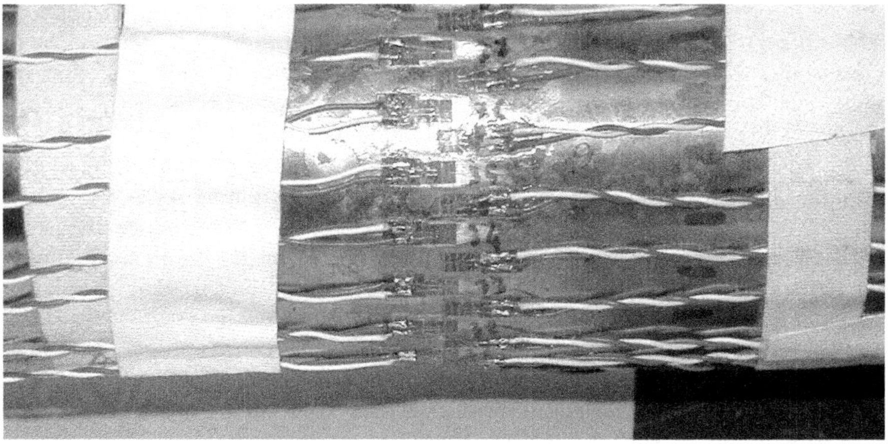

Fig. 4.2. Detail of strain-gauge positions close and across the crack.

The horizontal cracked specimen has been loaded with different stationary loads and has been rotated in different angular positions in order to excite the breathing of the crack (see figure 4.3).

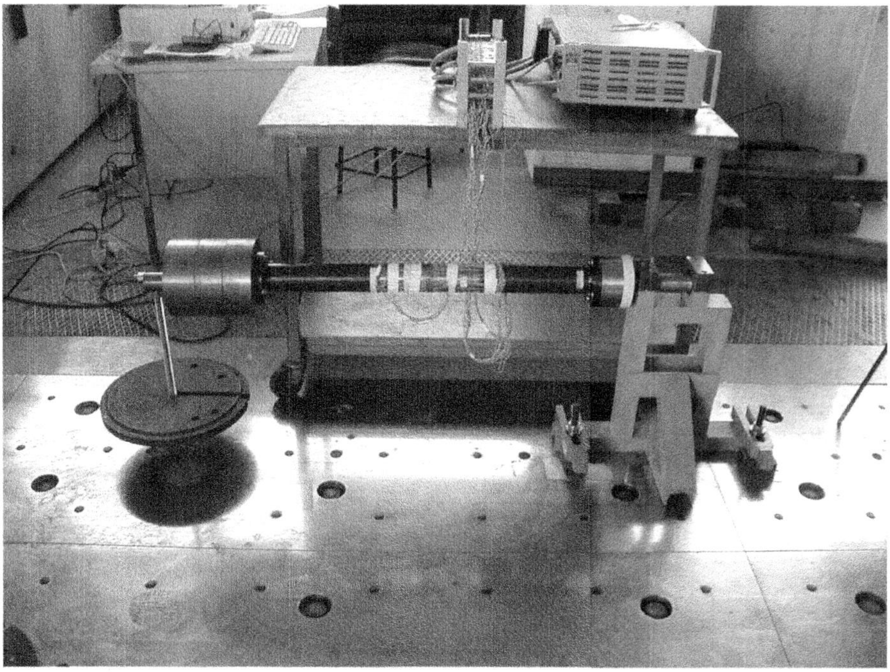

Fig. 4.3. Cracked shaft clamped at the stand.

Two different non trivial effects have been observed: the *crack closure effect* and the *local contact conditions of the crack lips* in closed crack configuration, which are here described.

a) Crack closure effect: Small loads, generating small bending moments in correspondence of the crack, were not able to open the crack as it resulted in all the different measuring points: the crack closure effect generates an internal bending moment that holds the crack closed. Only when the external bending moment overcomes the internal bending moment, then the lips of the crack start to open.

b) Local contact conditions of the crack lips: when the crack is closed, with an external bending moment that sums up to the internal bending moment, then the measured compressive strain is much higher than the theoretical strain calculated assuming a linear compressive stress distribution over the cracked section. This can be explained by assuming that the contact is not spread over all the cracked area when the crack is closed, but it occurs only on a smaller portion of the cracked surface, or on the crack lips only, determining higher strains associated also to stress intensity factors. This aspect is also related to the crack closure effect.

Despite the fact that crack closure effects have been studied by several researchers (see for instance [4.4]), their influence on the breathing behaviour of rotating shafts has never been modelled suitably to the authors' knowledge.

In the following of this section, the experimental results are presented and are compared to calculated results using the FLEX model described in section 5.4.3. More details are given in [4.5].

Despite the highly non-linear stress and strain distribution in the cracked area and the non-linear breathing behaviour, the overall load-strain behaviour results to be quite linear: the overall load vs. deflection law can therefore easily be represented by a linear model.

4.2.1 Description of the Experimental Apparatus

The dimensions of the cracked shaft are shown in figure 4.4. Specimen diameter was larger when the crack has been initiated by means of a small slot generated by electro-erosion and has been propagated roughly up to half way the shaft cross section by applying a constant bending load to the rotating shaft. The specimen was then machined and the diameter was reduced to the final one by turning, so that the initial slot was removed. The position of the crack is indicated as section C in figure 4.4.

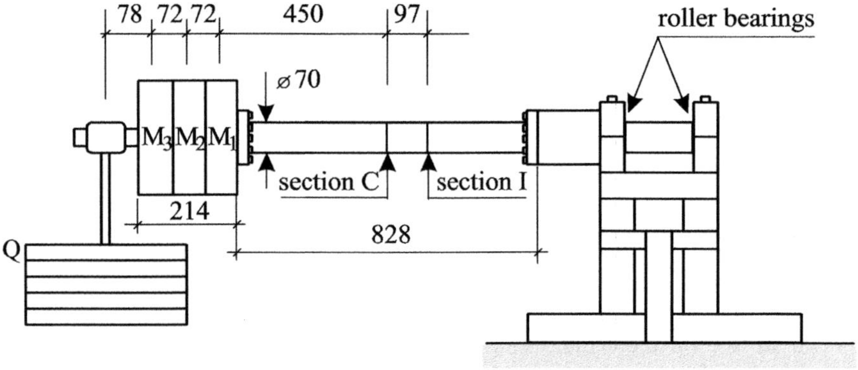

Fig. 4.4. Sketch of the test-rig (measures in mm).

The final cracked section has the shape shown in figure 4.5, as it resulted from ultrasonic test measurements. The cracked shaft has been clamped at one end but can be rotated around his axis by steps of 15° each. A vertical load has been applied at the other end and has been increased by steps. The loads, the resulting bending moments in correspondence of the crack and the theoretical stresses and strains are listed in table 4.1. Theoretical stresses and strains are calculated assuming no crack.

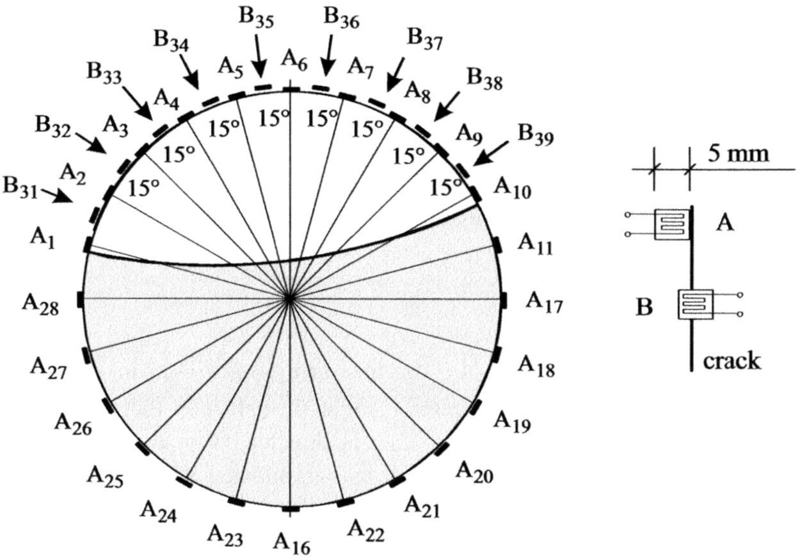

Fig. 4.5. Shape of the crack in section C (see figure 4.4) and strain-gauge positions.

Table 4.1. Loads, bending moments, theoretical stresses and strains in correspondence of the crack.

Load	Bending Moment in section C [Nm]	Max stress in section C [MPa]	Max strain in section C [με]
Load S	37	1.11	5.29
Load A	105	3.11	14.80
Load B	182	5.40	25.72
Load C	271	8.04	38.28
Load D	426	12.64	60.18
Load E	558	16.56	78.88
Load F	690	20.49	97.58
Load G	822	24.42	116.28
Load H	954	28.35	134.98
Load I	1087	32.27	153.68
Load J	1219	36.20	172.38
Load K	1351	40.13	191.08

Strain-gauges from A_1 to A_{11} are applied each 15° as close as possible to the crack, in section C (cracked) of the specimen, see figure 4.4. Strain-gauges from A_{12} to A_{15} are applied to section I (integer), which is sufficiently far away from the crack to be not influenced by its breathing behaviour and their measurements are used as reference signals. Strain-gauges from A_{16} to A_{28} are applied each 15° in correspondence of the crack but on the integer part opposite to the crack. Strain-gauges from B_{31} to B_{39} have been applied across the lips, as it is shown in figure 4.5.

4.2.2 Experimental Results

Typical results are shown in figure 4.6 for measuring point A_6, in which the maximum effect of the crack is expected. Angle 0° indicates that crack axis (which passes close to measuring point A_6) is directed vertically downwards (crack open), angle 180° indicates that crack axis is directed vertically upwards (crack closed). A flat zone in strain level (about 140 $\mu\varepsilon$) is clearly recognisable between 0° and 60° and between 300° and 360°, for a total range of angular positions of 120°. In this range of angular positions where the strain does not change, provided that sufficient load is applied, the crack is supposed to be definitely open. The strain for the open crack is not zero but a tensile strain of 140 $\mu\varepsilon$ is measured on the open crack lips.

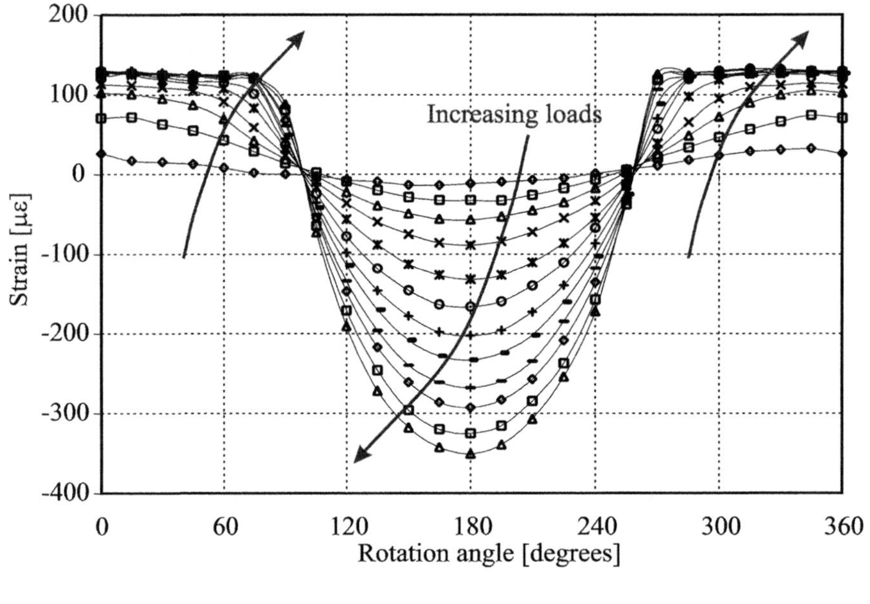

Fig. 4.6. Strain vs. rotation measured in correspondence of strain-gauge A_6.

This can be explained by the presence of an internal pre-stress on the crack lips due to the crack closure effect (see section 5.2.3.3), generating a compressive strain of 140 με, when the strain-gauges have been applied to the shaft with no external loads. Only the loads that generate a similar value of tensile stresses are able to open the crack. Similar results, but generally with higher values (up to 240 με) of compressive strains due to the crack closure effect, have been found in the other measuring points A_3, A_4, A_5 and A_7. The behaviour is different in positions that are closer to the crack ends, i.e. also closer to the crack tip, as in measuring point A_{10}.

When the crack is closed (at the angular position of 180°), the compressive stress of 350 με in correspondence of maximum load, is much higher with respect to the theoretical value of 190 με, calculated considering the load applied on a un-cracked specimen. This can be attributed to the fact that only a smaller part of the cracked surface is in contact, generating locally higher stresses and also stress intensity factors which would be absent if the crack faces were completely in contact each other over the complete crack area. This strain magnification is present for all the different loads, but its value is different for different measuring points and tends to unity (no strain magnification) close to the crack ends (in measuring points A_1 and A_{10}), as could be expected. The value of the stress concentration factors is represented in figure 4.7, along with the depth of the crack below the measuring points.

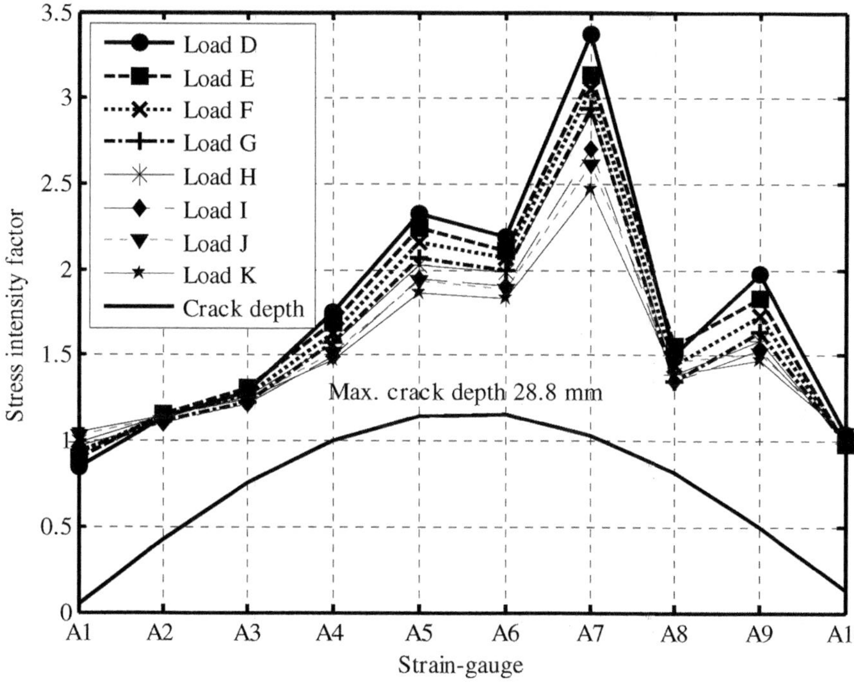

Fig. 4.7. Stress intensity factor versus number of strain-gauge (A_1 to A_{10}).

It is interesting to observe that these stress intensity factors are roughly independent from load.

In the measuring points close to the crack ends, when the loads generate tensile stresses, during the gradually opening of the crack, rather high stress intensity factors have been measured that are due to the closeness of the crack tip to the measuring points. In measuring point A_{16} (opposite to A_6), the maximum tensile strain is 240 $\mu\varepsilon$, which is somewhat higher than the theoretical value of 190 $\mu\varepsilon$: this might be caused by the fact that not all the cracked area is in contact with the opposite face and is therefore collaborating.

When the crack is completely open (angular position 0°) the compressive strain reaches 310 $\mu\varepsilon$ in measuring point A_{16} : this value cannot be compared with the theoretical value calculated assuming linear stress and strain distribution over the reduced section area, because the presence of the crack introduces a highly non-linear stress distribution, as described in section 5.3.1 and figure 5.23 where a 3D non-linear model of the cracked shaft is used.

Other interesting results are provided by the strain-gauges which were applied across the crack lips (B-series) and indicate the gradually opening of the crack (see figure 4.8).

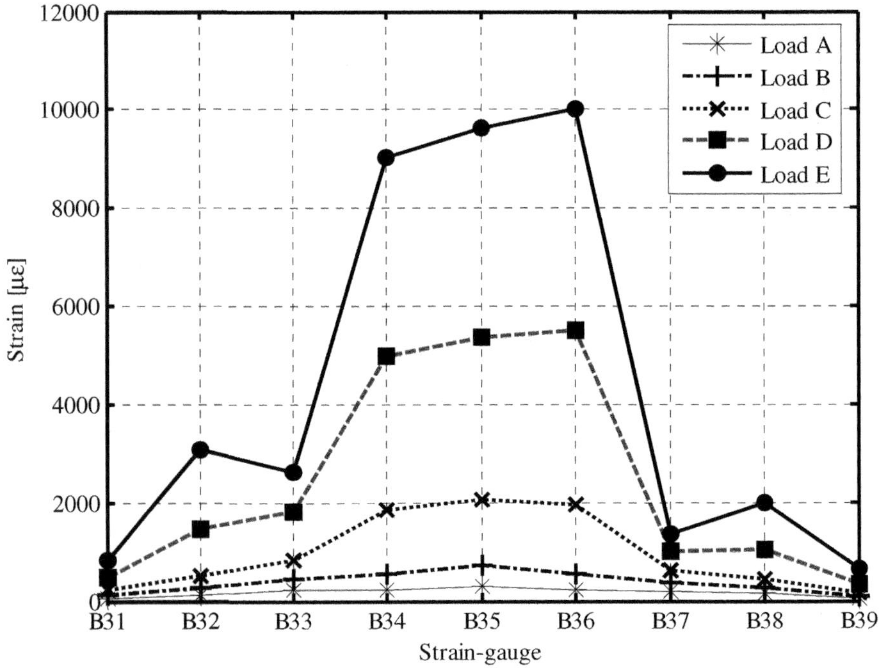

Fig. 4.8. The opening of crack lips as measured by the B-series gauges.

The measured behaviour has been also simulated, with the aid of a model of the crack. The FLEX model used for calculating the breathing behaviour and the reduced stiffness of the cracked shaft is described in the section 5.4.3. The crack closure effect has been simulated by an external bending moment that tends to hold the crack closed, generating a maximum of 140 $\mu\varepsilon$ of compressive strain. These strain values have been corrected with the stress intensity factors of figure 4.7. Now a comparison is possible between calculated results and experimental results, taking into account the crack closure effects and the stress intensity factors.

Figure 4.9 shows the comparison of calculated results obtained with the above specified model, with the experimental results in measuring point A_6 for the maximum load. It is surprising how good the simplified model is able to reproduce the experimental behaviour. This occurs in almost all measuring points. The quasi-linear breathing behaviour model can be considered completely validated with these experimental results. The results of the comparison in the diametrically opposite measuring point A_{16} are shown in figure 4.10.

The calculated maximum compressive strain is much higher than the measured one, because the model assumes unrealistic linear stress/strain distribution, as pointed out in section 5.3. A suitable reduction coefficient could be used to obtain a better fitting.

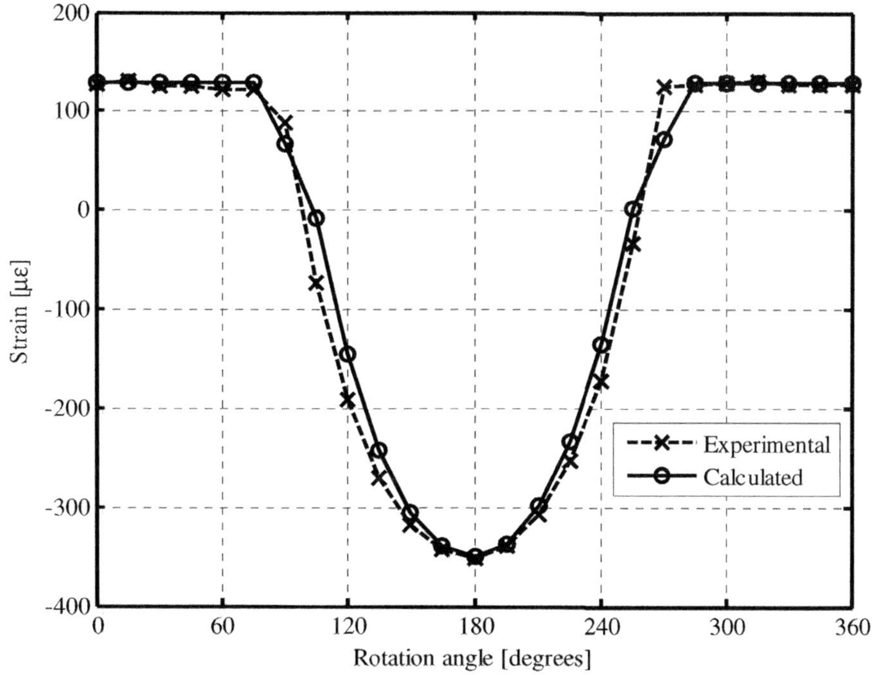

Fig. 4.9. Comparison of calculated and experimental result in point A_6 .

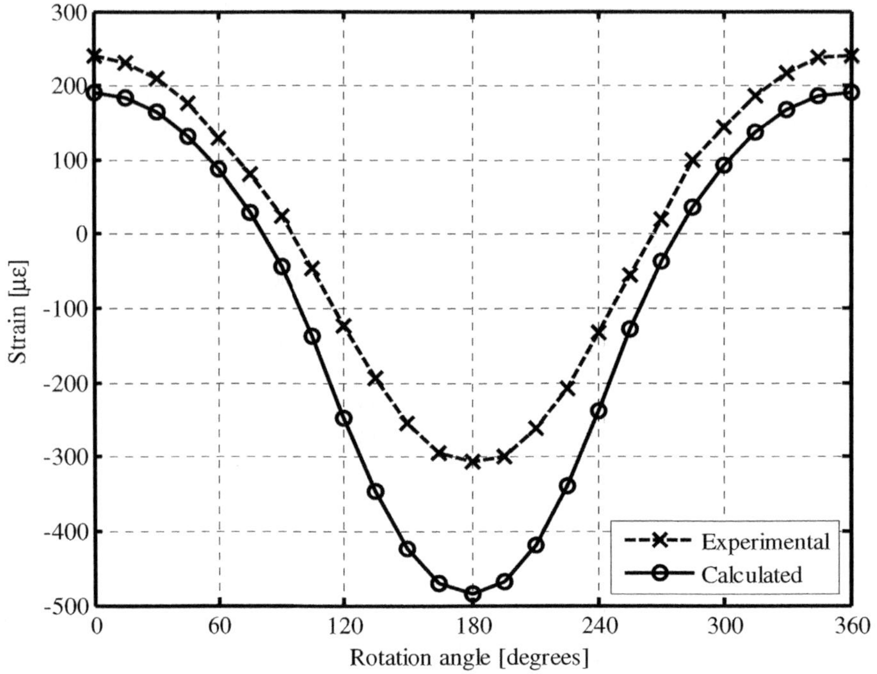

Fig. 4.10. Comparison of calculated and experimental results in point A_{16}.

4.3 Factory Test Results about Thermal Sensitivity on a Steam Turbine

The shaft of the cracked steam turbine considered in section 2.3.1 was disassembled from the turbo generator unit and was submitted to some tests in the factory. Temperature sensors have been applied in order to measure the internal temperature of the hollow rotor and that of the skin in the sections indicated in figure 4.11, while also some proximity probes have been used to measure vertical and horizontal displacements in the indicated measuring planes. Finally some strain-gauges have been applied close to the crack and in other positions as shown in figure 4.12.

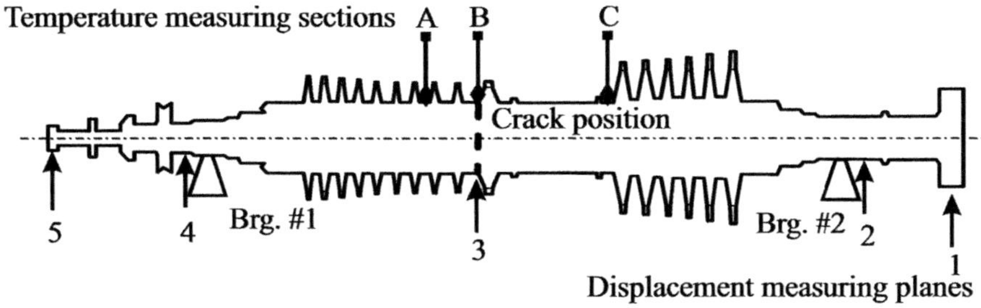

Fig. 4.11. Position of the temperature sensors and of the displacement measuring planes on the HP-IP steam turbine rotor in the factory.

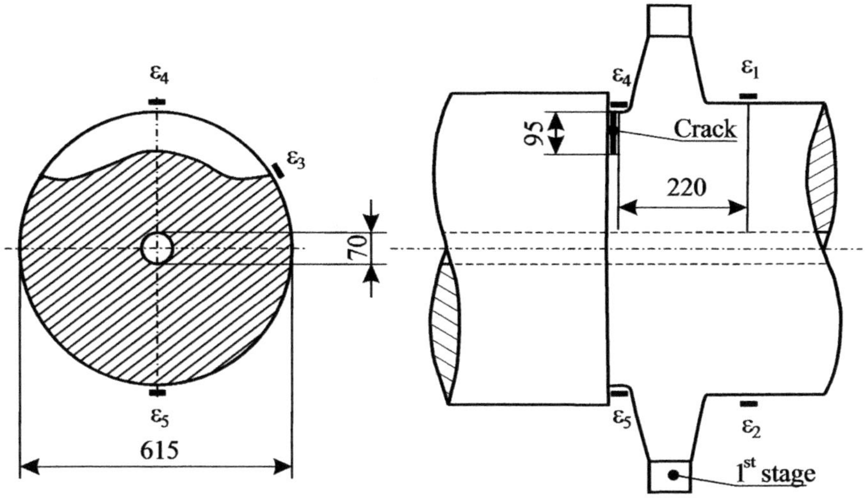

Fig. 4.12. Position of the strain-gauges in correspondence and close to the crack (measures in mm).

4.3.1 Tests at Low Rotating Speed

The shaft has been submitted to a heating transient, followed by a cooling transient, at low rotational speed. During the heating transient, a temperature gradient of the order of magnitude of 0.5°C/min was obtained. The calculated corresponding axial stresses arising in the shaft are given in figure 4.13 and figure 4.14. These stresses are high enough to affect consistently the breathing behaviour, compared to the maximum stress of 5 N/mm^2 due to weight bending.

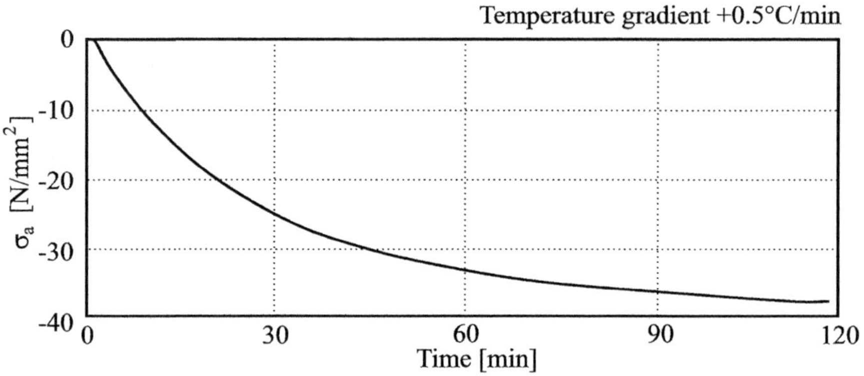

Fig. 4.13. Axial stress on shaft surface as function of time.

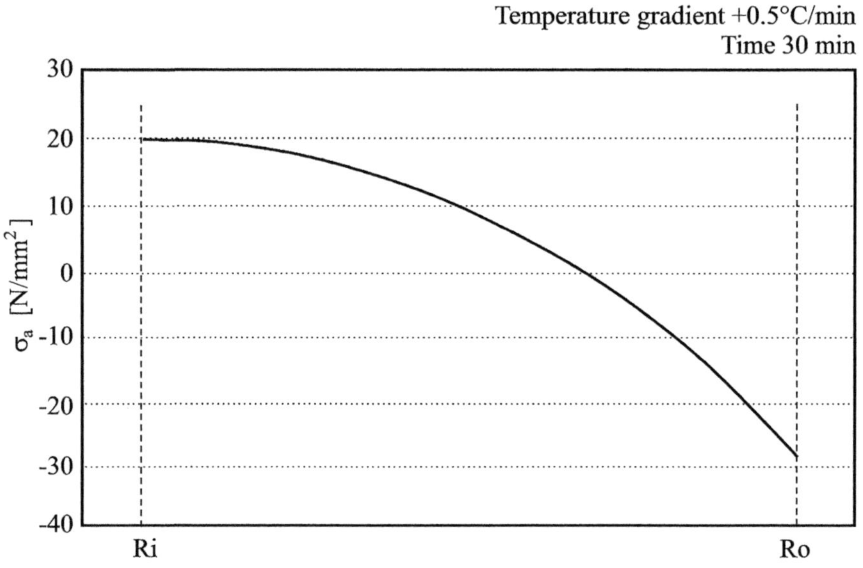

Fig. 4.14. Axial stress distribution over radius (Ro – rotor outer radius, Ri – rotor inner radius at the instant $t = 30$ min).

At constant temperature distribution, the crack is slightly open and the shaft has a bow. Then a situation, which is exactly opposite to that one of previous test specimen, has been measured in this case i.e. there is not the crack closure effect. Figure 4.15 shows the external and internal temperatures in section B of figure 4.11. Unfortunately the external temperature sensor has been lost after 165 min of operation. During the heating, the following periods can be identified:

a) high positive temperature gradient between outer and inner temperature: the crack is forced to close and the bow of the shaft changes;
b) the gradient is gradually vanishing: the crack is opening;
c) again positive gradient: crack is closing;
d) negative gradient: crack is opening quickly.

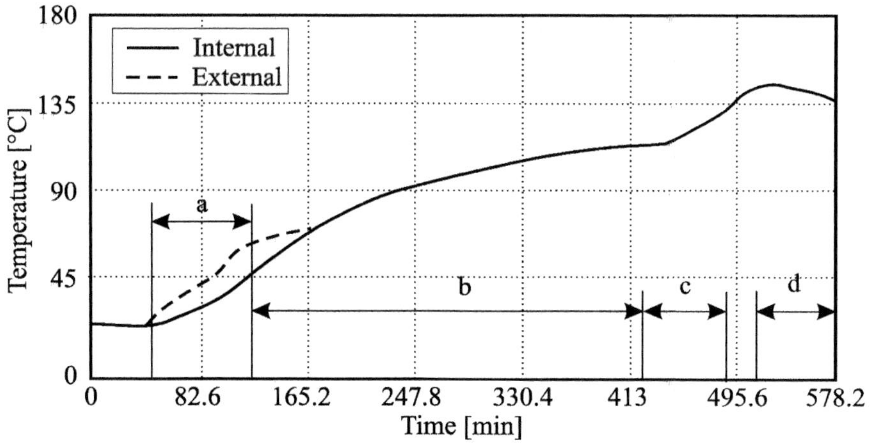

Fig. 4.15. Internal and external temperature vs. time in section B of figure 4.11.

Figure 4.16 shows the 1X amplitude of the strain measured in position ε_4, very close to the crack. When the crack is open, no measurement is given by the sensor. When the crack is closed, and it is kept closed by thermal stresses, the strain-gauge indicates a sinusoidal variation during the shaft slow rotation, with an amplitude of 290 με. When the temperature distribution becomes more even, the crack opens again. Also this signal has been lost when higher temperatures were reached.

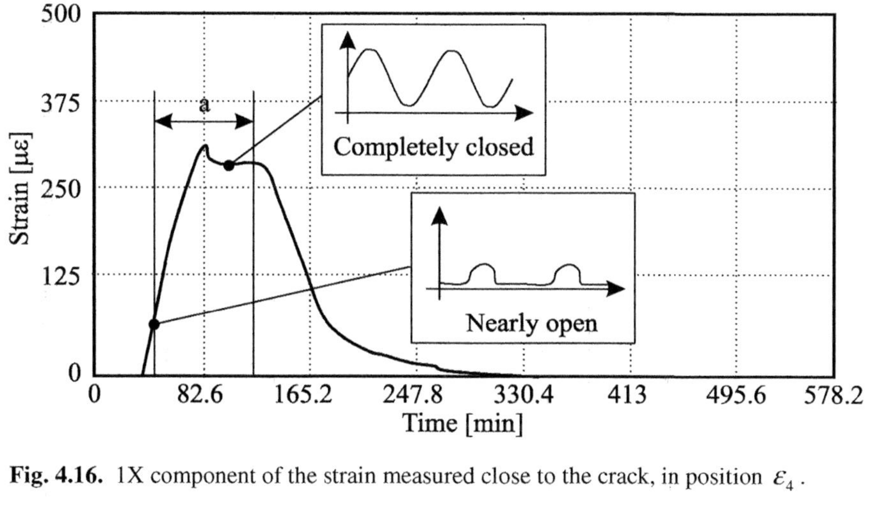

Fig. 4.16. 1X component of the strain measured close to the crack, in position ε_4.

Figure 4.17 shows the 1X displacement component in vertical direction, measured by a proximity probe in the measuring plane 3 of figure 4.11, i.e. in correspondence of the crack. The bow in steady state conditions is about 60 μm, during the heating the bow diminishes its amplitude (periods a and c), whilst during the cooling transient (period d) the bow increases consistently (up to 110 μm).

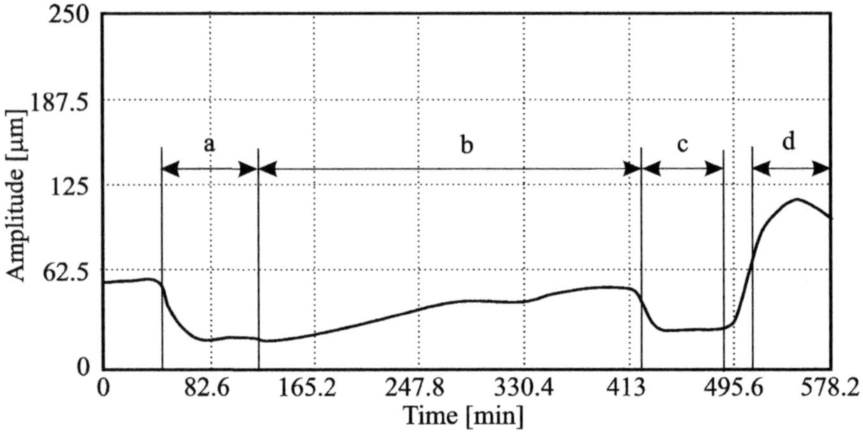

Fig. 4.17. Amplitude of the 1X vertical displacements measured in correspondence of the crack.

4.3.2 Tests during Speed Transients

The rotor has been tested also at higher rotating speeds, in slow run-up transients from 0 to 2000 rpm (so that the rotor first critical speed was passed), during heating and cooling transients.

Figure 4.18 shows the 1X vibration amplitudes, in vertical and horizontal direction, in correspondence of the crack (measuring plane 3 of figure 4.11), during the cooling transient (open crack). A bow of 115 μm is recognizable and the unbalance response compensates partially the bow response in correspondence of the critical speed (1750 rpm).

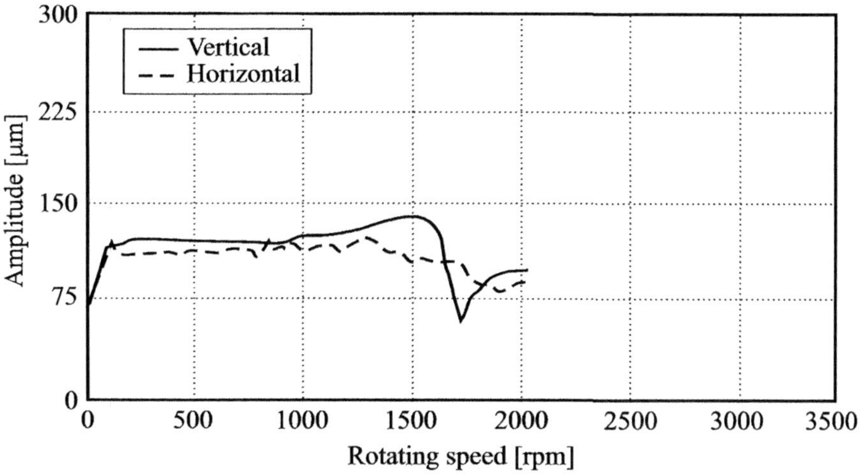

Fig. 4.18. Amplitude of the 1X vertical and horizontal displacements measured in correspondence of the crack during a cooling transient (open crack).

Figure 4.19 shows the same harmonic component, but during a heating transient. The bow is reduced and the classical resonance peak excited by the unbalance appears in vertical direction, in correspondence of the critical speed.

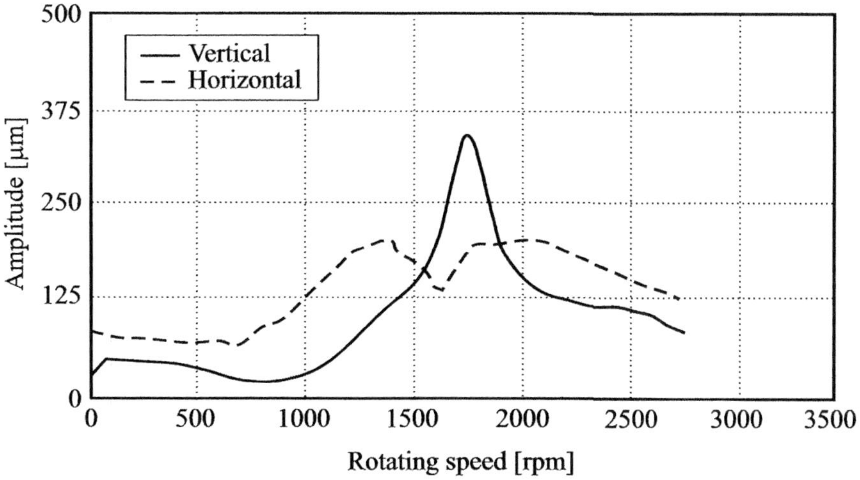

Fig. 4.19. Amplitude of the 1X vertical and horizontal displacements measured in correspondence of the crack during a heating transient (closed crack).

Figure 4.20 shows the 2X component with its resonance at half of the critical speed (875 rpm) in open crack configuration.

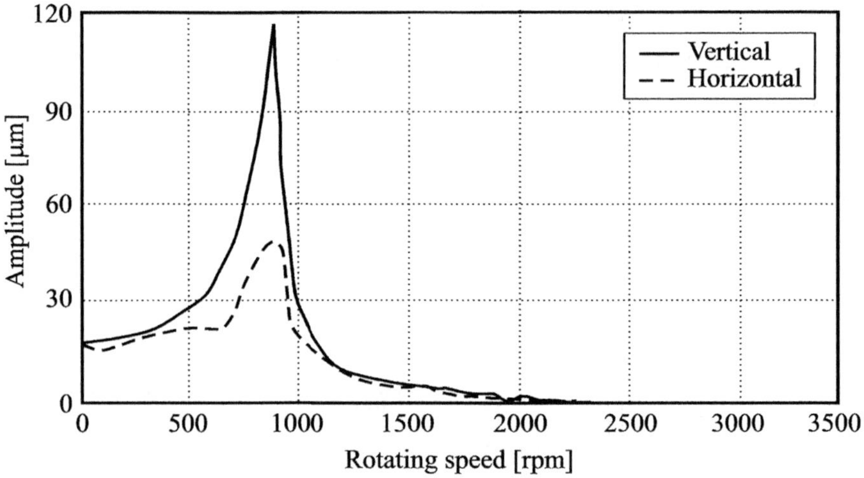

Fig. 4.20. Amplitude of the 2X vertical and horizontal displacements measured in correspondence of the crack during a cooling transient (open crack).

Figure 4.21 shows the same harmonic component, but in closed crack configuration during a heating transient: the 2X excitation is much smaller in this case than in that of figure 4.20.

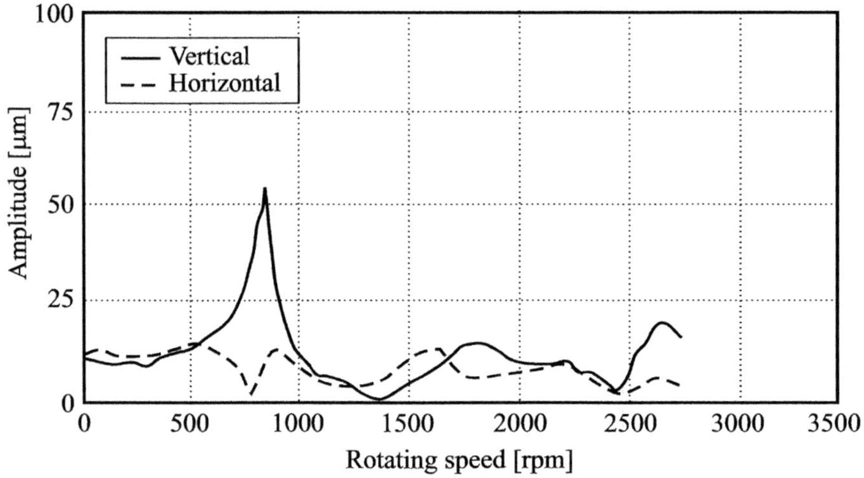

Fig. 4.21. Amplitude of the 2X vertical and horizontal displacements measured in correspondence of the crack during a heating transient (closed crack).

These tests, made on a real cracked steam turbine rotor, are unique to the authors' knowledge for the huge costs involved in performing them. They are also exceptional for the results shown about the thermal sensitivity of cracked shafts.

This kind of behaviour can be considered typical for actual cracked rotors and the knowledge of this aspect is extremely important when dealing with crack identification.

References

[4.1] Wauer J (1990) On the dynamics of cracked rotors: a literature survey. Applied Mechanics Review, 3, 13-17.

[4.2] Gasch R (1993) A survey of the dynamic behaviour of a simple rotating shaft with a transverse crack. Journal of Sound and Vibration, 160(2), 313-332.

[4.3] Dimarogonas AD (1996) Vibration of cracked structures: a state of the art review, Engineering Fracture Mechanics, 55(5), 831-857.

[4.4] Borri B, Carpinteri M, Chiaia B (1998) Contact, Closure and Friction Behaviour of Rough Crack Concrete Surface. Fracture mechanics of concrete structures, Freiburg, Vol. 3, 1635-1644.

[4.5] Bachschmid N, Tanzi E (2003) Stresses and Strains in Correspondence of a Transverse Crack in a Shaft: Effect of Crack Closure. Crack Path CP 2003 Int. Conference, September 2003, Parma, Italy.

Chapter 5
Crack Modelling

5.1 A Review about Crack Modelling

Cracked rotating shafts have been the object of studies and investigations since the 1970s, even if some failure cases are also reported during the 1950s (see section 1.2), therefore now we have a story lasting about 50 years. Important achievements have been made during these years, the knowledge of the dynamical behaviour of cracked rotors has allowed the presence of cracks to be recognized and cracked rotors to be stopped in time before catastrophic failures. Accurate inspections revealed that cracks had already propagated up to a depth of almost 50% of the diameter in many shafts (see for instance figure 2.15), which is obviously a very critical situation. Bearing in mind the fact that it is generally believed that propagating velocity increases exponentially, in many of these cases some further days of operation would have been sufficient to provoke a catastrophic failure, with the loss of the complete machine train and high risks for people and other equipment; also the loss of production of the unit, before the replacement with a new unit, is a comparable economic loss.

When a crack is discovered in time, the cracked rotor can be substituted by a spare rotor in few days or weeks, with affordable economic losses.

This situation explains the increasing interest in studying cracks in general and cracked shafts in particular. Starting from the 1980s up to nowadays researchers from everywhere in the world have contributed with papers addressing different topics related to cracks in rotating shafts.

The total number of papers is probably close to 1000 and is still increasing. Only very few papers present experimental results and generally these are related to very simple machine layouts or test-rigs, resembling the Jeffcott / de Laval rotor. Not always these results can be extended to real rotors.

Among the papers in literature, in the first phase bending vibrations only were studied, by utilizing the fracture mechanics approach for calculating the loss of stiffness due to the crack and simple models were used for the breathing mechanism. Breathing is typical for rotating cracked shafts, where the transition from closed crack situation to open crack situation is governed by the rotation of the shaft, as shown in section 4.2. It is then obvious that we will have full shaft stiffness in closed crack situation and a reduced stiffness in open crack situation.

Later on interest arose also for:

- models of cross coupling effects like excitation of torsional and axial vibrations;
- shafts with multiple cracks;
- slant or helicoidal cracks;

- more accurate models of the breathing mechanisms and of the crack behaviour;
- non-linear behaviour of light shafts with deep cracks.

Identification and diagnosis of cracks developed methods with increasing accuracy up to those presented in the recent years.

Excellent review papers have been written by Wauer [5.1] in 1990, who listed 162 papers, by Gasch [5.2] in 1993, who described the behaviour of cracked rotors and by Dimarogonas [5.3] in 1996, who included in a list of 346 papers not only about cracks in rotating shafts, but also about cracks in any kind of structure. These exhaustive reviews were followed more recently, in 2004, by a literature review about cracked shaft detection and diagnosis by Sabnavis et al. [5.4], in which 98 references are listed. Since 1996 for modelling and simulation and since 2004 for detection and diagnosis up to the present year 2009, the literature has been enriched consistently with further contributions at a rate of roughly 10-15 papers per year referred to cracks in rotating machinery.

A state of the art about cracked rotor behaviour, to which a group of authors that have been deeply involved in cracked rotor dynamics has contributed, was published in 2008 [5.5]. These papers include a very rich bibliography, with a total of 307 references.

Authors think that a complete exhaustive literature review was outside the scope of the book. Therefore only strictly necessary references are listed when they are related to a figure or the description of an event or to a specific topic.

This chapter describes the different aspects of crack modelling and cracked rotor modelling and is organized as follows:

- section 5.2 describes the fracture mechanics approach and gives some information on crack propagation;
- in order to analyse real rotor behaviour in a first step the breathing mechanism is accurately analysed and modelled in section 5.3. A simple model for the breathing simulation is proposed and results are compared to cumbersome 3D numerical results;
- in section 5.4, as a second step, the stiffness variation as function of the breathing mechanism is analysed and modelled. The fracture mechanics approach is adapted to circular cross section beams and results obtained with fracture mechanics approach are compared to 3D numerical results and to results obtained with an original simplified 1D model;
- section 5.5 shows all the results which are obtained, mainly by analytical derivation, for the simplest rotor model (Jeffcott / de Laval rotor) affected by a crack that is modelled in the simplest way by a hinge equipped with a spring. This way it is possible to represent the reduction in flexural stiffness of the shaft in open crack configuration (stiffness is given by the spring), and to have the full stiffness in closed crack configuration (hinge is closed);
- finally in section 5.6 the model of the cracked beam element is introduced in the finite element model of the rotating shaft suitable to represent accurately real industrial machines. A procedure for calculating the dynamic response of the cracked shaft is described in detail.

5.2 Fracture Mechanics Approach and Propagation of Cracks

Contributed by Laura Vergani and Mario Guagliano, Department of Mechanical Engineering, Politecnico di Milano.

5.2.1 Strain Energy Release Rate Approach

The first significant paper in the development of Fracture Mechanics was published by Inglis [5.6] in 1913. Inglis studied the problem of the stresses in a large plate with an elliptical hole, as shown in figure 5.1. He observed that there was an intensification of the stresses near the tip of the ellipse and estimated that the stress concentration was approximately equal to $2\frac{a}{b}$ or $2\sqrt{\frac{a}{\rho}}$, where a and b are the semi-major and semi-minor ellipse axes and ρ is the root radius at the ellipse tip.

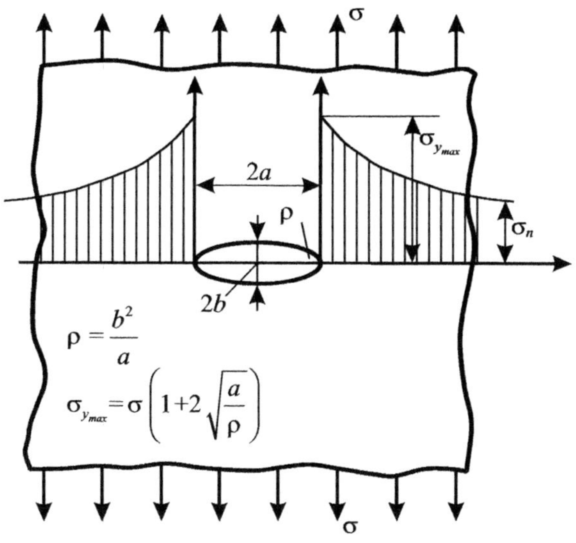

Fig. 5.1. Plate with an elliptical hole considered by Inglis.

In 1920, Griffith [5.7] proposed an energy balance approach to study the fracture phenomenon in cracked bodies, using Inglis work as foundation. He proposed that the reduction in strain energy of a body when the crack propagates could be equated to the increase in surface energy due to the increase of the crack surface area. Griffith assumed that the incipient fracture in ideally brittle material occurs when the elastic energy supplied at the crack tip is equal to, or greater than, the energy required to create new crack surfaces. He considered a model as that of figure 5.2 with a large plate with a central through crack, characterized by a length equal to $2a$.

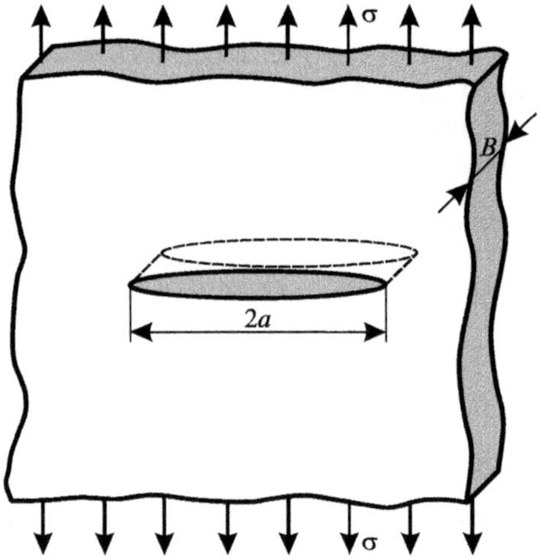

Fig. 5.2. Griffith crack model.

He developed a relationship between crack size a, surface energy 2γ connected with traction-free crack surfaces and applied stress σ:

$$\sigma^2 \pi a = 2\gamma E \tag{5.1}$$

Where E is the Young's modulus. Griffith's theory considered only brittle materials, in particular he conducted his experimentation on glass specimens. In 1948, Irwin [5.8] extended Griffith's model to metals by including the energy dissipated by local plastic flow. This modification was based on the idea that the resistance to the crack extension was due to the combined effect of the elastic surface and the plastic strain work.

Irwin's criterion was that the *strain energy release rate* G (SERR) must be larger than the critical work G_c, which is required to create new crack area.

Irwin defined the strain energy release rate G by considering a model of a cracked plate similar to the one of figure 5.2. If the material has a linear – elastic behaviour and if the load P increases, then the potential energy is stored in the plate, as shown in figure 5.3.

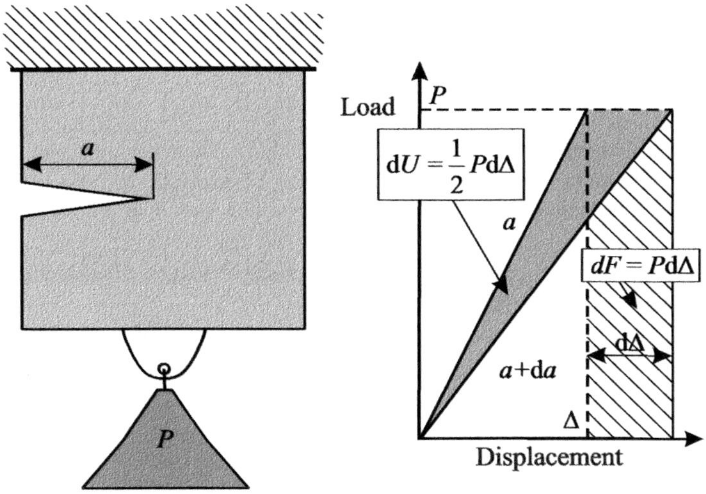

a) Load controlled crack estension

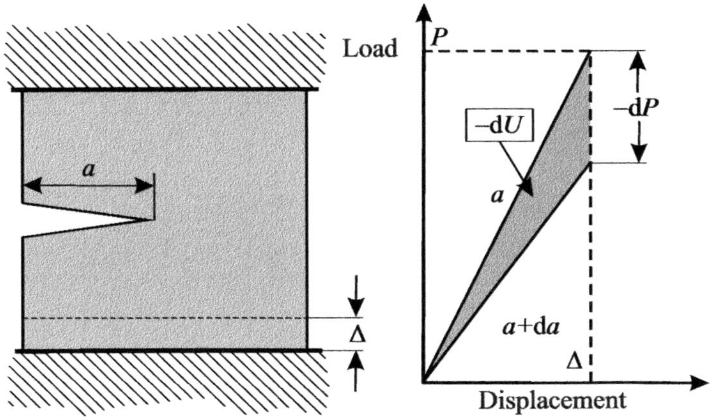

b) Displacement controlled crack estension

Fig. 5.3. Definition of the strain energy release rate G: a) load controlled crack extension; b) displacement controlled crack extension.

If the crack moves ahead by a small amount equal to $\mathrm{d}a$, while the displacement is held constant, the stiffness of the element decreases, as shown in figure 5.3b. The potential energy U decreases by an amount $\mathrm{d}U$. The rate of the change of potential energy with the increase of the crack area is defined as the strain energy release rate G:

$$G = -\frac{1}{B}\frac{\mathrm{d}U}{\mathrm{d}a} \tag{5.2}$$

where B is the thickness of the plate.

Two cases of crack extension are shown in figure 5.3: when the structure is load controlled and when the structure is displacement controlled. In both these cases, the rate of the change of potential energy with increase in crack area is equal to the strain energy rate.

The negative sign in eq. (5.2) causes G to have a positive value. G characterizes the energy per unit crack area required to extend the crack.

Irwin, successively, related his work to the stress field at the crack tip using Westergaard's work [5.9][5.10]. Irwin, in fact, showed that the stress field in the area of the crack tip is completely determined by a quantity K called the *stress intensity factor* (SIF).

The stresses are given by the following equations:

$$\begin{aligned}
\sigma_x &= \frac{K_I}{\sqrt{2\pi r}}\cos\left(\frac{\theta}{2}\right)\left[1-\sin\left(\frac{\theta}{2}\right)\sin\left(\frac{3\theta}{2}\right)\right] \\
\sigma_y &= \frac{K_I}{\sqrt{2\pi r}}\cos\left(\frac{\theta}{2}\right)\left[1+\sin\left(\frac{\theta}{2}\right)\sin\left(\frac{3\theta}{2}\right)\right] \\
\tau_{xy} &= \frac{K_I}{\sqrt{2\pi r}}\cos\left(\frac{\theta}{2}\right)\sin\left(\frac{\theta}{2}\right)\cos\left(\frac{3\theta}{2}\right)
\end{aligned} \tag{5.3}$$

where r and θ are the polar coordinates described in figure 5.4.

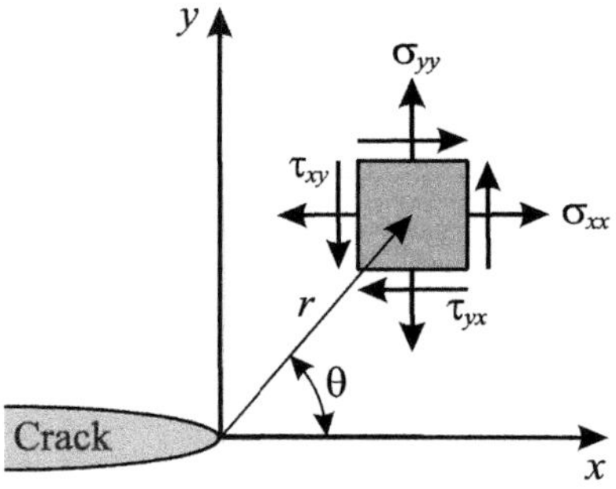

Fig. 5.4. Stress intensity zone and polar coordinates.

The subscripts of K are usually given in roman numerals, *I* (opening mode), *II* (in-plane shear or sliding mode) and *III* (out-plane shear or tearing mode), which refer to the modes of loading, as it is shown in figure 5.5.

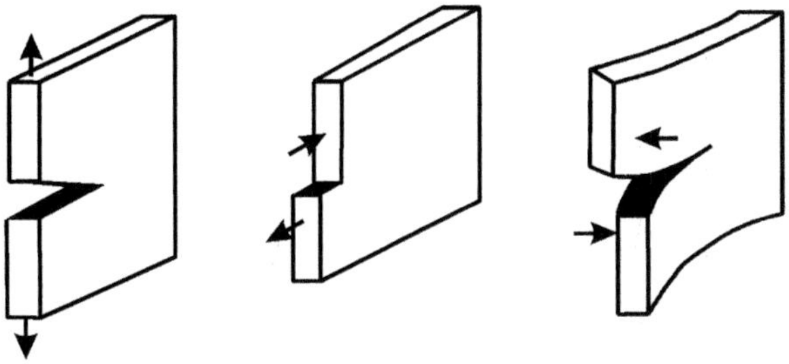

Mode *I*
opening mode

Mode *II*
in-plane shear

Mode *III*
out-plane shear

Fig. 5.5. Modes of crack loading.

For practical reason mode *I* is the most important one. In fact, experience shows that the greatest number of cracks results from opening mode *I*, while the other two modes are rare.

The relation between G and K is the following:

$$K_I = \sqrt{GE}\ \text{, in the case of plane stress} \tag{5.4}$$

$$K_I = \sqrt{\frac{GE}{(1-\nu^2)}}\ \text{, in the case of plane strain} \tag{5.5}$$

The stress intensity factor can be related to the applied stress in a general way:

$$K_I = \beta\left(\frac{a}{W},\frac{a}{d},\cdots\right)\sigma\sqrt{\pi a} \tag{5.6}$$

where β is a non dimensional function depending on cracked element geometry and on the external loads (a/W indicates the ratio of crack depth to the thickness of element and a/d indicates the ratio of crack depth to the diameter of a cylinder).

The knowledge of the value of the stress intensity factor K allows to verify if a cracked mechanical element is reliable or not. With this aim it is necessary to compare the stress intensity factor to the material toughness:

$$K_I < K_{I_c} \tag{5.7}$$

The toughness K_{I_c} is one of the most important parameters in fracture mechanics. In fact this is the critical value of stress intensity factor K_I at which fracture takes place.

The value of the toughness is experimentally determined by following the testing procedures of ASTM. The procedure is that a crack-notched specimen of suitable dimensions is progressively loaded until the crack becomes unstable, causing abrupt extension. The value of the critical load is used to calculate the toughness, that however is depending on the specimen thickness, as it is shown in figure 5.6.

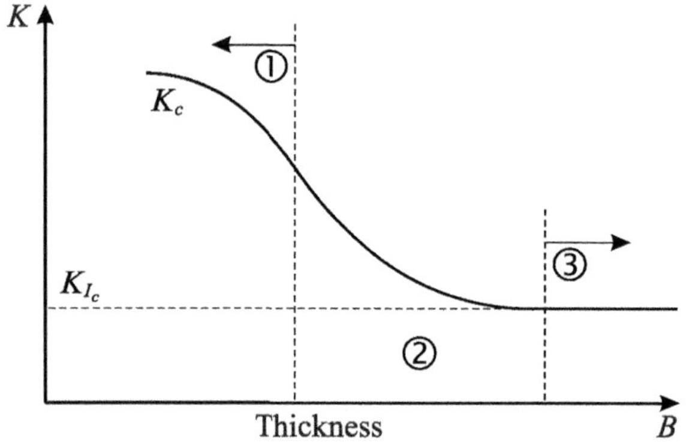

Fig. 5.6. Toughness as function of thickness; ① plane stress zone, ② transition zone, ③ plane strain zone.

In order to obtain the minimum value of the toughness K_{I_c}, the specimen thickness has to be larger than a minimum value $B_{\min}$ equal to:

$$B_{\min} = 2.5\left(\frac{K_{I_c}}{\sigma_y}\right)^2 \quad (5.8)$$

where σ_y is the yielding stress of the material.

If this condition is respected, the toughness value is that of the plane strain and is the minimum. Material toughness in general can be described as the ability to carry the load and to deform plastically in presence of a structural discontinuity as a sharp crack.

If the K_{I_c} value is known, it is possible to estimate, for an element containing a crack of size a, the corresponding nominal stress applied, by using eq. (5.7). If the crack size a is changed, the applied stress σ must also change to be consistent with the K_{I_c} value for a given material. In other words eq. (5.7) can be used to predict the size of the crack that a structural member can tolerate if the material toughness and the applied stress are known.

In literature, there are several handbooks that list the values of stress intensity factors for many cases and the relations of them when the geometries, the loading conditions and the crack shape vary. It is therefore easy to evaluate K_I value and to verify the cracked structure. On the contrary, when it is not possible to evaluate the stress intensity value from the literature, it is necessary to create a numerical model of the cracked element.

5.2.2 Finite Element Modelling of Cracked Elements

The behaviour of a cracked element in a structure or a machine can be simulated numerically by means of a finite element (FE) model of the crack that is included in the finite element model of the structure. This FE model allows the calculation of not only the deflections of loaded cracked elements, but also of the stress intensity factors, which can be compared to the values obtained using the SERR approach. Moreover, the obtained results allow checking the validity of the used model or mesh.

The most widespread technique to correctly simulate the stress singular field around the crack tip, which characterizes fracture mechanics problems, is the so called *one quarter point technique*. This technique consists in adequately distorting the usual 8 nodes (2D) and 20 nodes (3D) iso-parametric second order elements. In the following the attention will be concentrated on plain stress/strain 2D elements, while the results will be then extended to 3D.

5.2.2.1 Two Dimensional Elements

Consider an iso-parametric element with 8 nodes, distorted like the one shown in figure 5.7.

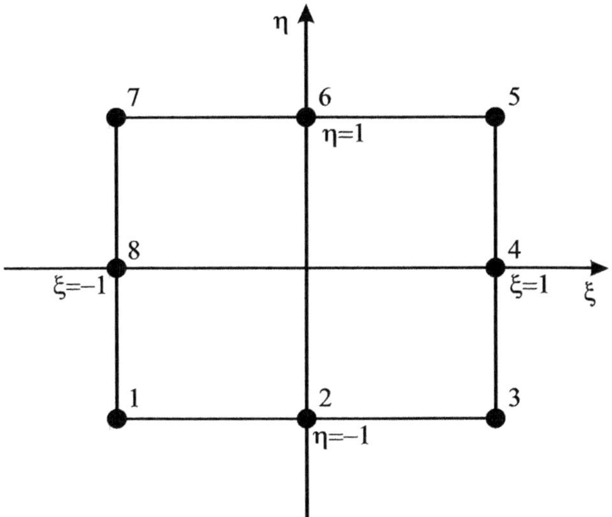

Fig. 5.7. 8 nodes iso-parametric element.

Its geometry is described by the following relations with respect of nodal coordinates:

$$x = \sum_{i=1}^{8} N_i(\xi,\eta)x_i \tag{5.9}$$

$$y = \sum_{i=1}^{8} N_i(\xi,\eta)\, y_i \tag{5.10}$$

with:

$$\begin{aligned} N_i(\xi,\eta) &= \\ &= \left[(1+\xi\xi_i)(1+\eta\eta_i)-(1-\xi^2)(1+\eta\eta_i)-(1-\eta^2)(1+\xi\xi_i)\right]\frac{\xi_i^2\eta_i^2}{4}+ \\ &+(1+\eta\eta_i)(1-\xi^2)\frac{\eta_i^2}{2}+(1+\xi\xi_i)(1-\eta^2)\frac{\xi_i^2}{2} \end{aligned} \tag{5.11}$$

where x and y are the global coordinates of a point in the continuum, ξ and η are the local iso-parametric coordinates, and N_i is the shape function (see e.g. [5.11]).

If node 1 of the element is considered to lie on the crack tip (see figure 5.8), it is possible to mathematically prove that if the side formed by the nodes 1-8-7 is collapsed to one point and if nodes 6 and 2 are moved from the middle of their side ($1/2h$) to a distance from node 1 (that is to say from the crack tip) equal to $1/4$ of the length h of the side, then the shape functions describing both the geometry and

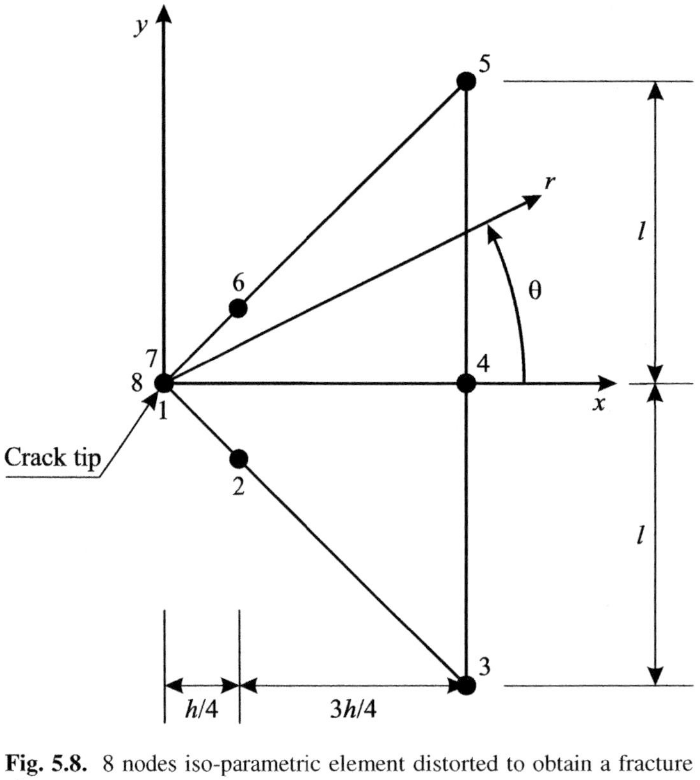

Fig. 5.8. 8 nodes iso-parametric element distorted to obtain a fracture mechanics singular element.

the displacement modes inside the element become singular in terms of strain and stresses with respect to the crack tip [5.12][5.13].

In this case, it is possible to demonstrate that:

$$\frac{\partial u}{\partial x}=\frac{A_0}{\sqrt{r}}+\frac{A_1}{r}+A_2 \tag{5.12}$$

$$\frac{\partial u}{\partial y}=\frac{B_0}{\sqrt{r}}+\frac{B_1}{r}+B_2 \tag{5.13}$$

$$\frac{\partial v}{\partial x}=\frac{C_0}{\sqrt{r}}+\frac{C_1}{r}+C_2 \tag{5.14}$$

$$\frac{\partial v}{\partial y}=\frac{D_0}{\sqrt{r}}+\frac{D_1}{r}+D_2 \tag{5.15}$$

where u and v are the displacement components in the x and y direction respectively, r is the distance from crack tip and all the capital letters represent constant quantities that depend on the geometry of the cracked element. The order of singularity of the shape function, that is the power of the function with respect to which stresses and strains tend to infinity approaching the crack tip, depends on the way the nodes that defines the collapsed side of the element are constrained.

If the displacement components of the nodes 1, 7 and 8 are constrained to be the same, then:

$$u_1=u_7=u_8 \tag{5.16}$$

$$v_1=v_7=v_8 \tag{5.17}$$

and it is possible to prove that the constants multiplying the terms in $1/r$ in eqs. (5.12)-(5.15) become null. The order of singularity with respect to the crack tip becomes $1/\sqrt{r}$, that is the one that correctly describes the stress and strain field in the close neighbourhood of the crack tip if a linear elastic stress-strain constitutive law of the material is considered (Linear Elastic Fracture Mechanics). On the contrary, if the displacement components remain free, the order of singularity is $1/r$, that is the order of singularity in the case of an elastic-perfect plastic material.

5.2.2.2 Three Dimensional Singular Elements

What has been introduced in the previous paragraph about 2D fracture mechanics problems and 2D iso-parametric singular elements can be extended for modelling 3D fracture mechanics problems. The condition that has to be respected to obtain elements able to respect the $1/\sqrt{r}$ singularity of the stress/strain field in the close neighbourhood of the crack tip is similar to that of 2D elements. The mid-sides nodes of the elements converging on the crack front have to move from their original position to a distance from the crack tip equal to $1/4$ of the length h of the element side. Like in 2D cases, both 20 nodes elements, with or without one side collapsed, and 15 nodes-$1/4$ points wedge elements can be used (see figure 5.9).

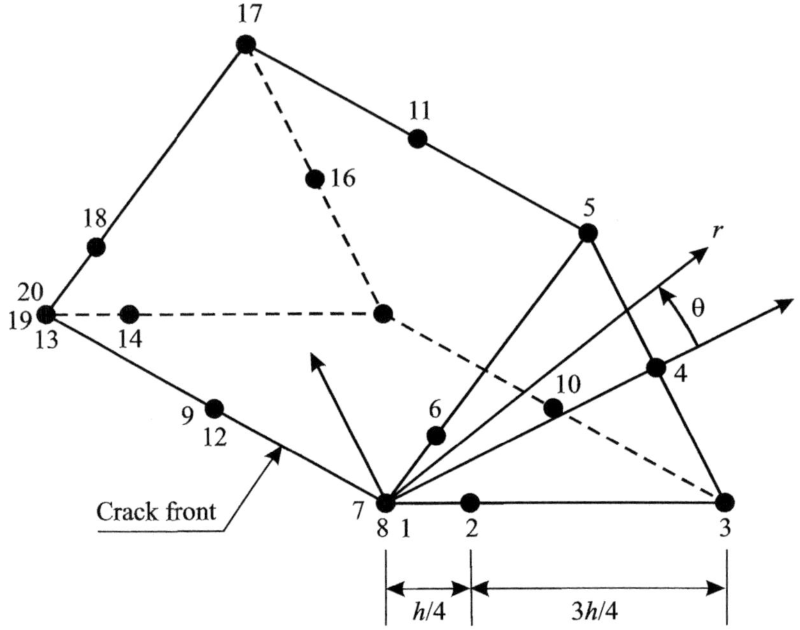

Fig. 5.9. 3D ¼ point singular element.

However, in the case of curvilinear crack fronts, the results can be improved by using some additional condition to the usual 20 nodes-$1/4$ points elements. For example, if one side collapsed elements are considered like that in figure 5.10, it is possible to demonstrate that the accuracy of the results increases if following conditions are imposed to the nodes of the elements defining the crack front [5.14]:

$$x_9 = x_{12} = a, \quad x_{10} = x_3 + a, \quad x_{11} = x_5 + a$$

$$y_9 = y_{12} = 0, \quad y_{10} = \frac{y_3 + y_{15}}{2}, \quad y_{11} = \frac{y_5 + y_{17}}{2} \tag{5.18}$$

$$z_9 = c_1 + \frac{c}{2}, \quad z_{10} = z_{11} = z_{12}$$

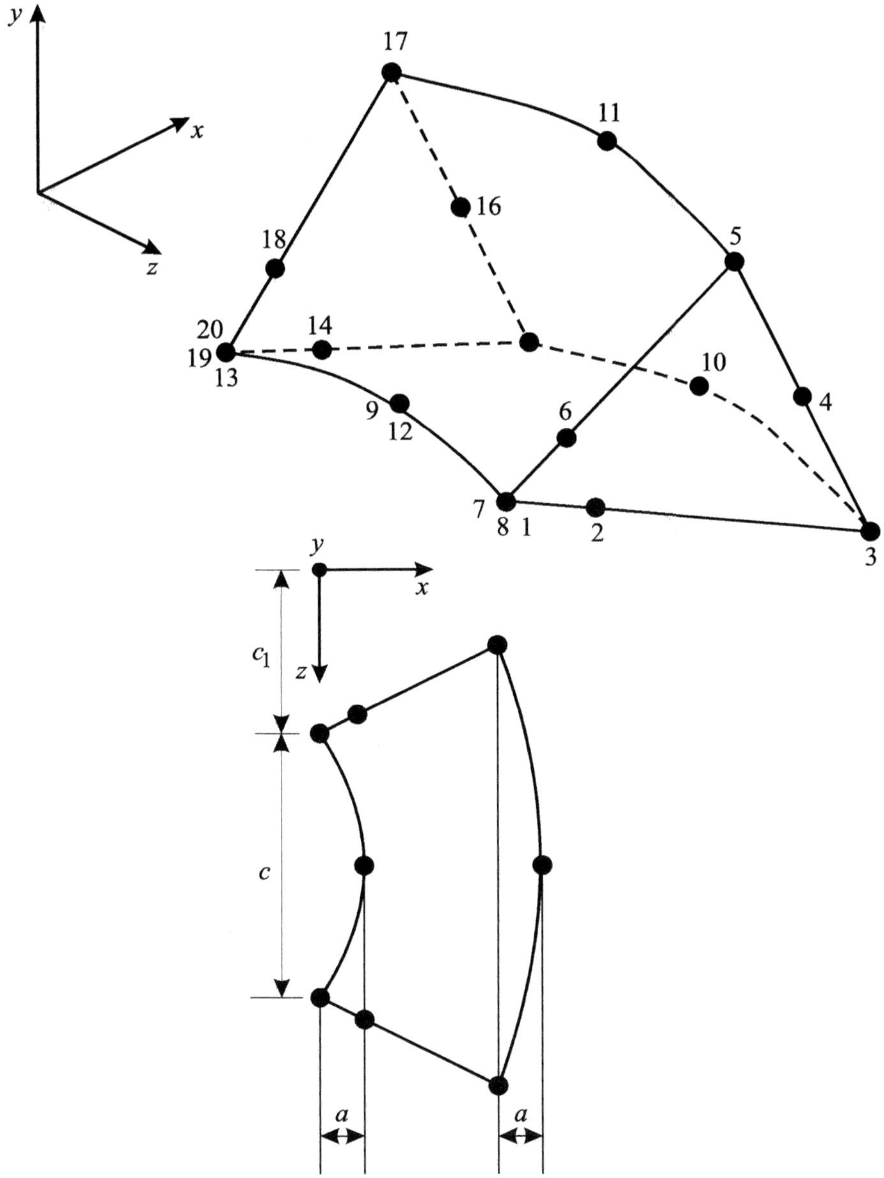

Fig. 5.10. ¼ point one side collapsed 20 nodes element defining a curvilinear crack front.

The correct FE simulation of a cracked element requires both the correct selection of the finite element type and the correct selection of the dimensions of the elements themselves, being the aim to obtain accurate and satisfactory results by minimizing the modelling and calculation effort.

When this problem has to be solved, an analysis of the convergence of the results has to be performed. It is necessary to develop different models with an increasing element density in the neighbourhood of the crack tip in order to assess how much the results vary by considering more refined meshes. When the differences of the results related to meshes with increasing density become negligible, then the mesh definition can be considered satisfactory.

Additional indications about the correct development of the FE model are pertinent to element dimensions and numbers, see figure 5.11.

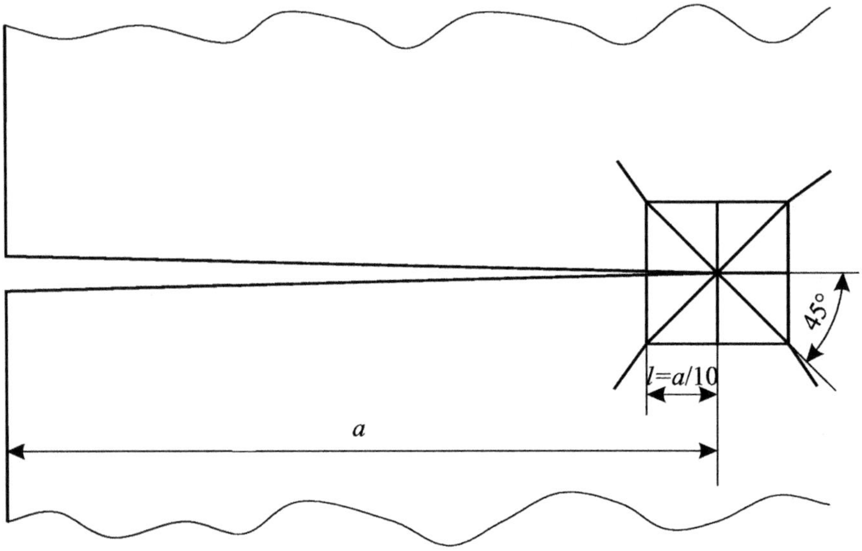

Fig. 5.11. Some indications to correctly model a cracked element.

As regards the dimensions of the elements converging to the crack tip, the length of their sides should be no longer than 1/10 of the crack depth a. The number of elements around the crack tip should not be less than 4 (8 in the case of mixed mode problems). A good model requires also that the angle of the converging sides of the elements defining the crack tip to be not less than 45° to avoid numerical errors due to excessive distortion of the element [5.15].

5.2.2.3 Methods to Calculate the Stress Intensity Factors from Finite Element Results

The final step of a FE analysis of a cracked element is the calculation of the stress intensity factor K_I (and K_{II}, K_{III} in mixed mode problems) from finite element results. The methods used with this aim can be divided in two groups, the first one based directly on the stress and displacement results (*direct methods*), the second one based on the elaboration of energy parameters related to crack propagation (*energy methods*).

Direct methods

The results obtained from a finite element analysis can be directly used to calculate the stress intensity factor.

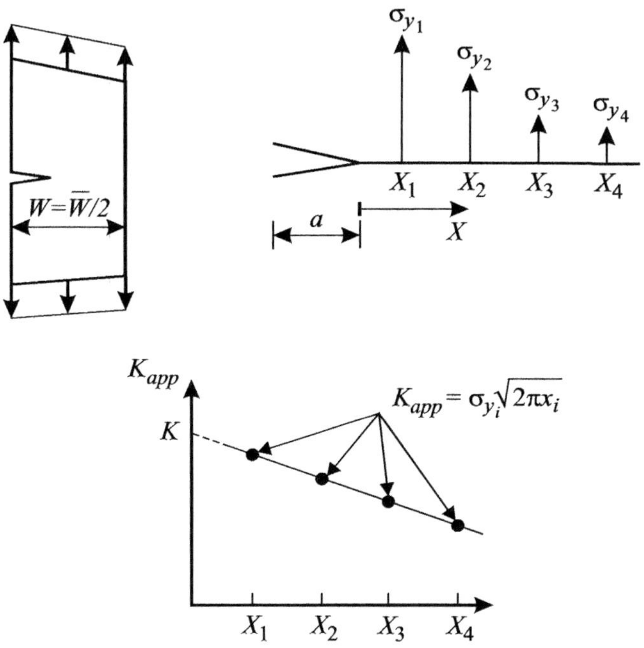

Fig. 5.12. Calculation of K_I by nodal stresses extrapolation [5.16].

According to the symbols of figure 5.12 and considering the nodes lying on the straight line $y = 0$ it is known that the stress field is described by the following relations:

$$\sigma_x = \frac{K_I}{\sqrt{2\pi x}} \tag{5.19}$$

$$\sigma_y = \frac{K_I}{\sqrt{2\pi x}} \tag{5.20}$$

$$\tau_{xy} = 0 \tag{5.21}$$

The calculation of the stress intensity factor is easy from eqs. (5.19)-(5.21). In fact, by substituting the stress values σ of the nodes close to the crack tip and the correspondent distance x of the nodes from the crack tip, it is possible to calculate the value of the stress intensity factor as (see figure 5.4 with $\theta = 0$):

$$K_I = \lim_{r \to 0} \sigma_y \sqrt{2\pi r} \tag{5.22}$$

The numeric results are more accurate if σ_y instead of σ_x is used, as indicated in eq. (5.22).

The same procedure can be used to calculate the value K_{II}:

$$K_{II} = \lim_{r \to 0} \tau_{xy} \sqrt{2\pi r} \tag{5.23}$$

The stress intensity factors can be calculated by using the displacements determined with the finite element analysis. In this case it is better to use the displacements of the crack faces (negative values of x, $y = 0$). This is due to the fact that the value of the displacements of these nodes are larger and, consequently, the results are less sensitive to the numerical errors. In addition, the calculation of displacements is more accurate than that of stresses, due to the fact that these latter are obtained numerically by deriving the displacements. The following relations are obtained:

$$K_I = 2G\sqrt{\frac{2\pi}{r}}(\kappa+1)v \quad K_{II} = 2G\sqrt{\frac{2\pi}{r}}(\kappa+1)u \tag{5.24}$$

with:

$\kappa = (3-4\nu)$ for plane strain cases;

$\kappa = \dfrac{3-\nu}{1+\nu}$ for plane stress cases.

In 3D problems with curvilinear crack fronts [5.17] it is possible to use the same procedures if a local normal, binormal and tangential coordinate system is defined (see figure 5.13).

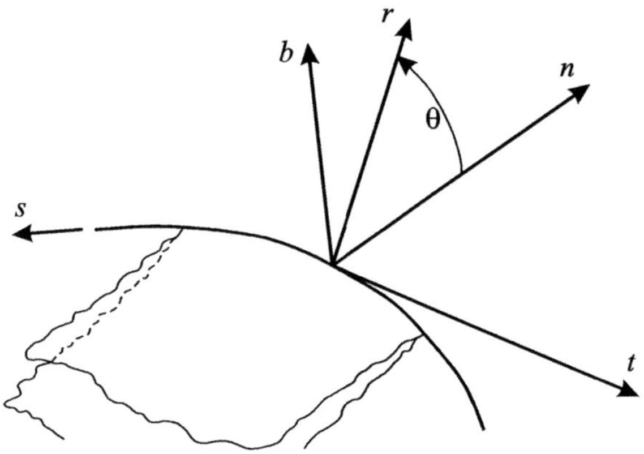

Fig. 5.13. Normal, bi-normal and tangential local coordinate system (*n*-*b*-*t*).

It is possible to calculate the stress intensity factors as:

$$K_I = \lim_{r \to 0} \sigma_b \sqrt{2\pi r} \tag{5.25}$$

$$K_{II} = \lim_{r \to 0} \tau_{nb} \sqrt{2\pi r} \tag{5.26}$$

$$K_{III} = \lim_{r \to 0} \tau_{tb} \sqrt{2\pi r} \tag{5.27}$$

One of the advantages of direct methods is the possibility to distinguish the contributions related to the different fracture modes and to use them with a general finite element code, without special routines. The main drawback is that they require very accurate finite element results to give precise results, especially for analysing 3D problems.

Energy methods

The second type of approach that can be used for calculating the stress intensity factors is based on the calculation of some energy parameter related to the propagation of a crack. Among these procedures, the most widely used one will be reported here. It is called the *Virtual Crack Extension* technique and consists in calculating the energy released by the increasing of the crack depth of a virtual value [5.18].

The procedure starts from the calculation of the elastic potential energy U associated with a finite element model of a cracked element. The elastic potential energy can be written as:

$$U = \frac{1}{2}\mathbf{u}^{\mathrm{T}}[\mathbf{K}]\mathbf{u} - \mathbf{u}^{\mathrm{T}}\mathbf{F} \tag{5.28}$$

where $\mathbf{u}$ is the displacement vector, $[\mathbf{K}]$ is the stiffness matrix of the system and $\mathbf{F}$ is the nodal forces vector.

By deriving this expression with respect to the crack depth a and by maintaining the load constant, the energy G released for unit propagation depth of the crack can be calculated. The following expression is obtained:

$$-\frac{\partial U}{\partial a} = G = -\frac{\partial \mathbf{u}^{\mathrm{T}}}{\partial a}\left([\mathbf{K}]\mathbf{u} - \mathbf{F}\right) - \frac{1}{2}\mathbf{u}^{\mathrm{T}}\frac{\partial [\mathbf{K}]}{\partial a}\mathbf{u} + \mathbf{u}^{\mathrm{T}}\frac{\partial \mathbf{F}}{\partial a} \tag{5.29}$$

It is obvious that the first and the third term on the r.h.s. of eq. (5.29) are null, so:

$$-\frac{\partial U}{\partial a} = G = -\frac{1}{2}\mathbf{u}^{\mathrm{T}}\frac{\partial [\mathbf{K}]}{\partial a}\mathbf{u} \tag{5.30}$$

The matrix $\partial[\mathbf{K}]/\partial a$ is the variation of the stiffness matrix for an unit propagation depth of the crack and it can be said that G is proportional to this variation. To obtain the values of G, it is necessary to calculate the values of $[\mathbf{K}]$ for two different depths of the crack without performing two complete finite element analyses that consider two different crack depths.

Generally the calculation of $\partial[\mathbf{K}]/\partial a$ is obtained by developing a mesh focussed on the crack tip with different hulls of elements converging on it. This way it is possible to define different contours Γ_i ($\Gamma_0, \Gamma_1, \ldots$, see figure 5.14) and to move all the nodes inside the different hulls of elements by a quantity Δa, while all the other nodes do not change their original position.

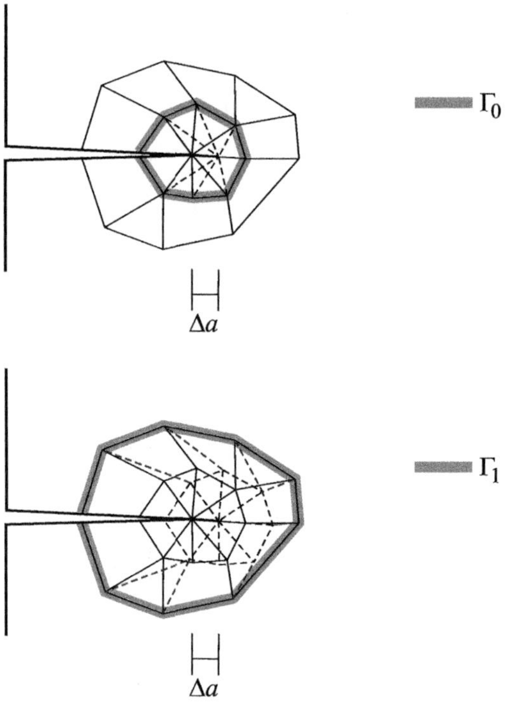

Fig. 5.14. Definition of the contours $\Gamma_0, \Gamma_1, \ldots$ used to calculate G.

The variation of $[\mathbf{K}]$ is due only to the different positions of the nodes inside the hulls and the calculation is fast. By considering the displacement vector calculated with the FE analysis and $\partial[\mathbf{K}]/\partial a$ it is possible to obtain G. If the values of G obtained by considering different hulls Γ_i are constant, then this result can be considered an index of the accuracy of the analysis.

This method is really more accurate than the previous ones, because it is not related to local quantities like stresses or displacements but to global ones. The disadvantage is that it is not possible to directly separate the contributions of the different propagation modes in mixed mode problems.

5.2.3 *Fatigue Crack Propagation*

The damage of materials induced by the cyclic repetition of stress cycles is called *fatigue* and is one of the most dangerous mechanisms of damage of structural materials. In fact fatigue development depends on many factors: surface roughness, residual stresses, dimensions of the components, effects of accidental overload, presence of notches, presence of a multiaxial stress, presence of fretting. These are only some of the key factors that strongly influence the fatigue strength of materials, which prevented till now an universally accepted definition of general design criteria able to take into account all these factors.

Fatigue damage is formed by three phases: i) fatigue crack initiation, ii) fatigue crack propagation and iii) final fracture. In the first phase surface microscopic damage due to repeated sliding of crystal grains takes place, inducing intrusion and extrusions of the surface (PSB – persistent slip bands) that, cycle after cycle, lead to the formation of a crack that starts to propagate. In the second phase the crack begins to propagate in the maximum shear stress direction but soon, given the difficulty in propagating by shear stress due to the friction among the crystal grains, it deviates its trajectory and propagates by normal stresses [5.19].

Generally components and structural elements working with a propagating crack are not allowed in mechanical design. Anyway, modern fatigue design criteria admit a propagating crack in a structural element if it is possible to predict its evolution, that means if it is possible to determine crack growth rate and the number of load cycles needed to reach the critical dimension causing the final sudden fracture of the component.

This type of approach, usually called *fail-safe* or *damage tolerant design* [5.20], needs also the definition of a plan of non-destructive inspections by ultrasonic or x-ray or some other techniques, to assess the real dimension reached by the crack during operational service. Also the residual life needs to be determined after each inspection and a plan for the removal and substitution of the damaged component must be ready.

From another point of view, the mechanical behaviour of a cracked element and its evolution during fatigue crack propagation can be used and compared with a non-cracked one to monitor the development of the crack and to estimate the residual life of the cracked element.

Considering rotating elements and, in particular, large rotors and shafts, the presence of a fatigue crack influences the dynamic behaviour of the element, inducing changes in characteristic vibrations and in the natural frequencies of the rotor. In other words, if it is possible to relate the dynamic behaviour of a cracked rotor to the dimensions (surface length and depth), of the propagating crack, then the residual life of the rotor can be estimated and it is possible to determine when it needs to be substituted.

This approach can be applied only when a suitable law of fatigue crack growth rate is known. The crack growth rate, which is the derivative of the crack depth a with respect to the number of fatigue cycles N (i.e. $\mathrm{d}a/\mathrm{d}N$), depends mainly on the variation of the stress intensity factors K_i during one cycle.

Since fatigue crack propagation is influenced generally only by Mode I, the index I is omitted in the following and only K is used. At a first glance this is surprising, since fatigue crack growth is a mechanism that involves plastic deformation at the crack tip while the stress intensity factor K is a parameter used in Linear Elastic Fracture Mechanics (LEFM). But yielding is usually reached only in a little zone around the crack tip (small scale yielding), thus not preventing the application of LEFM.

The stress intensity factor can vary during one cycle between a maximum K_{max} related to maximum positive tensile stress for open crack with maximum load and a minimum K_{min} that can be 0 for closed and un-loaded crack, or can be related to the maximum negative compressive stress in case of always open crack loaded by alternate loads. When the load is a symmetrical bending load due to weight and rotation of the shaft, in case of always open crack we get the maximum variation of ΔK , which is equal to $2K$.

5.2.3.1 Paris' law

Figure 5.15 shows a graph of the fatigue crack growth rate ($\mathrm{d}a/\mathrm{d}N$) as function of the cyclic variation of the stress intensity factor, $\Delta K = K_{max} - K_{min}$ .Three zones are depicted: in the first one (Zone I in figure 5.15), the threshold value ΔK_{th} is defined and the crack growth rate increases rapidly with larger ΔK_I , with a slope that is decreasing with a tendency to the constant slope m in the second zone. If the applied ΔK_I is lower than the threshold value no propagation will take place. Otherwise, the crack will propagate and will reach Zone II, characterized by a linear relation between $\log(\mathrm{d}a/\mathrm{d}N)$ and $\log \Delta K_I$. The last Zone, III, is characterized by an increment of the fatigue crack growth rate with an increasing slope and finishes with the fracture of the component, when $\Delta K = \Delta K_c$.

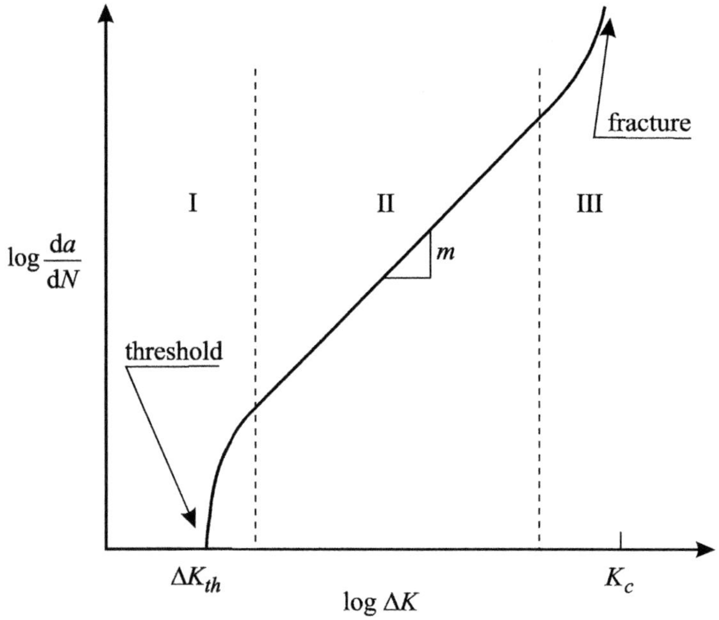

Fig. 5.15. Fatigue crack growth rate ($\mathrm{d}a/\mathrm{d}N$) versus stress intensity factor variation (ΔK_I).

Zone II is the most important in damage tolerant design and is also called *Paris' zone*, by the name of P.C. Paris that defined the law that is still used today to predict the propagation of a fatigue crack [5.21]:

$$\frac{\mathrm{d}a}{\mathrm{d}N} = C\,\Delta K^{n} \tag{5.31}$$

where C and n depend on the material. This law permits to determine the fatigue crack growth rate in a complex system by simply knowing C and n, the values of which can be determined by means of tests on simple specimens and by calculating the value of ΔK, with handbooks or numerical calculations. The trend of the crack depth a as function of the number of cycles N can be calculated by integrating Paris' equation. However, as soon as the crack has propagated deeper, the value of K changes and a new – and higher – value of crack propagation velocity will be found. Residual life can then be estimated considering the step by step increasing propagation velocities.

Anyway, Paris' equation does not consider all the effects affecting the fatigue crack growth, first of all the ratio R between the minimum and the maximum stress of a fatigue cycle, and can be applied only to the Zone II of the diagram of figure 5.15. Besides, residual stresses due to some treatment or induced by overloads can strongly influence the crack growth rate. Also the interaction between cycles with variable amplitude can have an influence on fatigue crack growth rate. To overcome these limitations other laws have been developed with the aim of taking into account the factors affecting the fatigue growth rate, but basically all the laws are derived from Paris' one.

5.2.3.2 Example of Propagation Speed Calculation

The crack propagation speed has been calculated for the steam turbine shaft described in section 2.3.1. The necessary data are reported in table 5.1.

Table 5.1. Data for propagation speed calculation.

Parameter	Value
Parameter C for 30CrMoV steel	$C = 0.65 \cdot 10^{-18}$ (when stresses are expressed in $\mathrm{dN/mm^2}$ and dimensions in mm)
Parameter n	$n = 6.7$
Depth of the crack a	$a = 135$ mm
Diameter in correspondence of the crack d	$d = 615$ mm
Stress intensity factor K	$K = 15.69\ \mathrm{dN/mm^{-3/2}}$
Stress intensity factor variation (for open crack and symmetrical alternate bending stress) ΔK	$\Delta K = 2K = 31.38\ \mathrm{dN/mm^{-3/2}}$
fatigue crack growth rate $\mathrm{d}a/\mathrm{d}N$	$\mathrm{d}a/\mathrm{d}N = 0.69 \cdot 10^{-8}$ mm/cycle

At the speed of 3000 rpm (i.e. 3000 cycles per minute) the propagation is equal to 0.69 mm every 555 hours of operation, that means 23 days of continuous operation.

5.2.3.3 Crack Closure Effect

Another important factor affecting the crack growth rate is the so called *crack closure* effect. In fact, it was observed that, when a tensile or bending load is applied to a cracked element, its stiffness does not vary till a threshold load is reached. This means that the crack remains closed unless the threshold load is exceeded (see also section 4.2.2, where the crack closure effect has been measured). The logical consequence of this fact is that the effective fatigue cycle of a cracked element could not be the nominal one:

$$\Delta K = K_{max} - K_{min} \tag{5.32}$$

while it is more likely given by the following equation:

$$\Delta K_{eff} = K_{max} - K_{op} \tag{5.33}$$

where K_{op} is the value of the stress intensity factor which causes the opening of the crack (see figure 5.16).

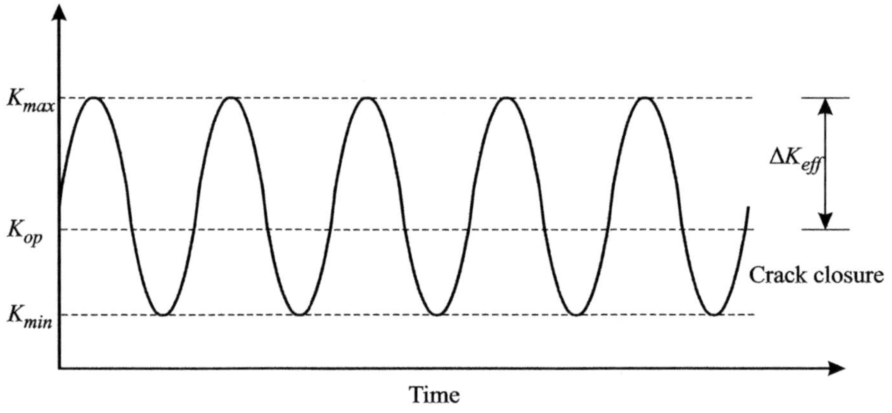

Fig. 5.16. Definition of ΔK_{eff} .

There are different causes of the crack closure: the plastic deformation in the crack tip zone, the surface roughness of the crack faces, the presence of oxides on the crack faces.

Elber [5.22] developed an equation to consider this effect that is a modification of Paris' equation:

$$\frac{\mathrm{d}a}{\mathrm{d}N} = C\,\Delta K_{eff}^{n} = C\left(U\,\Delta K\right)^{n} \tag{5.34}$$

with

$$U = \frac{\Delta K_{eff}}{\Delta K} = \frac{K_{max} - K_{op}}{K_{max} - K_{min}} \tag{5.35}$$

and C and n depend on the material type.

A more accurate calculation of propagation velocity, which takes also into account the effect of the ratio R of the minimum stress to the maximum stress, is given in [5.23] and [5.24]:

$$\frac{\mathrm{d}a}{\mathrm{d}N} = \frac{C\left(1-f\right)^{n}\Delta K^{n}}{\left(1-R\right)^{n}} \frac{\left(1-\dfrac{\Delta K_{th}}{\Delta K}\right)}{\left(1-\dfrac{\Delta K}{(1-R)K_{c}}\right)} \tag{5.36}$$

where f is function of the crack opening type, K_c is the toughness of the material, C and n depend on the material type.

5.3 Modelling the Breathing Behaviour and Its Thermal Sensitivity

Rotating heavy shafts with horizontal axis and affected by cracks have two peculiar characteristics with respect to cracked stationary structures:

1. crack related vibrations are excited naturally by the rotation of the shaft, while the vibrations are excited by external forces in stationary structures;
2. the crack gradually opens and closes periodically, in other words breaths during the revolution of the shaft (independently from vibration), while the crack opens and closes abruptly, as a function of the applied forces, in stationary structures.

Since the vibrations that are excited by the crack are one of the most expressive symptoms of a crack in rotating shafts and since the vibration characteristics (amplitudes of the different harmonic components) depend, apart from the depth the shape and the position of the crack, only on the breathing mechanism, this mechanism need to be accurately modelled if accurate prediction and simulation of vibration components are desired.

The breathing mechanism is the result of the stress and strain distribution around the cracked area, which is due to:

1. static loads, like the weight, the bearing reaction forces and so on;
2. dynamical loads, like the unbalance and the vibration induced inertia force distribution.

Accurate modelling of the breathing mechanism has been generally disregarded. When the static loads overcome the dynamical ones, the breathing is governed by the angular position of the shaft with respect to the stationary load direction and the crack opens and closes again completely once per revolution. The transition from closed crack (*full*) stiffness to the open crack (*weak*) stiffness has been generally considered abrupt [5.25] (see section 5.5) or represented by a given trigonometric function [5.26].

The reason for this is that the SERR approach for calculating the additional flexibility due to the crack is valid only for the fully open crack, but cannot be extended to other intermediate situations. Darpe et al. [5.27] and Papadopoulos [5.28] used the SERR approach and calculated the breathing by evaluating on the rectilinear crack tip the point where the crack is starting to close, assuming that the closed part of the crack surface is delimited by a boundary, the *crack closure line* (CCL), represented by a segment, orthogonal to the crack tip, that can be drawn from that point on the crack tip. A result is shown in figure 5.17.

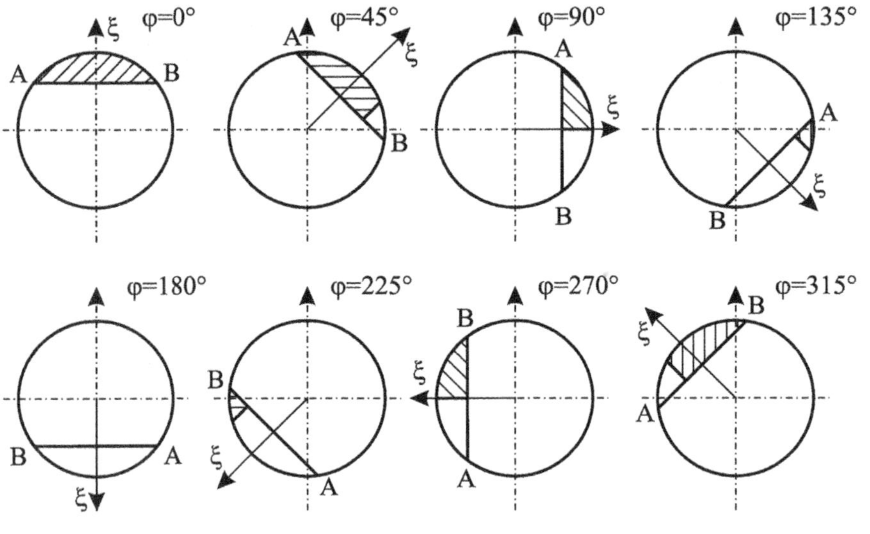

Fig. 5.17. Breathing mechanism model with CCL (crack closure line) orthogonal to the rectilinear tip, as required by the SERR approach (from [5.27]).

The same approach has been used also by Wu et al. [5.29] in time step calculations where breathing was determined by vibrations also. Researchers of EDF [5.30] have also developed a model in the time domain, in which the true breathing mechanism, or better directly the compliance of the cracked element, is evaluated by means of a 3D model as a function of the applied forces in each time step. Therefore the true breathing is accurately respected in this approach.

The true breathing mechanism can be calculated by means of 3D finite element models, in a non-linear approach: contact conditions on the cracked surface require iterative calculations. The calculation is cumbersome and time consuming, and also the required mesh is rather refined for taking into account the geometry of the crack. A mesh example is given in figure 5.18: a bending load is applied and the calculation is repeated for all different angular positions of the cracked shaft specimen.

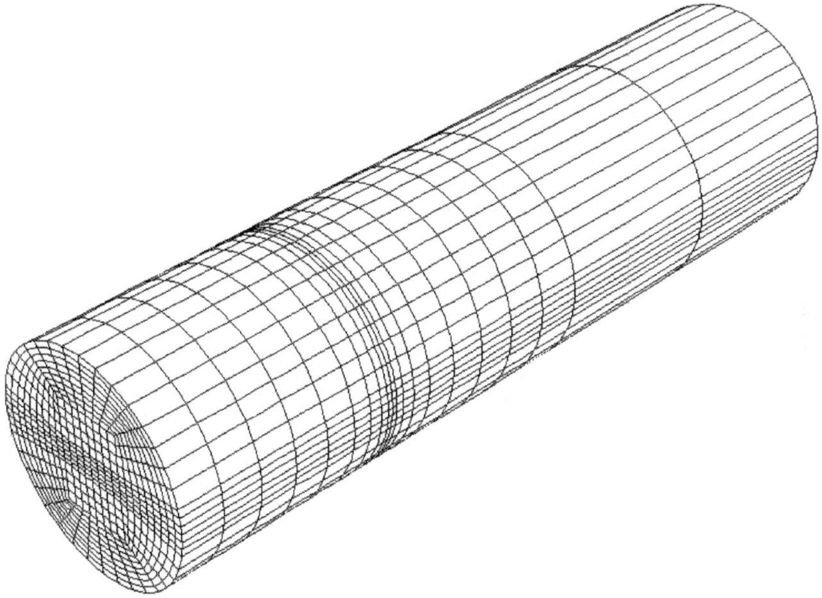

Fig. 5.18. 3D mesh of the cracked specimen.

Figure 5.19 shows some of the results obtained with the 3D non-linear approach. As can be seen there are some relevant differences with respect to the model used by Darpe et al. [5.27] in the 3D model the crack opens more slowly at the beginning, but increases its opening speed so that at 90° it is more open and at 135° it is already completely open.

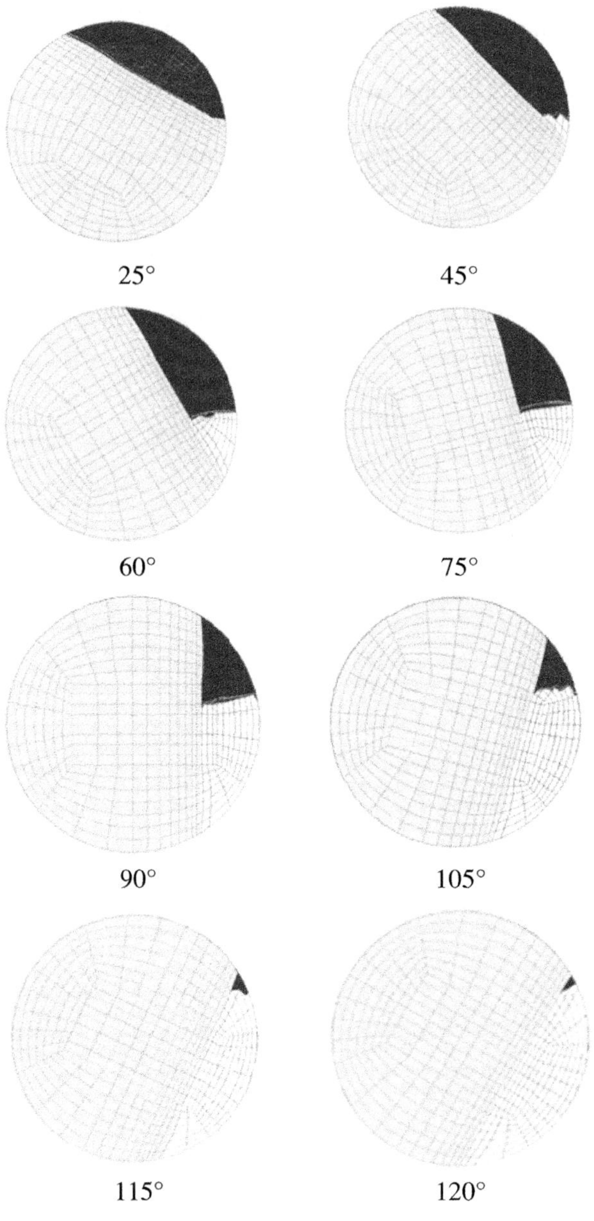

Fig. 5.19. Open and closed areas in different angular positions: from 125° to 180° the crack is completely open.

3D non-linear finite element calculations allow the breathing mechanism to be predicted accurately, when the loads are known, but are extremely cumbersome, costly and time consuming (due to the need of a refined mesh, as shown in section 5.2.2, in the crack region and to the non-linear contact conditions).

5.3.1 The Breathing Crack Simplified Model

A simplified model, which assumes linear stress and strain distributions has been developed by the authors, to calculate the breathing mechanism, and proved to be very accurate. The determination of the breathing mechanism is a non-linear iterative procedure. The breathing mechanism is affected also by transient thermal stresses that can arise in rotating shafts during a change in operating conditions and by pre-stresses that can develop during the crack propagation. These pre-stresses can open further the crack or can tend to hold the crack more closed, influencing the breathing mechanism.

The thermal stresses resulting from a thermal transient are calculated according to the following simple model that considers an infinite cylinder and axial symmetrical temperature and stresses distributions.

To determine the temperature distribution, the equation of the thermal exchange is used in the case of axial-symmetry and of an infinite cylinder:

$$\frac{\rho c_p}{k}\frac{\partial T}{\partial t} = \frac{1}{r}\frac{\partial}{\partial r} r \frac{\partial T}{\partial r} \tag{5.37}$$

where T is the temperature, t time, r radial distance from axis, ρ, c_p and k are respectively the density, the specific heat and the heat exchange coefficient of the shaft material.

Temperature gradient $\partial T/\partial t$ is imposed on the external surface as boundary condition. This way it is possible to determine the temperature distribution in each point of the section as a function of the time, integrating the equation by finite differences.

The axial stress distribution corresponding to the temperature distribution is given by:

$$\sigma = \frac{\alpha E}{1-\nu}\left(T_m - T(r)\right) \tag{5.38}$$

where α is the linear thermal expansion, E the Young's modulus, ν the Poisson's coefficient, $T(r)$ the temperature at the radial distance r and T_m the average section temperature. The effect of the centrifugal forces on the axial stresses has been neglected.

The proposed model assumes:

- the initial position of the main inertia axes of the supposedly partially open crack surface as known,
- a linear axial stress distribution due to the bending load, to which eventually also thermal stresses are superposed.

Then the compressive and tensile stresses are defined: cracked surfaces where tensile stresses should appear are "open" areas, where compressive stresses appear are "closed" areas (figure 5.20).

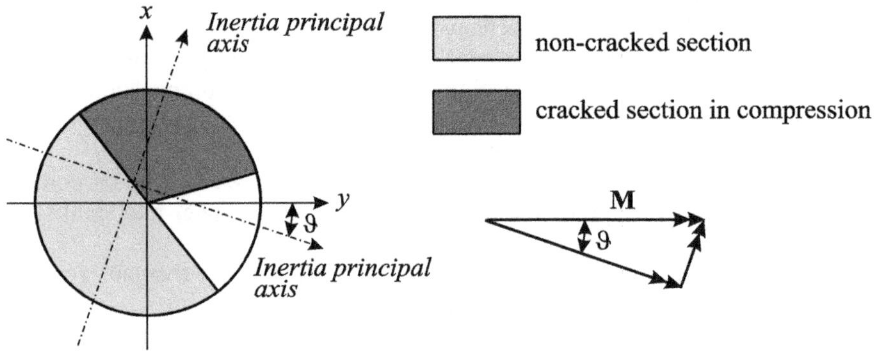

Fig. 5.20. Main axes of inertia and the bending moment decomposition in a generic position of the cracked section.

Indeed open and closed areas define also the actual main inertia axes position. Therefore the procedure has to be repeated iteratively, until the position of the main axes remains stable. In order to apply this procedure the cracked section has been subdivided in small elementary areas, as shown in figure 5.21. On each one of these elementary areas, the stresses are calculated in order to check if the area is in the closed or open part of the crack.

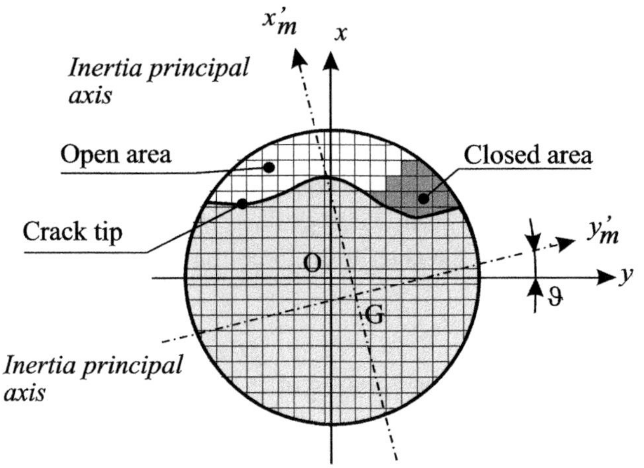

Fig. 5.21. Transforming the continuous crack surface in a discrete number of elementary areas.

The calculation of the breathing mechanism is performed according following steps:

1. Initially the main axes are considered to be coincident with the rotating crack axes ($\vartheta = 0$); the stresses due to bending moments are calculated (with the assumed main axes) and the thermal stresses (if present) are then added in each point.
2. Now the stress distribution is known over the cross section and the sign of the stress can be checked in each point of the cracked area: '+' means tension and therefore we have no contact forces in this point (the crack area element is "open"), '–' means compression and therefore we have contact forces (the crack area element is "closed"). The open and closed area sections have been determined.
3. The surface gravity centre G of the total area (formed by the un-cracked area plus the closed cracked area) can be calculated.
4. The second moments of area can now be calculated with respect to the reference system with origin in G and the angular position ϑ of the main axis of inertia can be found.
5. Now the procedure from 1 to 4 is repeated with the new value of ϑ, until ϑ converges to a stable value;
6. At this point the position of the main axis, the second moments of area and the bending moments with respect to the main axes are known. The second moment of area J_x, J_y and J_{xy} with respect to the fixed reference frame (xOy) and the components of the moments due to the thermal stress distribution with respect to the same reference frame are calculated. This will be repeated for each angular position of the shaft.

The actual axial stress distribution in the cracked area is obviously strongly non-linear as can be seen in figure 5.22, where the axial stress distribution in a cylindrical beam, affected by a 50% deep crack and loaded by a bending moment suitable to open the crack, is shown.

If the cracked cross section is considered, the stress distribution over the diameter, in direction orthogonal to the crack tip, can be calculated in the two different conditions: closed crack (linear distribution) and open crack (strongly non-linear distribution). The results calculated on a 25 mm diameter cylindrical beam are shown in figure 5.23.

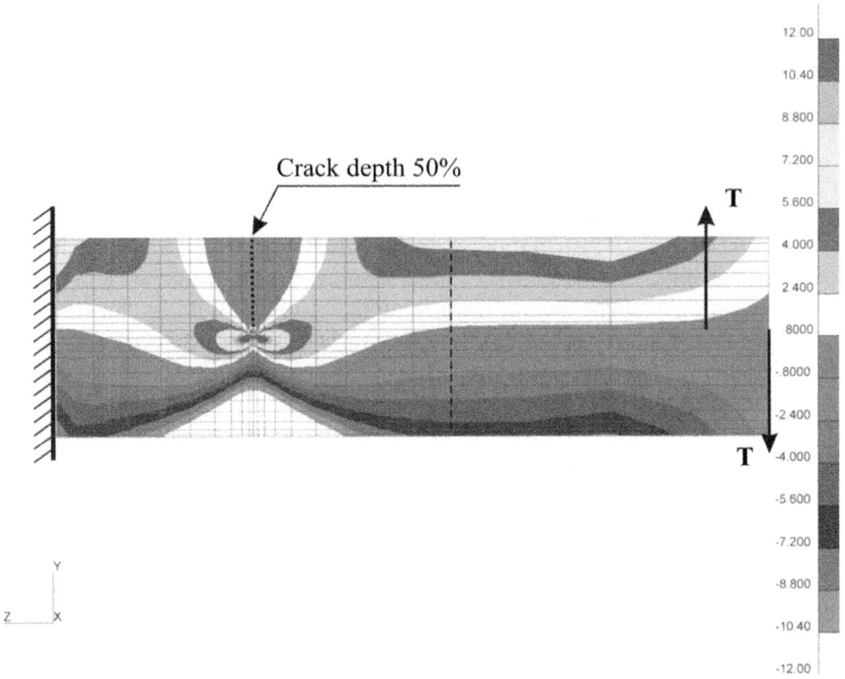

Fig. 5.22. Axial stress distribution in a cylindrical beam loaded by a bending moment.

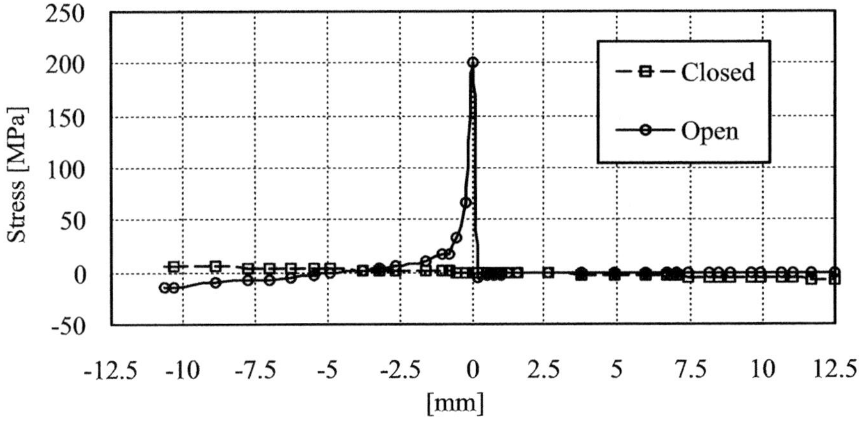

Fig. 5.23. Stresses in open and closed crack configuration along the diameter of a 25 mm beam with a crack 50% deep.

Nevertheless with the simplified model, which assumes linear stress distribution in all different configurations, very accurate results have been found. Breathing mechanism calculation is summarized in the flow-chart of figure 5.24.

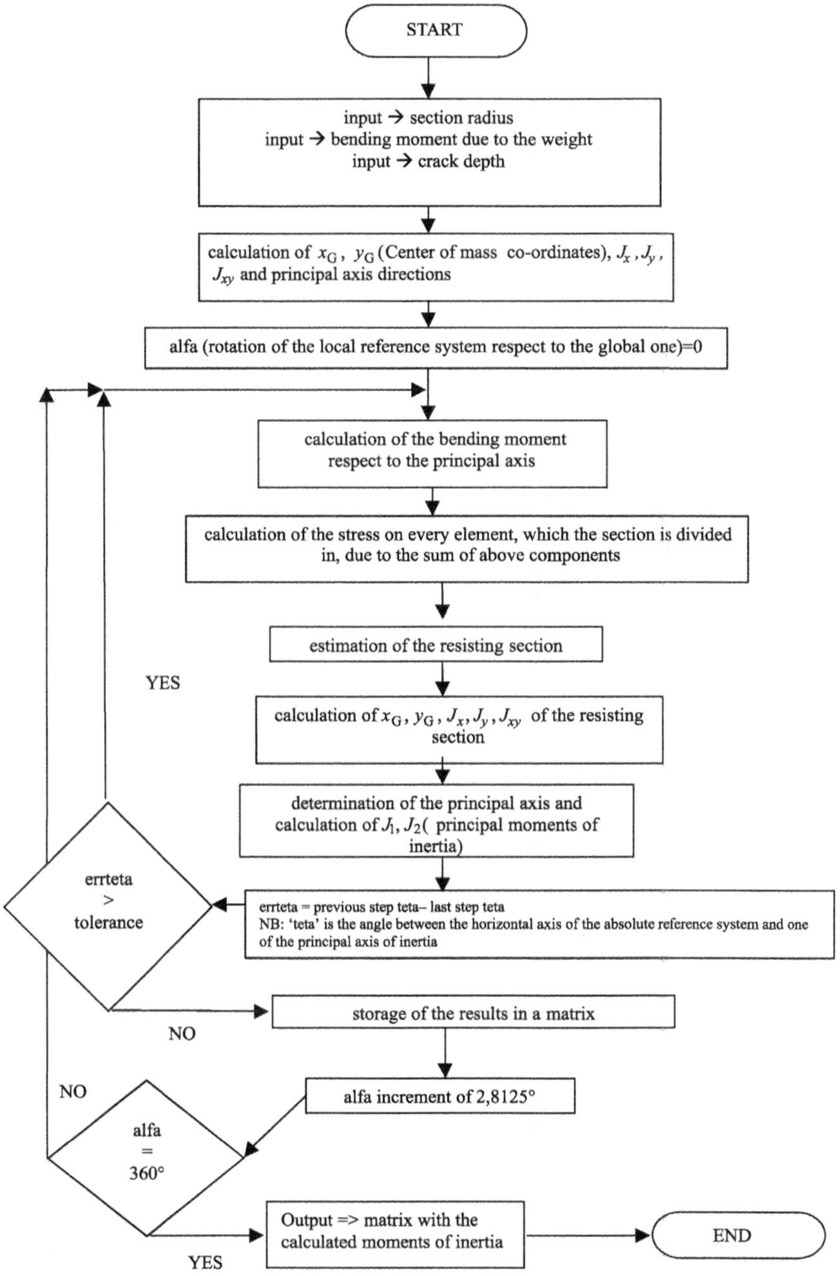

Fig. 5.24. Flow-chart of the procedure for the breathing mechanism iterative calculation.

The breathing mechanism calculated with the described simplified approach, has been validated by numerical results obtained with a 3D model of a cracked cylindrical beam, clamped at one end and loaded mechanically at the other end with a rotating load. Also temperature gradients have been imposed to the outer surface of the cylindrical specimen.

Generally an excellent agreement has been found between the simplified linear model and the 3D non-linear model. Figure 5.25 shows the comparison of 3D results, in the position where the rotor (or the load) is rotated by 75°, for the 25% and the 50% deep crack, with the simplified model results: a very good agreement has been found as also in all the other positions that are not shown for the sake of brevity. The dark areas of the crack are closed, the white areas are open.

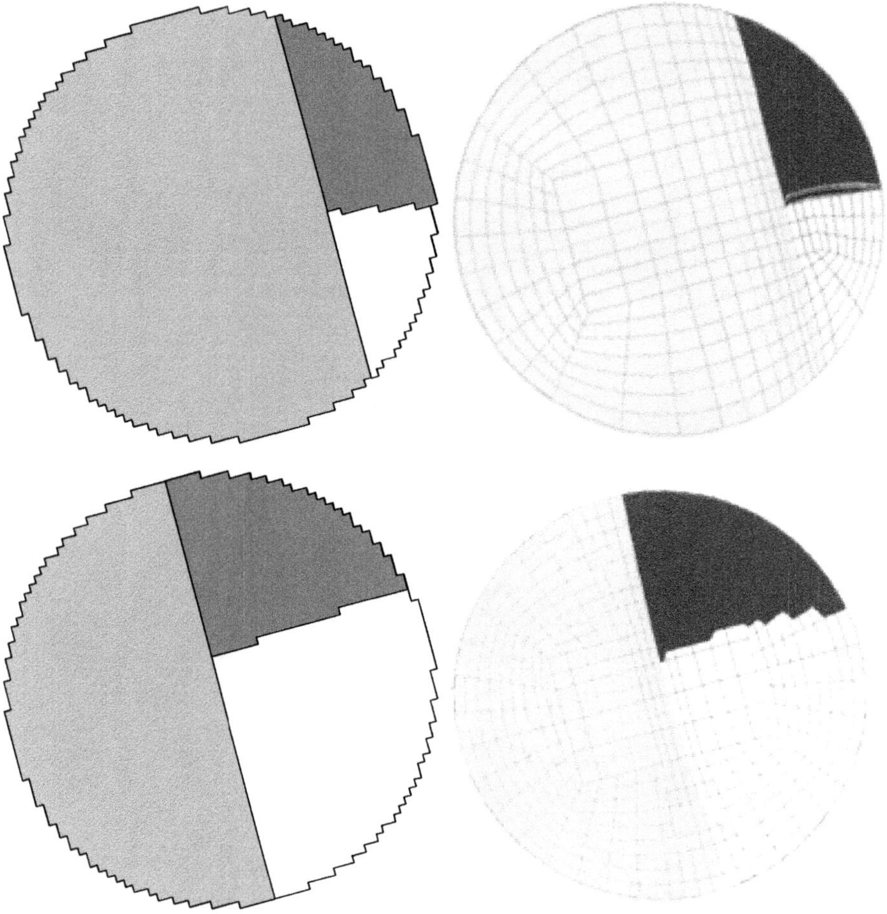

Fig. 5.25. Comparison between the results of the simplified model (left) and the 3D FE model (right) for two different crack depths (25% and 50%), rotation angle of 75°.

When a thermal transient is superposed to the mechanical loading, then the agreement is also found to be good in general. This is shown in figure 5.26 where the crack breathing in some angular positions of the load with respect to the crack is represented in case of negative thermal transient, applied to a 50% deep crack.

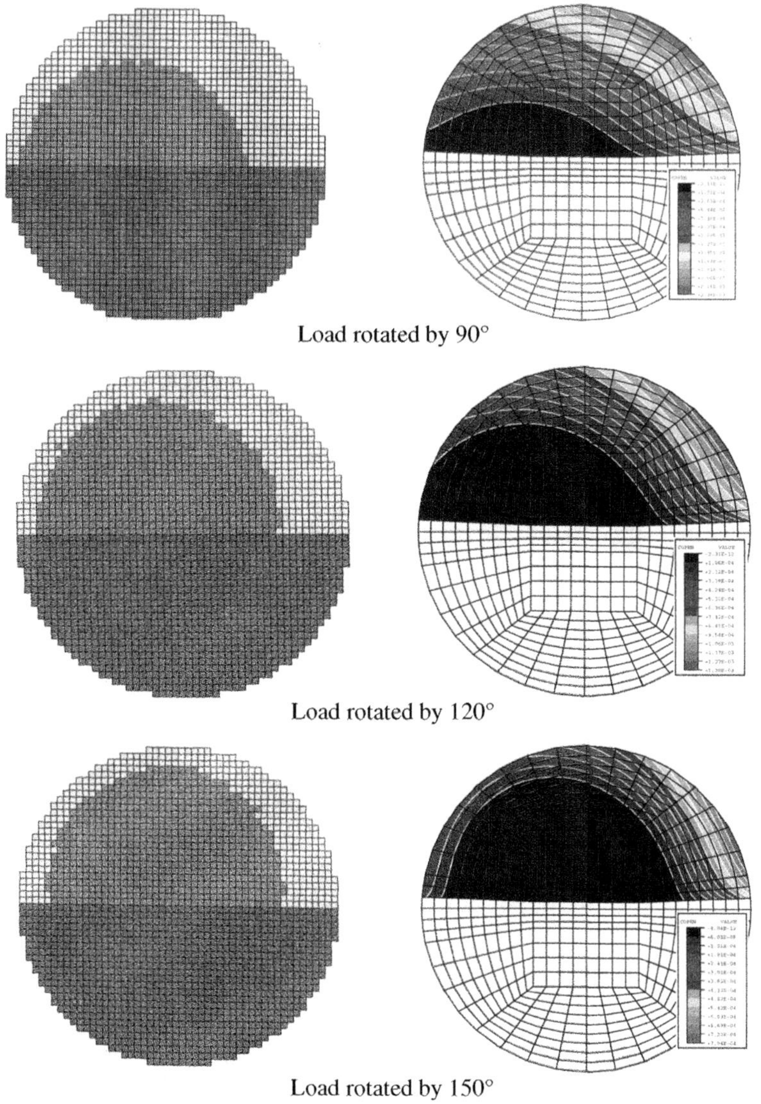

Fig. 5.26. Thermal and mechanical load – Negative gradient – Angular positions 90°, 120°, 150° and 180°, results of simplified model (left) and 3D model (right).

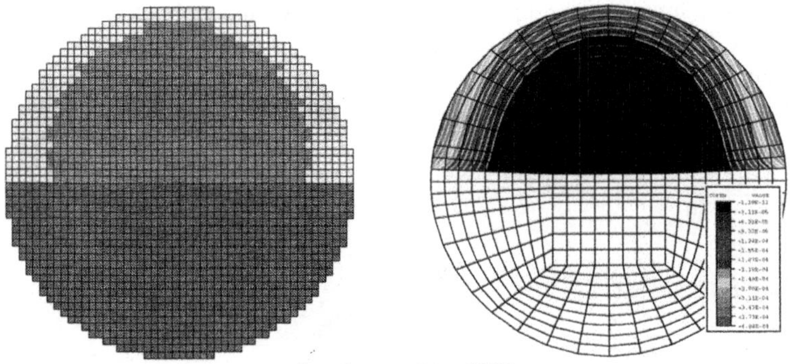

Load rotated by 180°

Fig. 5.26. (*continued*)

Breathing mechanism calculated by means of the model has also been compared experimentally with the results obtained using the crack specimen considered in section 4.2. The crack in the shaft specimen had been propagated by rotation of the specimen loaded with huge bending moment: in these conditions residual stresses may develop at the crack tip region, which try to maintain the crack closed causing the crack closure effect. The crack remains closed as long as the stresses deriving from external loads (bending moment) do not overcome the residual internal stresses, as described in section 5.2.3.3. The cracked specimen has been equipped with a series of strain-gauges, a static bending moment has been applied and the specimen has been rotated in different angular positions. In the position of completely open crack, the strain-gauge applied in correspondence of the crack axis to one of the crack lips as close as possible to the crack itself (see figure 4.2), indicates 120 $\mu\varepsilon$ of elongation instead of 0: this means that in closed configuration and no load, a compression of 120 $\mu\varepsilon$ occurs due to this crack closure effect.

The diagram of the strain measured by this strain-gauge as a function of the angular position, is shown in figure 5.27 for the maximum load of figure 4.4. The crack starts to close as soon as the angle of rotation reaches 75° and opens again at 270°. This has been simulated by means of the breathing model applied to the shape of the crack shown in figure 5.28. The crack closure effect has been simulated by superposing a constant strain value to the strain given by the applied load. A stress concentration factor has also been considered to fit the maximum measured strain with the calculated one (in closed crack configuration). The strain values calculated using these two corrective factors fit very well the measured data. It is interesting to note that the angular positions where the crack midpoint closes and opens again are the same in the model and in the test. Strain-gauges in other positions along the crack (e.g. closer to one crack end) indicated earlier closing and opening, according to the breathing model. This confirms the results obtained by both the 3D non-linear model of the crack and the extremely simple proposed 1D model.

Finally it is worth to say that 3D non-linear breathing mechanism calculation requires some hours of workstation calculation, whilst the proposed method requires only few minutes on a PC.

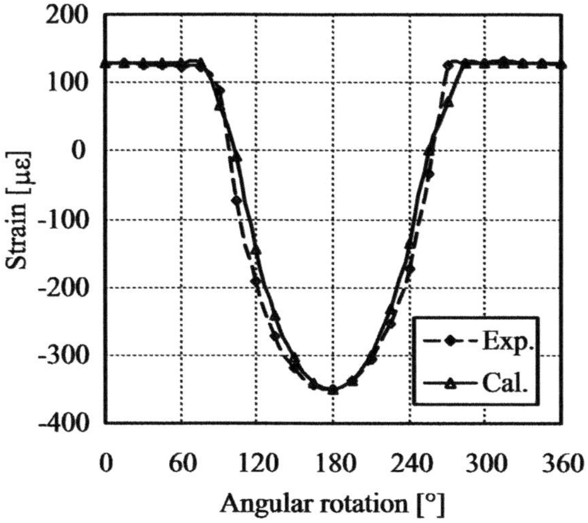

Fig. 5.27. Strain measured in correspondence of the crack (in position A6 of figure 4.5) in a cracked specimen loaded by static bending moment and rotated step by step in different angular positions.

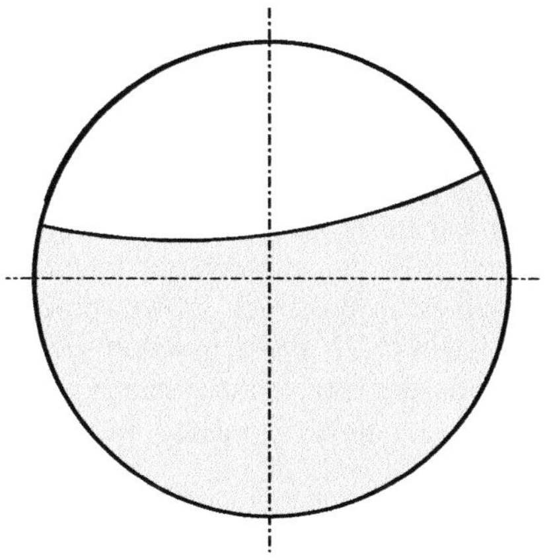

Fig. 5.28. Shape of the crack.

It has been shown that breathing mechanism in rotating shafts can be accurately reproduced by 3D non-linear finite element models and that a simple approximated model which has been developed by the authors and which can be used for calculating the variation of the compliance or of the stiffness of a cracked shaft during one revolution, is also able to simulate accurately the breathing behaviour.

5.4 Modelling the Crack

Modelling the crack, with the aim to model its effects on the static behaviour of a shaft, means to define the additional compliance, sometimes called flexibility, due to the crack in its different configurations, or to define the loss of stiffness of the element containing the crack.

This can be done using 3D finite element models, as described in section 5.2.2, by applying suitable loads and evaluating deflections. Due to the breathing behaviour, also contact conditions on crack faces must be considered, making the problem non-linear and the calculation effort much higher.

Another approach, based on fracture mechanics and on the strain energy release rate evaluation (which will be called *SERR* approach) has been used by the majority of the researchers. This approach has the advantage of a straightforward calculation of the compliance of a fully open crack, but has the drawback that breathing is generally not taken into account, except with rather approximated approaches.

The authors have developed an original approach based on the breathing behaviour evaluation described in section 5.3, which utilizes results of the 3D models for tuning one of the parameters (the equivalent length of the element): this approach will be called *FLEX* approach.

These three approaches as long as another approach, which uses also results from 3D models, are described in this section. A brief history of the approach based on fracture mechanics, as well as some comparison with other approaches is reported in [5.31].

5.4.1 Approach Based on Fracture Mechanics (SERR Approach)

The effects caused by the local flexibility of the shaft due to the presence of a crack can be considered with a *nodal model*. With this approach, the stiffness matrix of a crack finite element that is included in the FE model of the rotor is obtained by inverting a *compliance matrix*, the terms of which are evaluated on the basis of three important notions in fracture mechanics: the elastic strain energy release rate (SERR), the stress intensity factors (SIF) and the contour integrals. This method considers the energy balance involved in the crack growth process and is based on the works of Griffith [5.7] and Irwin [5.9], as discussed in section 5.2.1. The strain energy release rate G, which is the total elastic energy made available per unit increase in the crack surface area, can be expressed, in a more general way with respect to eq. (5.2) of section 5.2.1 as:

$$G = \frac{\partial U}{\partial A}(P_i, A) \tag{5.39}$$

where U is the total strain energy in the cracked body with a crack area A under the load P_i. A crack in a solid can be stressed in three different modes: opening mode, sliding mode and tearing mode. They can be denoted as mode *I*, *II* and *III* respectively, as shown in figure 5.5 of section 5.2.1. The stress intensity factors K, which are proportional to the discontinuity of the displacements of the crack sides, characterize the intensity of the singularity of the stress field. Each crack opening mode is characterized by one of these factors (K_I, K_{II}, K_{III}). They can be defined in terms of the stress components or in terms of the displacement discontinuities. By using the stress components the factors K_k, as shown in section 5.2.2, can be defined as follows:

$$K_k = \lim_{r \to 0} \left(\sigma_{ij} \sqrt{2\pi r} \right) \tag{5.40}$$

where the terms σ_{ij} are the stresses involved in the opening mode k and r is the radial coordinate in the plane perpendicular to the crack face as shown in figure 5.29.

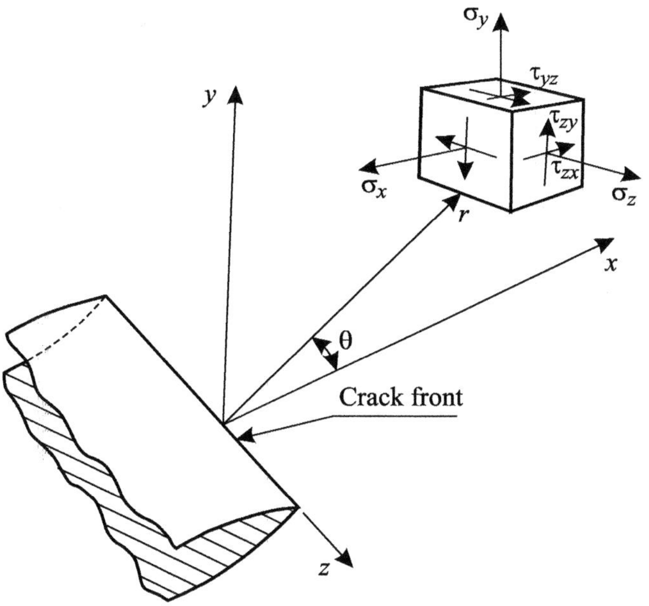

Fig. 5.29. Three dimensional crack tip coordinates and stress systems.

In the case of a rotor that is affected by a crack, whose local depth is denoted with η, with η varying between 0 and the total depth a, the factors K_k can be expressed as:

$$K_k = \sqrt{\pi \eta}\, F_k\, D_k(\sigma_{ij}) \tag{5.41}$$

The functions D_k depend on the mode k and on the stresses σ_{ij}, while the functions F_k are defined for beams with rectangular cross sections and depend on the ratio a/h between the crack depth a and the width h of the beam cross section (see figure 5.30).

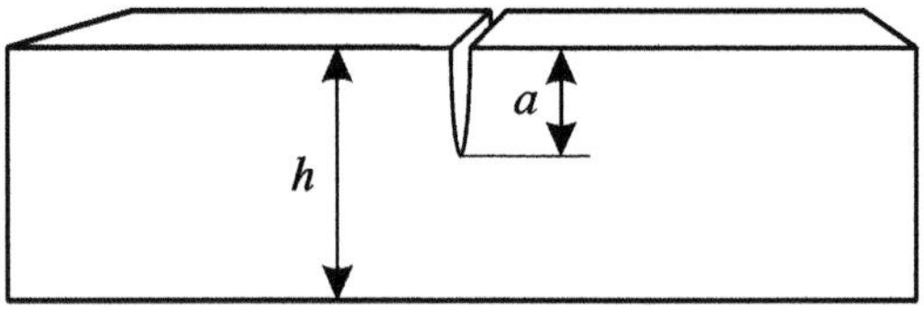

Fig. 5.30. Cracked beam with a rectangular cross section.

Then, the total energy release rate G can be expressed in terms of the stress intensity factors as:

$$G = \frac{1-\nu^2}{E}\left[\left(\sum_{i=1}^{6} K_{I\,i}\right)^2 + \left(\sum_{i=1}^{6} K_{II\,i}\right)^2\right] + \frac{1+\nu}{E}\left(\sum_{i=1}^{6} K_{III\,i}\right)^2 \tag{5.42}$$

where $K_{k\,i}$ is the stress intensity factor generated by load P_i or moment M_i in the crack opening mode k. The additional displacement u_i in direction i, due to the crack with a constant depth a and unit length, can be expressed in relation to the total deformation energy and to the load in the same direction i as:

$$u_{ia} = \frac{\partial}{\partial P_i}\left(\int_0^a G(\eta)d\eta\right) \tag{5.43}$$

The terms of the compliance matrix can be obtained by a second derivation. In the case of a crack with variable depth $a(\xi)$ along axis z and non unit width equal to $2b$, these terms are expressed as:

$$C_{ij} = \frac{\partial^2}{\partial P_i \partial P_j}\left(\int_0^a \int_{-b}^{b} G(\xi,\eta)d\xi d\eta\right) \tag{5.44}$$

Since the functions F_k and the stress intensity factors are only determined for beams with rectangular cross sections, their application in the case of usual rotors requires subdividing the circular cross section into a large number of inter-parallel rectangular sections (figure 5.31).

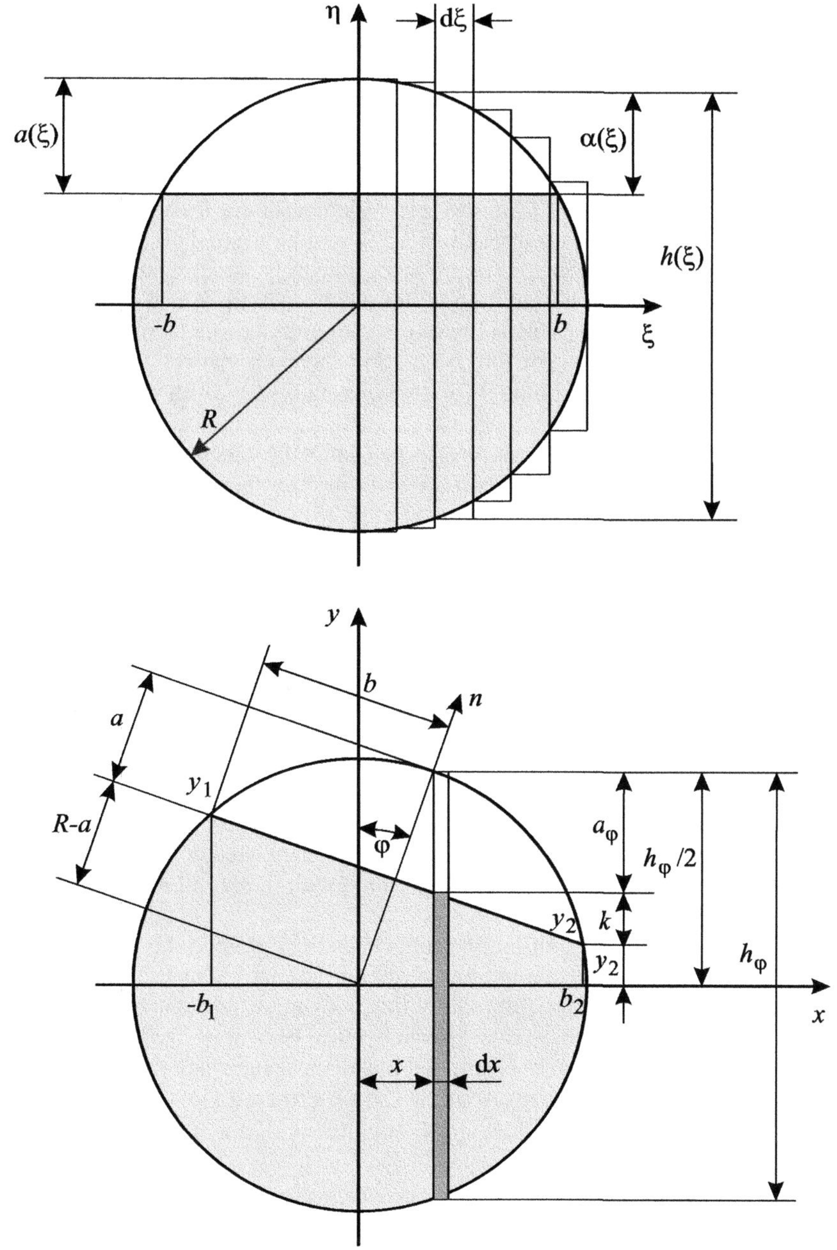

Fig. 5.31. Discretization of a circular cross section and SERR approach in rotated position.

The SERR approach, which has been applied to shafts with open cracks by Dimarogonas et al. [5.32] is relatively simple and the nodal type model allows the local character of the crack to be considered. When the cracked shaft is rotated, the calculation of the compliances is a little more complicated for taking account of the actual geometry in the rotated position as shown in the figure. When the crack is partially closed, due to the breathing, then for avoiding errors in the SERR approach, a crack closure line (CCL), the line which separates the open part from closed part, which has been introduced in [5.27], can be used. This line must be orthogonal to the crack tip, and the compliance integrals are computed with respect to the rotating reference frame, only along the crack tip which is open, see also figure 5.17 of section 5.3. This is obviously a limitation: true breathing cannot be modelled. A similar approach for calculating the compliances of a rotating cracked shaft, with a simplified model for considering also breathing behaviour is described in [5.33].

Thermal stresses and crack closure effects cannot be taken into account by these approaches. For cracks deeper than 50% of diameter the SERR approach has a singularity and cannot be used (as discussed in [5.28]).

Once the complete additional compliance matrix due to the crack has been calculated, the local stiffness matrix can be obtained by inversion: but where the compliance is zero (closed crack) there the stiffness becomes infinite and numerical problems arise. This can be avoided by inverting not the additional compliance only, but joining a piece of shaft with known compliance, and calculating the total 6×6 compliance, where all diagonal terms are different from 0. This can then be inverted and processed, as will be shown also in the following, to obtain the 12×12 stiffness matrix of the element. This element can then be introduced in the shaft finite element model.

It is important to emphasize that, in the nodal type model, the strain state of the various inter parallel rectangular sections, in which the cracked area can be subdivided, is assumed to be plane. The results obtained with the 3D FEM allowed this hypothesis to be checked.

Figure 5.32 shows the distribution of strains and stresses along the tip of a 25% depth crack. In the vicinity of the external edge of the crack, the strain increases and the stress decreases rapidly: fortunately this does not affect much the overall nodal type flexibility. A further aspect, which must be emphasized, is that the material behaviour can be no more elastic but plastic near the crack tip.

Figure 5.33 shows the stress distribution vs. the dimensionless distance from the crack tip (r/a). These results have been obtained with the 3D FEM by applying a bilinear elasto plastic behaviour as well as considering a linear elastic behaviour. The small area close to the crack tip, in which plastic behaviour can occur, does not affect the overall elastic behaviour of the cracked section.

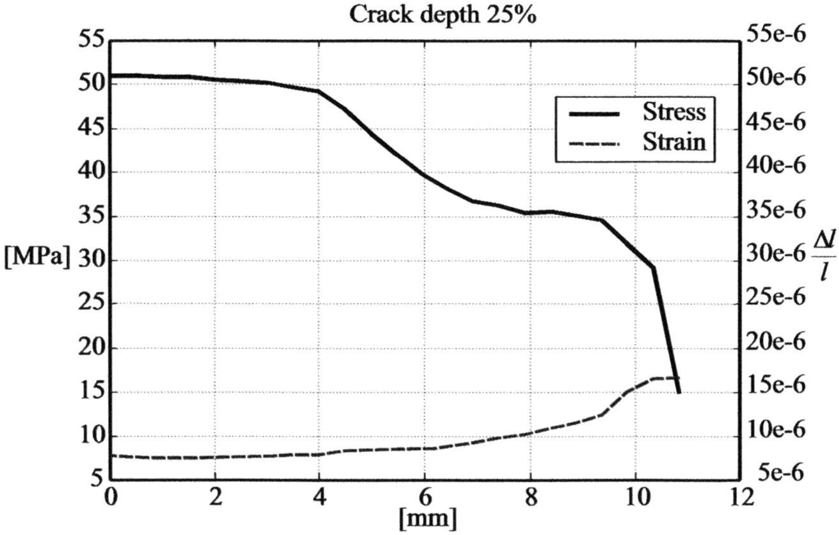

Fig. 5.32. Distribution of strains and stresses along the tip of a crack in a cylindrical specimen of diameter 25 mm, loaded by a bending moment of 100 Nm.

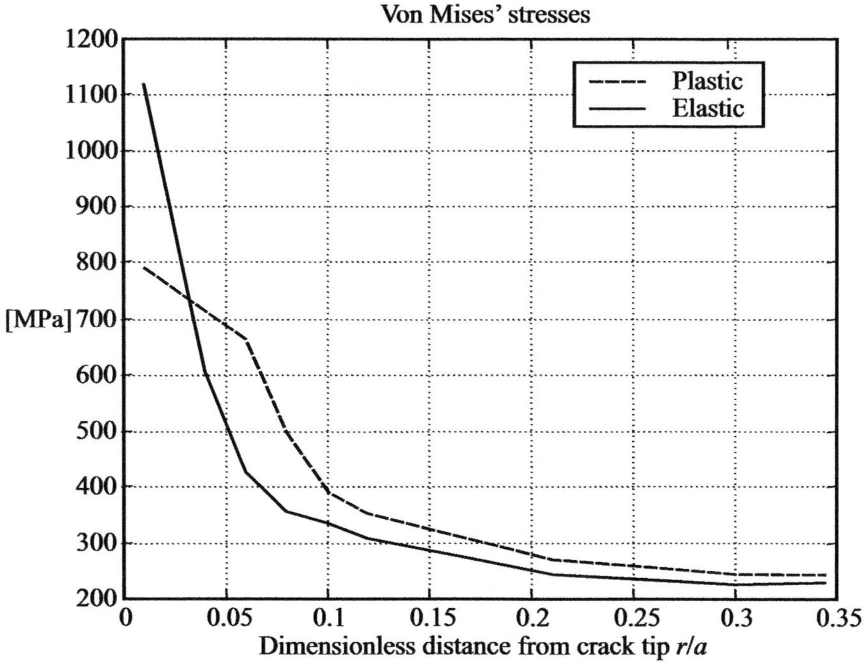

Fig. 5.33. Stress distributions evaluated by considering elastic and elastic-plastic behaviour of the material.

Since the SERR approach combined with the SIF had been used by almost all authors (as it is shown in Wauer [5.1], Gasch [5.2], and Dimarogonas [5.3]) for the calculations of the cracked beam bending behaviour, several calculations according to this approach and for different crack depths have been made. This approach allows the additional flexibility introduced by the crack to be calculated, when the crack is open. Nothing can be said when the crack is partially open, due to the breathing mechanism, except when the approximated CCL is used for simulating the breathing. Very deep cracks (more than 50% deep) as well as multiple cracks on the same cross section cannot be dealt with this approach.

For making a comparison with the other models, the breathing mechanism was assumed known (from FE model or from FLEX simplified model) and the SERR approach was applied to the cracked cross section, with its open and closed portions, in order to calculate the beam bending stiffness. The extension of this approach to the breathing crack is affected by some errors due to the fact that the crack tip is supposed to be formed by the boundary between the cracked areas and the un-cracked areas for the regions in which the breathing crack is "open", which is correct, and by the boundary between the "closed" cracked areas and the "open" cracked areas, which is not correct because on this boundary no stress intensity factors will appear.

The cracked cross section is not anymore planar, but is distorted. This is not taken into account by the fracture mechanics approach. Moreover the fracture mechanics approach does not consider any friction on the cracked area and this also seems to be unrealistic. If torsion is present, the contribution of friction forces on the cracked area can be taken into account only by the non-linear 3D calculation and by the FLEX simplified model in an approximate way. Nevertheless the results obtained with this approach are very accurate as regards the additional flexibility introduced by the crack, for a completely open crack, as well as for providing stress intensity factory at the crack tip, which are extremely important for evaluating the propagation mechanism. The additional flexibility can be easily transformed in local crack stiffness.

5.4.2 Approach Based on Energy Balance (EDF)

R&D – AMA Department of Électricité de France (EDF) has developed an original method that utilizes 3D modelling of the crack and derives the crack local stiffness by means of an elastic energy approach for each angular position of the actual bending moment (as it results from static and dynamic loads) with respect to the crack. The method is described in [5.30] in detail.

A 3D cracked beam element is considered, the length and the diameter of which are respectively equal to $2L$ and $D = L$, see figure 5.34.

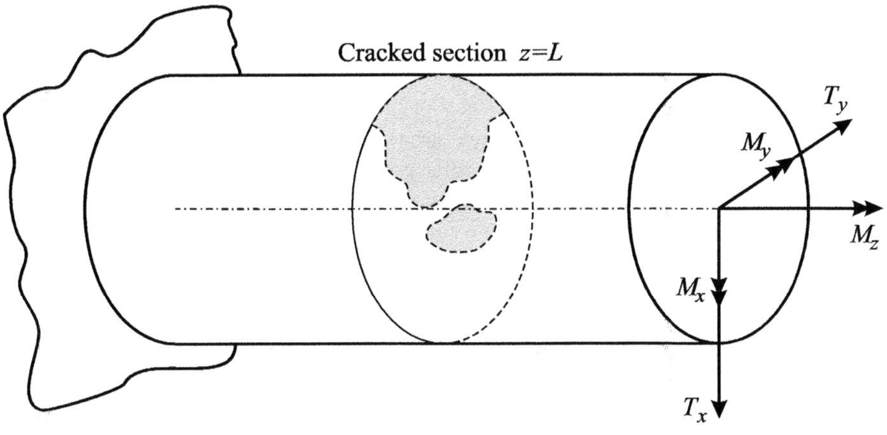

Fig. 5.34. Overall geometry of the 3D general problem under bending loads, torque and shear forces.

The beam is clamped at $z=0$ and moments $\mathbf{M}=(M_x,M_y)$ are applied at $z=2L$. The crack is located at $z=L$, and its profile is not relevant. A monodimensional equivalent system is constituted of two beams of length L, corresponding to the crack-free parts of the cracked beam element, connected by a discrete non-linear spring which represents the lumped crack model. The aim is to determine the stiffness of the nodal beam element that represents the crack in the 1D beam model, starting from the flexibility introduced by the crack and calculated on the 3D beam element. Applied load is a bending moment and flexibility is evaluated as relative angular deflection of the crack faces. That means that flexibility is 0 in closed crack configuration and is maximum in open crack configuration. Moreover the relative angular deflection of the crack faces develops in a direction which is different from that one of the applied bending moment due to the anisotropy of the cracked beam.

The calculation of the lumped flexibility is based on a 3D energetic approach. The following main hypotheses are introduced:

a) no surface traction loads are applied on the lateral surface of the beam;
b) no volume forces are applied on the beam;
c) small perturbations theory;
d) the effect of the crack are negligible at $z=2L$ and the 3D sections in $z=0$ and $z=2L$ are not deformed;
e) the initial displacement gap between the crack lips without any load is zero;
f) unilateral contact conditions without friction are applied on the crack lips.

The unilateral contact conditions applied to the crack lips mean:

- the crack lips cannot penetrate each other;
- the normal stress is only compressive;

- either the crack is open and the normal stress is zero or the crack is closed and the normal stress is strictly negative.

Then calculation follows these steps:

1) the elastic energy of the crack area is calculated by means of the 3D model, for several different angular positions Φ, with respect to the crack tip, of a reference bending moment applied to the cracked beam;
2) the local flexibility (the relative rotation of the crack lips) due to the crack, which is obviously function of the angle Φ, is calculated from the energy. Figure 5.35 shows an example of the flexibility $s(\Phi)$;
3) the lumped bending 2×2 stiffness matrix $[\mathbf{K}(\varphi)]$ is calculated from flexibility $s(\Phi)$ and from eq. (5.45) that defines the direction φ of the angular deflection of the cracked element, which is different from Φ:

$$\varphi = \arctan\frac{\theta_y}{\theta_x} \tag{5.45}$$

where θ_y and θ_x are the relative rotations of the crack surface along y and x axes.

4) the different values of stiffness matrix $[\mathbf{K}(\varphi)]$ as function of angle φ are stored and will be used in the code for calculating the dynamic behaviour of cracked shafts.

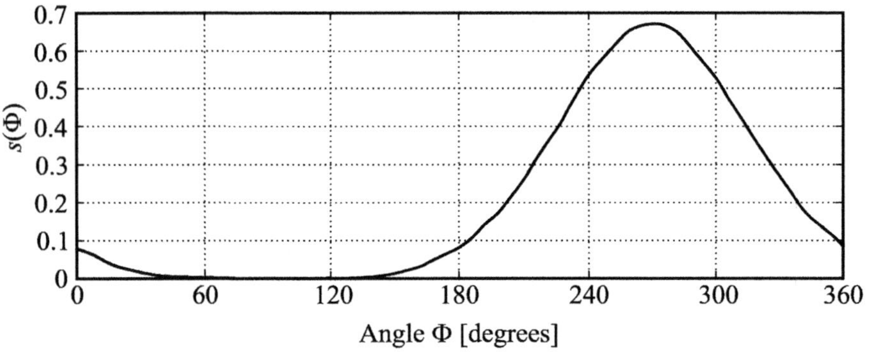

Fig. 5.35. Flexibility as function of the direction of the applied end-moment for a 50% cracked element.

Figure 5.35 shows the flexibility $s(\Phi)$ as a function of the angle Φ between the applied moment and the crack tip, for a 50% diameter cracked beam element, obtained using a 36000 degrees of freedom finite element model, with 320 nodes placed in the cracked area subjected to contact conditions. All 3D non-linear calculations have been carried out by EDF R&D software Code_Aster©, [5.35].

The conclusions obtained by means of the energy variation principles allow responding to the following needs:

- to study any type and dimension of crack;
- to reduce the problem of great number 3D calculation;
- to represent the crack breathing mechanism;
- once the shaftline is modelled, the non-linear system including the crack is solved by means of a time transient resolution using an explicit integration scheme. In fact, at each time step, the 1D software calculates θ_y, θ_x and φ by means of eq. (5.45) and gets the related values of $[\mathbf{K}(\varphi)]$ from the table where they are stored.

These values are inserted in the stiffness matrix of the rotor system and the dynamic behaviour equations are solved in the time domain by a time step integration procedure. Advantages are that breathing is accurately represented, and furthermore also non-linear behaviour is automatically taken into account by means of the time step procedure; disadvantage is the huge cost in terms of computation time.

5.4.3 Approach Based on "Equivalent Beam" (FLEX)

This approach has been developed in different steps by considering the equivalent beam of figure 5.36, in which the cross section is formed by the un-cracked part of the section plus the closed part of the cracked section. The length of the equivalent beam l_c will be defined later as function of the depth of the crack.

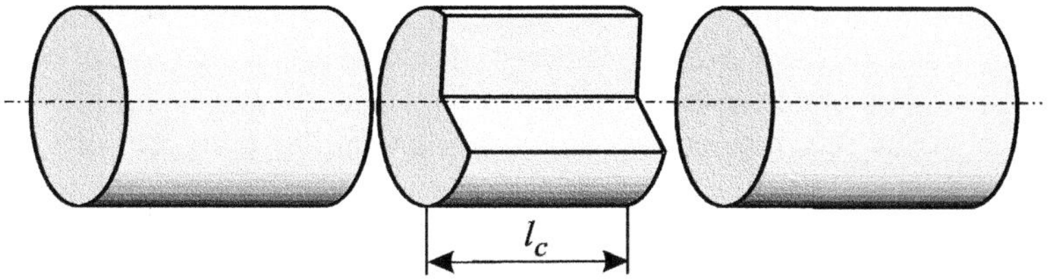

Fig. 5.36. Equivalent cracked beam element: its configuration changes with its angular position.

In the first step, the flexural behaviour has been defined by building the 4×4 stiffness matrix of the equivalent beam clamped at one end, as shown in figure 5.37. Only horizontal and vertical displacements and angular deflections around horizontal and vertical axes are considered, i.e. 4 degrees of freedom per node are considered. This is made by using standard Timoshenko beam approach and introducing the different second moments of area of the cross section. The extension to the 8×8 stiffness matrix of the free-free beam is trivial as it is derived from equilibrium conditions of the beam.

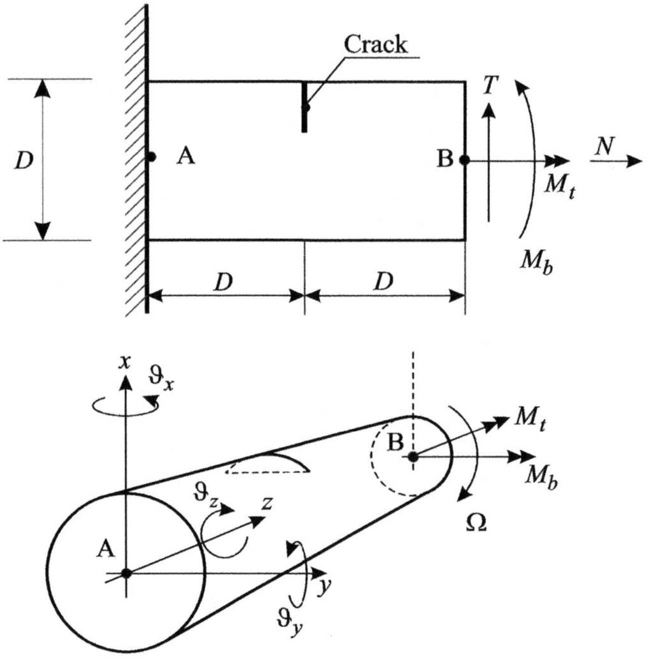

Fig. 5.37. Cracked beam boundary conditions and reference frame.

The primary interest is focused on the flexural bending behaviour of cracked shafts because the lateral vibrations in general are the only vibrations that are measured in industrial machines. Therefore in the first step only the above specified degrees of freedom have been considered. The derivation of the stiffness matrix was then rather straightforward.

In a second step, the complete behaviour according to all the 6 degrees of freedom per node, has been analysed in order to model also all coupling effects between bending behaviour and torsion and axial behaviour. It was not possible to use the same approach as for the 4×4 stiffness matrix. The easiest way was to switch from stiffness approach to flexibility approach, in order to introduce in the flexibility matrix the coupling coefficients that could be estimated by means of expected physical behaviour. The procedure was then the following. The clamped beam must be used for defining flexibility. The 4×4 flexibility matrix was obviously obtained by inversion of the corresponding stiffness matrix. Then the 6×6 flexibility has been built by adding rows and columns and introducing direct torsion and axial flexibilities and estimated coupling coefficients. Finally the flexibility matrix has been inverted to obtain the 6×6 stiffness matrix of the clamped beam and lastly the 12×12 stiffness matrix of the free-free beam has been obtained applying usual equilibrium conditions.

These steps are described more in detail in the following.

Once the breathing mechanism and the second moments of area have been defined for the different angular positions, as described in section 5.3, the stiffness matrix of an equivalent cracked beam element of suitable length l_c can be calculated, assuming a Timoshenko beam with constant section and second moments of area along l_c, as shown in figure 5.36, for each different angular position. The neighbour beam elements are normal circular section beams. Similarly to what will be shown for the simple Jeffcott / de Laval rotor in section 5.5, the angular position dependent stiffness matrix $[\mathbf{K}_c(\Omega t)]$ results periodical, has one constant term (the mean stiffness) $[\mathbf{K}_m]$ and several harmonic components $[\Delta\mathbf{K}_i]$ of which only the first three components (with $i = 1$, 2 or 3) are considered in the following calculations.

The stiffness matrix (square, symmetrical, 8×8 elements) is given by:

$$[\mathbf{K}_c(\Omega t)]\begin{Bmatrix}\mathbf{x}_1\\ \mathbf{x}_2\end{Bmatrix} = \begin{bmatrix} a & c & p & q & -a & c & -p & q \\ & e & -q & r & -c & f & q & s \\ & & b & -d & -p & -q & -b & -d \\ & & & h & -q & s & d & g \\ & & & & a & -c & p & -q \\ & & & & & e & q & r \\ & & & & & & b & d \\ & & & & & & & h \end{bmatrix}\begin{Bmatrix} x_1 \\ \vartheta_{y1} \\ y_1 \\ \vartheta_{x1} \\ x_2 \\ \vartheta_{y2} \\ y_2 \\ \vartheta_{x2} \end{Bmatrix} \tag{5.46}$$

where the coefficients are defined as:

$$\begin{gathered} a=\frac{12J_y E}{(1+\phi)l_c^3},\ b=\frac{12J_x E}{(1+\phi)l_c^3},\ c=\frac{6J_y E}{(1+\phi)l_c^2},\ d=\frac{6J_x E}{(1+\phi)l_c^2},\ e=\frac{(4+\phi)J_y E}{(1+\phi)l_c}, \\ f=\frac{(2-\phi)J_y E}{(1+\phi)l_c},\ g=\frac{(2-\phi)J_x E}{(1+\phi)l_c},\ h=\frac{(4+\phi)J_x E}{(1+\phi)l_c},\ p=\frac{12J_{xy} E}{(1+\phi)l_c^3}, \\ q=\frac{6J_{xy} E}{(1+\phi)l_c^2},\ r=\frac{(4+\phi)J_{xy} E}{(1+\phi)l_c},\ s=\frac{(2-\phi)\,J_{xy} E}{(1+\phi)l_c} \\ \phi=\frac{12EJ}{GSl_c^2} \end{gathered} \tag{5.47}$$

The parameter ϕ accounts for the shear effects, E and G are respectively the Young's modulus and the shear modulus, S is the cross section area, J are the second moments of area. These second moments of area depend on the breathing, on the open and closed parts of the crack, which are function of the angular position, and are therefore periodical, as also the stiffness matrix $[\mathbf{K}_c(\Omega t)]$. Performing a Fourier's expansion of these periodical quantities, it is possible to extract a constant component ($[\mathbf{K}_m]$ function of the mean values of the second moments J_m) and several harmonic components ($[\Delta\mathbf{K}_i]$ function of the corresponding ΔJ_i harmonic components of the second moment of area variations).

In order to extend the model to 6 degrees of freedom per node, considering also the axial and torsional coordinates, for expressing the direct and the cross-coupling coefficients related to the additional degrees of freedom, the stiffness approach has been left and the more suitable flexibility approach has been adopted. Following procedure has then been used, because the different additional stiffness coefficients could not be derived directly from geometrical cross section properties of the equivalent beam: bending flexibility matrix has been obtained from above stiffness matrix, the flexibility matrix has been enlarged to accommodate the additional degrees of freedom, and finally the complete stiffness matrix has been built by inversion of the corresponding flexibility. It should be reminded that differently from stiffness the flexibility cannot be defined for a free-free beam, but only for beams with constraints.

Figure 5.37 shows the cracked beam used for the stiffness and flexibility matrix calculation, along with the definition of the reference frame for the 6 degrees of freedom. The beam is considered clamped at one end (A) and the external loads are applied to the other end (B). Here the displacements according to the 6 degrees of freedom are evaluated. First the flexibility matrix is calculated.

The 4×4 stiffness matrix in node B $[\mathbf{K}_{\mathrm{BB}}]$ of the clamped beam is obtained by partitioning the 8×8 stiffness matrix of the free-free beam defined by eq. (5.46). Partitioning is shown in eq. (5.48), where also the 4×4 flexibility matrix $[\mathbf{H}_{\mathrm{BB}}]$ is shown as obtained by inversion of $[\mathbf{K}_{\mathrm{BB}}]$. Index A refers to the left end of the beam, which is clamped for the flexibility calculation, index B refers to the right hand side end of the beam, where loads are applied and deflections are calculated. In this flexibility matrix then the coefficients related to the axial and torsion degrees of freedom are introduced.

$$\begin{aligned}
\left[\mathbf{K}_c(\Omega t)\right]_{8\times8} &= \begin{bmatrix} [\mathbf{K}_{\mathrm{AA}}]_{4\times4} & [\mathbf{K}_{\mathrm{AB}}]_{4\times4} \\ [\mathbf{K}_{\mathrm{BA}}]_{4\times4} & [\mathbf{K}_{\mathrm{BB}}]_{4\times4} \end{bmatrix} \\
[\mathbf{H}_{\mathrm{BB}}]_{4\times4} &= \left([\mathbf{K}_{\mathrm{BB}}]_{4\times4}\right)^{-1}
\end{aligned} \tag{5.48}$$

The flexibility matrix $[\mathbf{H}_{\mathrm{BB}}]_{6\times6}$ is obtained from $[\mathbf{H}_{\mathrm{BB}}]_{4\times4}$ by adding rows and columns corresponding to the axial and torsion degrees of freedom and introducing direct flexibility coefficients and lateral-axial and lateral-torsion coupling coefficients.

The direct flexibility coefficient for the axial degree of freedom h_{55} is evaluated by considering a resisting area equal to the un-cracked sectional area plus the closed portion of the cracked area, combined to a suitable axial length l_{ax}:

$$h_{55} = \frac{EA}{l_{ax}} \tag{5.49}$$

This model remains valid as far as the axial vibration does not interfere with the breathing behaviour, which means that the stresses in the cracked section due to the axial vibrations should be negligible with respect to the bending stresses caused by the weight and other stationary forces.

The direct flexibility coefficient for the torsion degree of freedom h_{66} is evaluated by considering the polar second moment of area of the resisting area equal to the un-cracked sectional area plus the closed portion of the cracked area, provided that no micro-slipping occurs on the crack, combined to a suitable torsion length l_{tors}:

$$h_{66} = \frac{GJ_p}{l_{tors}} \tag{5.50}$$

In this simplified crack model the true tangential stress distribution is completely disregarded and linear stress distribution is assumed in radial direction, for the sake of simplicity.

The breathing behaviour model (described in section 5.3) does not consider the effect of tangential stresses or torsional loads.

Numerical simulations with 3D finite element models have shown that the influence of torque on the breathing behaviour for horizontal heavy shafts affected by transverse cracks is negligible.

In the numerical applications where torsion is involved following assumptions have been made for taking account of possible micro-slipping conditions: the closed portion of the crack contributes entirely to the polar moment if the tangential stress is less than the value of the friction coefficient times the normal axial stress, otherwise a reduction factor is introduced in the resisting area to account for micro-slipping conditions.

The values of friction coefficient as well as the corresponding values of the reduction factors have been assumed equal to 0.2 and sometimes 0.4 for accounting for rough crack surfaces.

The cross coupling coefficients are evaluated by considering the resultant shear force and/or axial force simply applied to the resisting area centre instead of the circular shaft centre: the distance between the two centres is responsible for the arising of torsion and/or bending moments, which generate torsion displacements (due to shear forces) and lateral displacements (due to axial forces). This approach is obviously roughly approximated, but the comparison with 3D non linear calculation results shows very good agreement, as long as single cracks with regular shapes (rectilinear or elliptical) are considered.

The cross coupling coefficients have following expressions:

- vertical displacement x and torsion moment M_t:

$$h_{16} = h_{61} = \frac{l_{tors}}{GJ_p} y_G \tag{5.51}$$

- horizontal displacement y and torsion moment M_t:

$$h_{36} = h_{63} = \frac{l_{tors}}{GJ_p} x_G \tag{5.52}$$

- vertical displacement x and axial force N :

$$h_{15} = h_{51} = \sqrt{x_G^2 + y_G^2} h_{12} \tag{5.53}$$

- rotation ϑ_y and axial force N :

$$h_{25} = h_{52} = \sqrt{x_G^2 + y_G^2} h_{22} \tag{5.54}$$

- horizontal displacement y and axial force N :

$$h_{35} = h_{53} = y_G h_{34} \tag{5.55}$$

- rotation ϑ_x and axial force N :

$$h_{45} = h_{54} = y_G h_{44} \tag{5.56}$$

where x_G and y_G are the resisting area centre coordinates.

Flexibility matrix with coupling terms between lateral displacements – rotations and axial force – torsion moment is then given by:

$$[\mathbf{H}_{BB}]_{6\times6} = \begin{bmatrix} h_{11} & h_{12} & h_{13} & h_{14} & h_{15} & h_{16} \\ h_{21} & h_{22} & h_{23} & h_{24} & h_{25} & 0 \\ h_{31} & h_{32} & h_{33} & h_{34} & h_{35} & h_{36} \\ h_{41} & h_{42} & h_{43} & h_{44} & h_{45} & 0 \\ h_{51} & h_{52} & h_{53} & h_{54} & h_{55} & 0 \\ h_{61} & 0 & h_{63} & 0 & 0 & h_{66} \end{bmatrix} \tag{5.57}$$

All the elements of the flexibility matrix as well as those of the stiffness matrix are dependent on the angular shaft position. The calculation of the stiffness matrix of the clamped beam element is obtained by inversion of the flexibility matrix:

$$[\mathbf{K}_{BB}]_{6\times6} = \left([\mathbf{H}_{BB}]_{6\times6}\right)^{-1} \tag{5.58}$$

Matrix $[\mathbf{K}_c(\Omega t)]_{12\times12}$ is the complete stiffness matrix of the cracked free-free beam, which can then be introduced in the stiffness matrix of the complete cracked shaft. Matrix $[\mathbf{K}_c(\Omega t)]_{12\times12}$ is obtained as usual by means of symmetry and equilibrium conditions.

This can be obtained introducing a transformation matrix $[\mathbf{T}]$ with which symmetry and hemi-symmetry conditions as well as equilibrium conditions can be expressed.

$$[\mathbf{T}]=\begin{bmatrix} 1 & 0 & 0 & 0 & 0 & 0 & -1 & l & 0 & 0 & 0 & 0 \\ 0 & 1 & 0 & 0 & 0 & 0 & 0 & -1 & 0 & 0 & 0 & 0 \\ 0 & 0 & 1 & 0 & 0 & 0 & 0 & 0 & -1 & -l & 0 & 0 \\ 0 & 0 & 0 & 1 & 0 & 0 & 0 & 0 & 0 & -1 & 0 & 0 \\ 0 & 0 & 0 & 0 & 1 & 0 & 0 & 0 & 0 & 0 & -1 & 0 \\ 0 & 0 & 0 & 0 & 0 & 1 & 0 & 0 & 0 & 0 & 0 & -1 \end{bmatrix} \tag{5.59}$$

$$[\mathbf{K}_c(\Omega t)]_{12\times12} = [\mathbf{T}]^{\mathrm{T}}[\mathbf{K}_{\mathrm{BB}}]_{6\times6}[\mathbf{T}] \tag{5.60}$$

Now the length of the equivalent beam l_c has to be defined. Length l_c represents in some way the length to which the influence of the discontinuity introduced by the crack is extended, which obviously depends on the crack depth. This cannot be found in a simple way, because axial stresses due to bending, start from strongly non-linear distribution in correspondence of the crack and along the beam gradually assume a quasi-linear distribution over the cross section as the distance from crack increases. Therefore the length of the equivalent beam l_c has been defined empirically as the length with which the best agreement (in terms of additional crack induced deflections) between 3D calculation results and the on l_c value based FLEX model results have been obtained. This length, which performs the best fitting between 3D and 1D FLEX, depends only on the depth of the crack and is considered constant and independent on how much the crack is open or closed. The length for bending behaviour called l_{flex} is in general different from length for torsion behaviour l_{tors}, as well as for axial behaviour l_{ax}. However, this last parameter has little influence because axial excitation is assumed to be negligible, otherwise it would affect the breathing behaviour.

The dimensionless values of l_{flex} and l_{tors} (referred to the depth a of the crack) are shown in figure 5.38 and figure 5.39 as function of the dimensionless depth p (referred to the diameter of the shaft D), for transverse cracks with rectilinear tip. Only negligible differences have been found in case of elliptical crack tips.

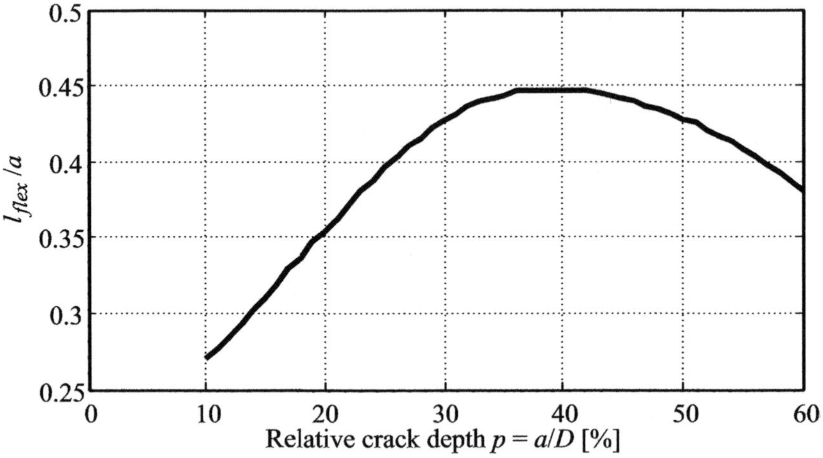

Fig. 5.38. Dimensionless length l_{flex}/a as function of dimensionless depth $p = a/D$.

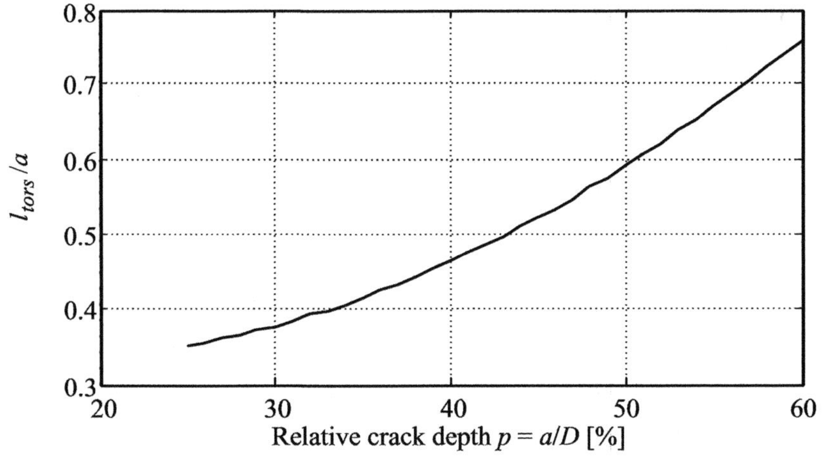

Fig. 5.39. Dimensionless length l_{tors}/a as function of dimensionless depth $p = a/D$.

It should finally be reminded that modelling the crack in a shaft requires to modify and adjust the finite elements of the shaft in order to fit the cracked element with its length l_c in the model of the shaft. If bending vibrations have to be calculated l_c is equal to l_{flex}, when torsion vibrations are analysed, l_c has to be taken equal to l_{tors} . The FE model has to be modified in order to accommodate the element with the desired length l_c . It is obviously possible to incorporate the equivalent beam in the middle of a longer beam by means of a static condensation. When the full degrees of freedom are used it is convenient to apply the static condensation technique to define the stiffness matrix of the longer beam for incorporating the equivalent beams of both lengths l_{flex} and l_{tors}.

5.4.4 The 3D Model

The 3D non-linear model allows obviously calculating also deflections and strains, by taking into account the breathing behaviour. Also the model developed by EDF), see section 5.4.2, uses a FE approach, but in this case the calculations were performed by using ABAQUS.

Figure 5.40 shows one of the meshes that have been used by the authors for the cracked test beam, that is that relative to 50% crack depth. In this case roughly 11000 elements have been used for the analysis of the cracked cylindrical beam. The mesh in the neighbourhood of the crack has been made according to the specifications described in section 5.2.2. The mesh has been chosen rather dense because not only deformations of the cracked specimen, but also stress intensity factors in correspondence of the crack tip have been calculated numerically and compared with those calculated by means of the classical fracture mechanics approach. This comparison allowed evaluating the accuracy of the model as regards its capability of representing real crack behaviour in the region close to the crack. The elastic limit was never exceeded in the simulations. Other comparisons, with other meshes for different crack depths, made with same criteria, are not reported for the sake of brevity, because they show similar results.

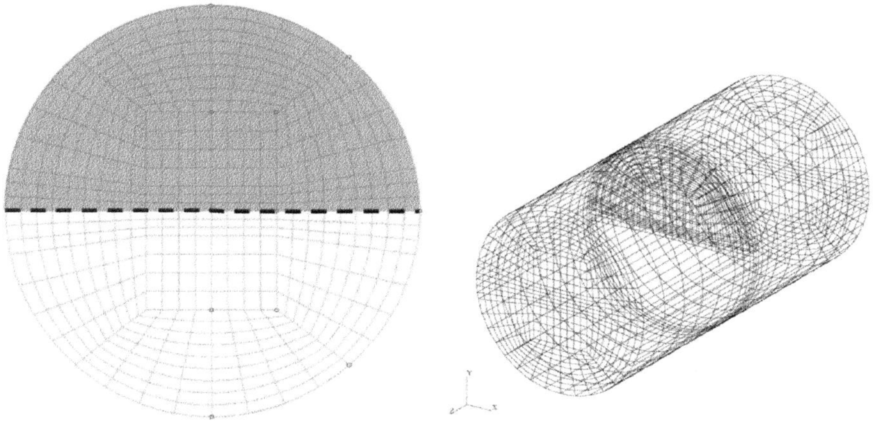

Fig. 5.40. Mesh of the section and isometric view of the model with a crack of 50% of the diameter.

The contact model in the cracked surface is obviously non-linear. Also a friction coefficient ($\mu = 0.2$) has been introduced in order to account for micro-slip conditions in the cracked area, due to shear forces and torsion. In order to avoid local deformations due to the application of loads, the model has been extended to a higher length where the load is applied to the specimen (see figure 5.22). This way no local deformations due to external load application are present in the cracked area and in the "measuring" section, where the deflections are evaluated, indicated by the dashed line in figure 5.22. The results obtained with this model will be called simply 3D results.

A typical deformation shape with an opening crack is shown in figure 5.41.

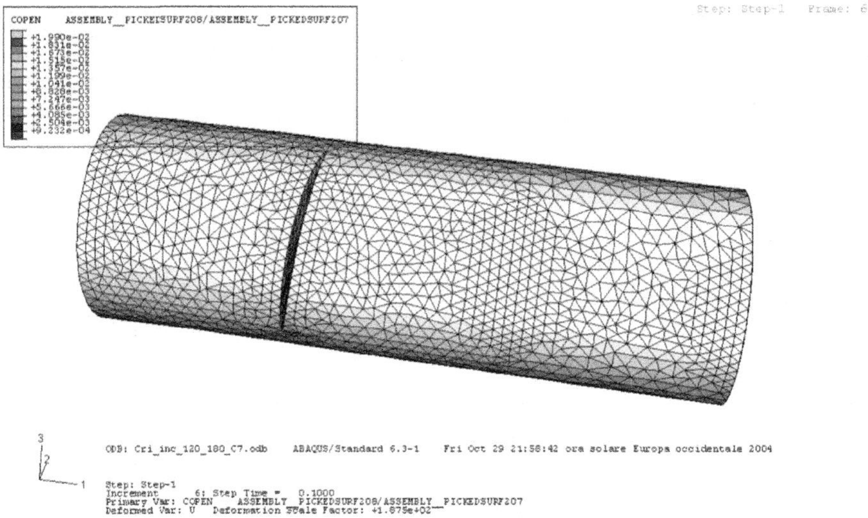

Fig. 5.41. Typical deformation shape of the cracked specimen in crack opening situation.

5.4.5 Comparison of Results Obtained with the Models

Three different models (3D, FLEX and SERR) have been compared by calculating deflections of a test beam in different load conditions. The test beam is a cylinder with a diameter of 25 mm and a length of 50 mm, clamped at one end, and with rotating loads applied to the other end. Figure 5.42 to figure 5.45 show some of the obtained results, for a crack with rectilinear tip and relative depth of 50% of the diameter. The considered loads are not only bending moments, but bending plus torsion loads or bending plus shear loads, as it occurs in real machines where torsion and shear loads are generally combined to bending.

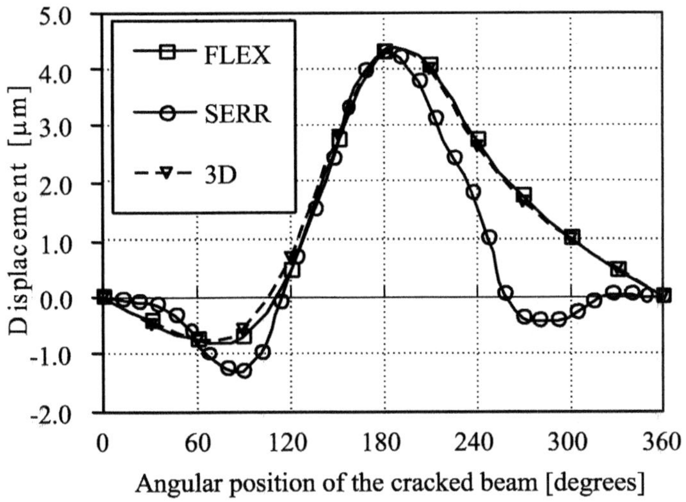

Fig. 5.42. Bending and torsion, 50% crack depth, x displacement as function of cracked beam angular position.

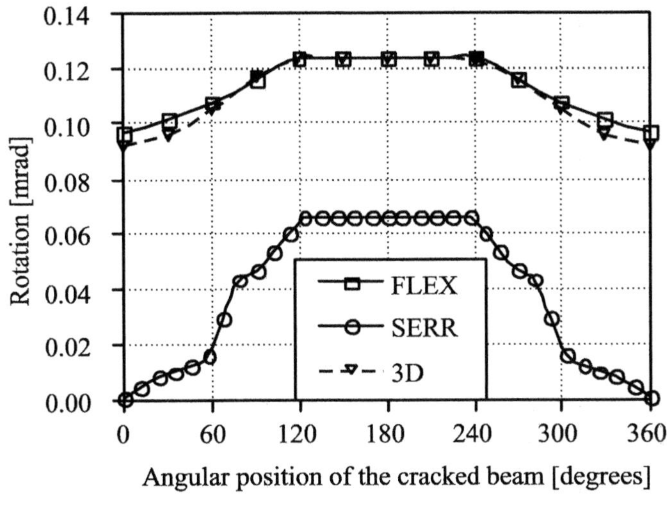

Fig. 5.43. Bending and torsion, 50% crack depth, θ_z torsional deflection as function of cracked beam angular position.

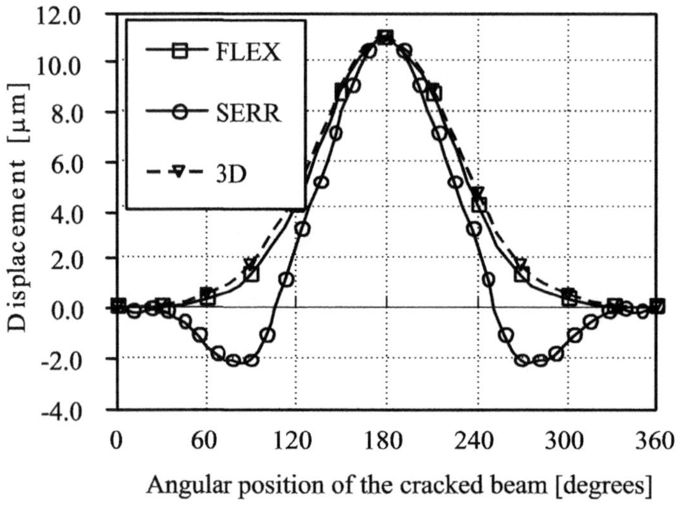

Fig. 5.44. Bending and shear, 50% crack depth, x displacement as function of cracked beam angular position.

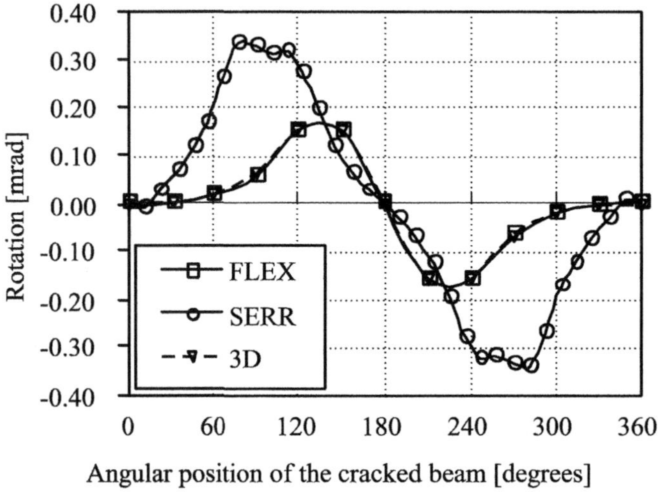

Fig. 5.45. Bending and shear, 50% crack depth, θ_x angular deflection as function of cracked beam angular position.

All displacements calculated for fixed beam and rotating load and referred to the fixed reference frame have been processed in order to represent displacements according to the same fixed reference, but caused by fixed load and rotating beam.

In order to highlight the effect of the crack only, the corresponding displacements of the un-cracked beam have been subtracted in all the figures. As can be seen, the FLEX model seems to be more accurate with respect to the modified

SERR model, assuming as reference the 3D calculation. In particular, the effects of tangential stresses due to torsion or shear forces cannot be correctly represented by the modified SERR model.

In regards to the bending-torsion coupling coefficients, a comparison between FLEX and 3D, made for the beam with different crack depths loaded by bending and torsion (the case with 60% relative depth and rectilinear tip is shown in figure 5.46 to figure 5.51), validates the FLEX approach.

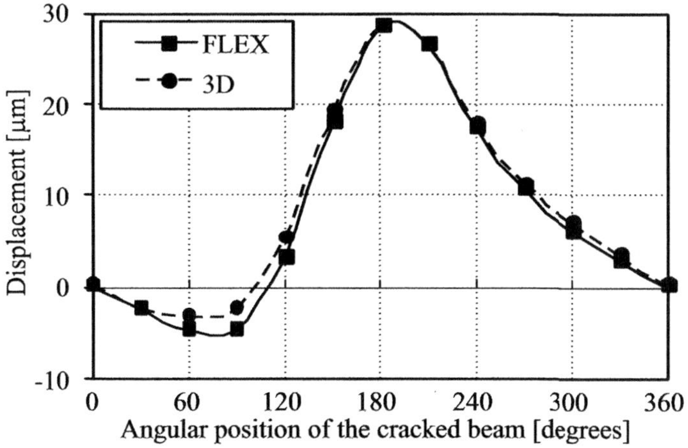

Fig. 5.46. Comparison between 3D calculation and FLEX simplified model, 60% crack depth, vertical displacement x as function of cracked beam angular position.

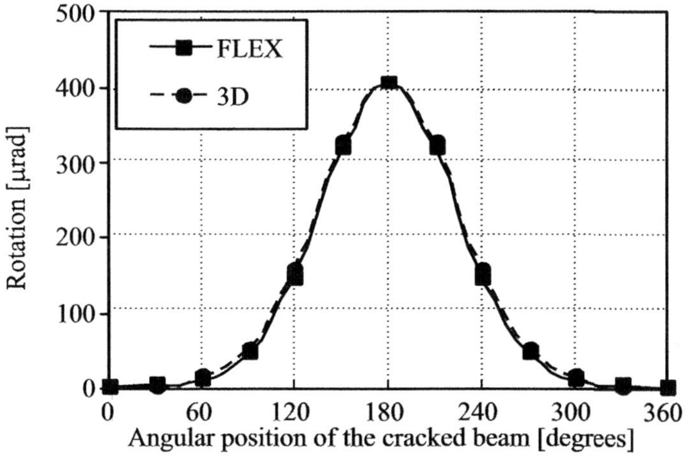

Fig. 5.47. Comparison between 3D calculation and FLEX simplified model, 60% crack depth, angular deflection about horizontal axis θ_y as function of cracked beam angular position.

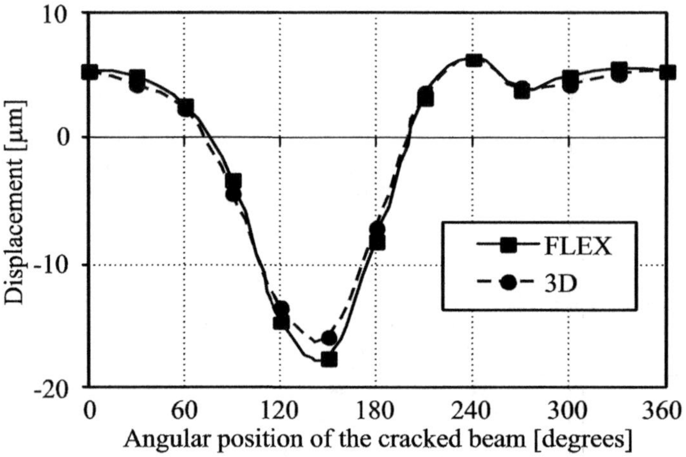

Fig. 5.48. Comparison between 3D calculation and FLEX simplified model, 60% crack depth, horizontal displacement y as function of cracked beam angular position.

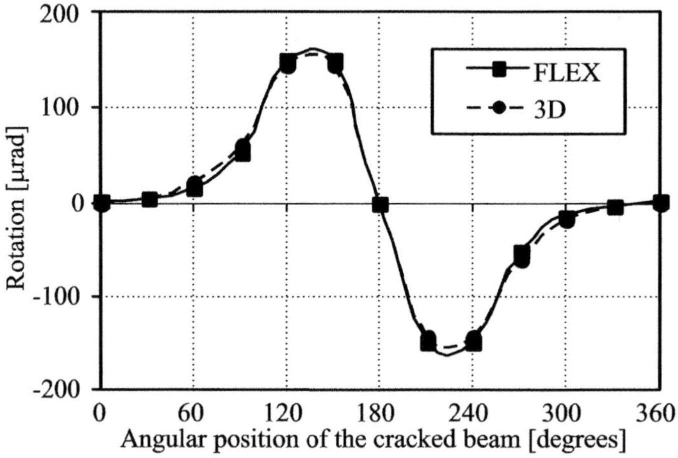

Fig. 5.49. Comparison between 3D calculation and FLEX simplified model, 60% crack depth, angular deflection about vertical axis θ_x as function of cracked beam angular position.

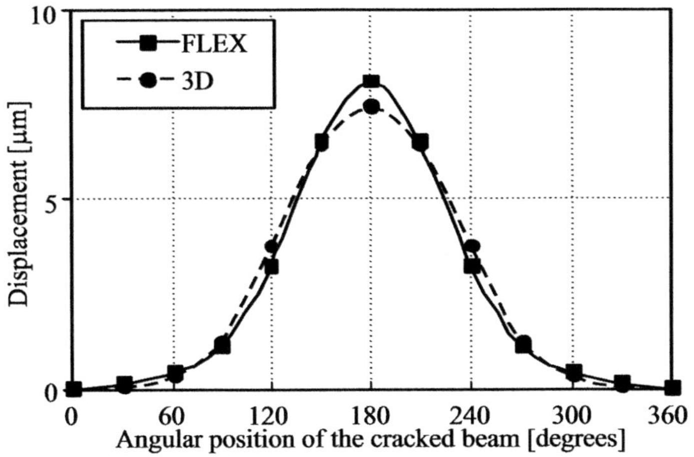

Fig. 5.50. Comparison between 3D calculation and FLEX simplified model, 60% crack depth, axial displacement z as function of cracked beam angular position.

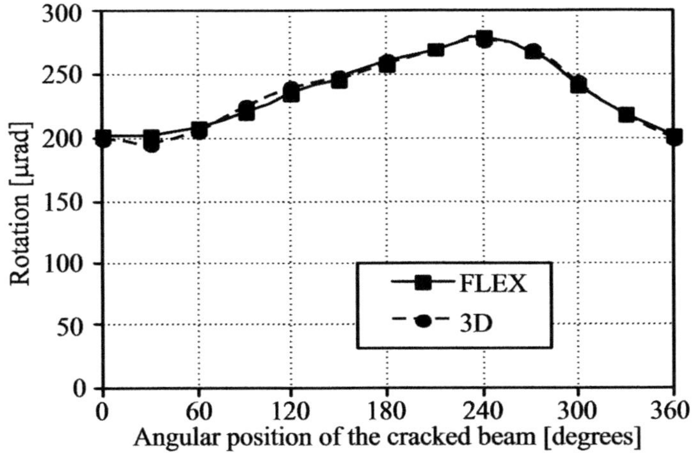

Fig. 5.51. Comparison between 3D calculation and FLEX simplified model, 60% crack depth, torsional deflection about axis of rotation θ_z as function of cracked beam angular position.

As can be seen from the above comparisons, once the equivalent length l_c (i.e. l_{flex} or l_{tors} depending on the circumstances) has been tuned, the fitting of the FLEX simplified model results is extraordinary good: the FLEX simplified model, despite some rough approximation that has been used, is able to represent the real additional flexibility caused by the breathing crack.

The periodical reduction, during one revolution of the shaft, of the torsion deflection from the maximum of 280 μrad to the minimum of 200 μrad shown in figure 5.51 is due to the contribution of the friction forces when the crack is closed: it is surprising to realize how close are the results of the FLEX simplified linear approximated approach to those of the non-linear 3D calculation. Calculation time for obtaining the results shown here is few seconds on a PC for the FLEX simplified model and several hours on a workstation for the 3D non-linear model.

A comparison of the results obtained by the FLEX model and experimental test is reported in section 4.2.2, but also other statical tests have been made in order to validate the FLEX crack model.

The cracked specimen used in section 4.2 was mounted on the bearings of the rotating fatigue test machine (as shown in figure 5.52). Two loads P of 20000 N have been applied in order to have in correspondence of the crack only bending moment and no shear force.

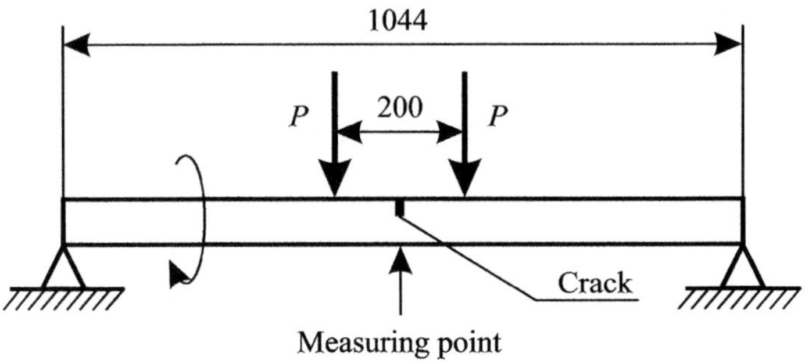

Fig. 5.52. Cracked test specimen.

The deflection in the different angular positions during slow rotation have been measured with a displacement sensor. The shape of the crack, shown in figure 5.53, has been introduced in the model, breathing has been calculated and, with the resulting stiffness matrix (from FLEX model), the deflections have been calculated and compared to the measured ones.

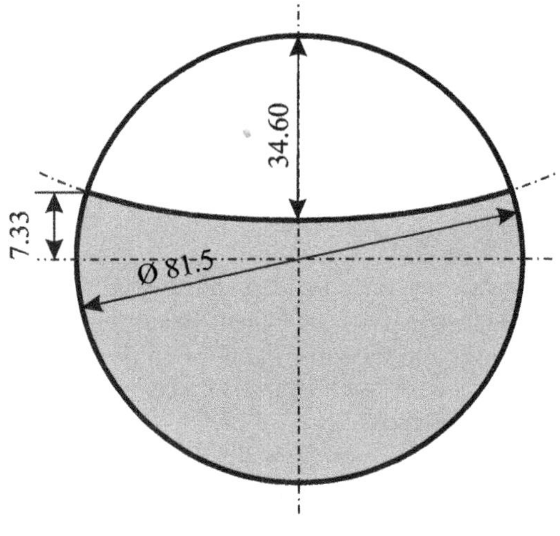

Fig. 5.53. Crack shape as per ultrasonic inspection.

The comparison is shown in figure 5.54: the excellent agreement confirms the breathing model and the crack model. It must be pointed out that also the model developed by EDF produced results in excellent agreement with the measured data.

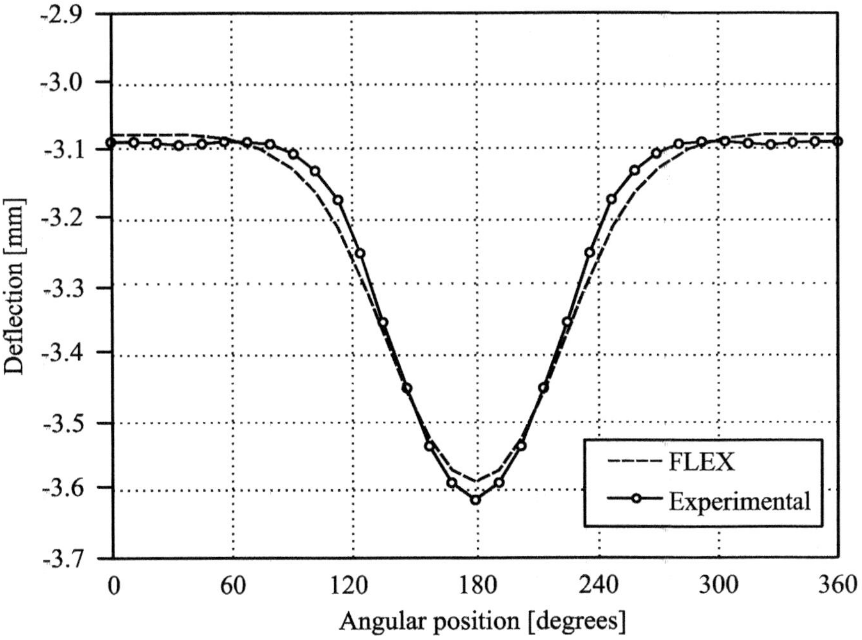

Fig. 5.54. Comparison between measured and calculated deflections.

5.5 Basic Modelling of a Cracked Rotor

Partially contributed by Robert Gasch, Institut für Luft-und Raumfahrt, Technical University of Berlin.

The aim of this section is to introduce a very simple model for a horizontal cracked rotor for a basic understanding, neglecting any effect of the coupling of lateral, torsional and axial vibrations. Therefore, a horizontal Jeffcott / de Laval rotor with a transverse crack is considered. The reader must be warned against the fact that this model can hardly reproduce the behaviour of a real rotating machine. Similar models have been proposed also in [5.36][5.37] and have the purpose to show the main issues of the dynamical response of a cracked rotor.

The deflection line of a shaft with a crack in the tension zone is given by the superposition of two parts (see figure 5.55): the deflection line of the un-cracked shaft and the additional deflection caused by the local compliance of the crack. This additional part cannot be found from the elastic beam theory, because, for the beam theory, a crack is the weakening of the bending stiffness EI on a zero length. Only 3D calculation (or 2D approximation) is able to determine this additional weakening of the stiffness [5.38][5.39][5.40] as shown also in section 5.2.2.

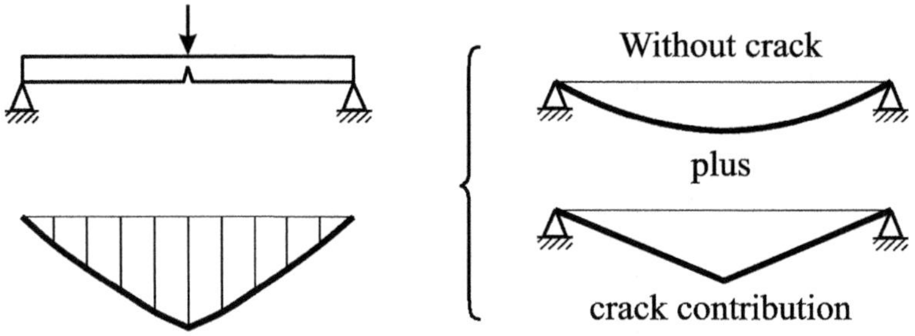

Fig. 5.55. Deflection line: contributions of the un-cracked shaft and of the local crack compliance.

In fact, since a cracked cylindrical shaft is no longer symmetric and isotropic, the crack produces the coupling of lateral, axial and torsional vibrations (see chapter 7). However for the sake of simplicity and to understand the basics of the dynamical behaviour of a cracked rotor, these effects are neglected and the interest is focused only on lateral vibrations.

The crack compliance coefficients for a general cracked beam element with 6 degrees of freedom at each end are reported for instance in [5.38]. Since a Jeffcott / de Laval rotor with a crack close to the disk is used to introduce the basics, only the main compliance H_{22} and the cross compliance H_{11} coefficients are considered, as shown in figure 5.56. These compliance coefficients are considered in a dimensionless form. They have to be divided by ER^3, where E is Young's modulus and R the radius of the shaft to obtain the physical compliances:

$$H_{22} = \frac{\hat{H}_{22}}{ER^3} \tag{5.61}$$

$$H_{11} = \frac{\hat{H}_{11}}{ER^3} \tag{5.62}$$

Figure 5.56 shows the compliances versus relative crack depth a/R. For small cracks ($a/R < 0.5$) the cross compliance coefficient H_{11} is much smaller than the main one H_{22}, so the cross compliance is neglected as a first approximation. This has the advantage that crack compliance is represented by only one parameter and the understanding of the analytical results in the following will be eased.

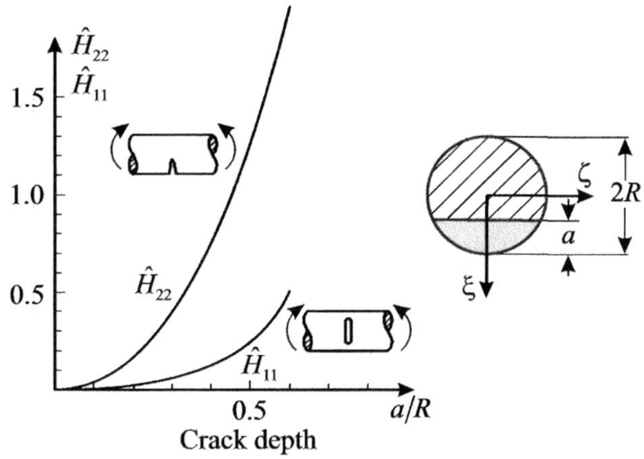

a/R	0	0.1	0.2	0.3	0.4	0.5	0.6
$\hat{H}_{11}$	0	0.001	0.009	0.038	0.107	0.245	0.493
$\hat{H}_{22}$	0	0.028	0.144	0.374	0.734	1.248	1.947

Fig. 5.56. Main and cross compliance coefficients of the open crack from [5.38].

Thus the deflection in the disc position, with an open crack nearby, can be written as:

$$\begin{Bmatrix} \xi \\ \eta \end{Bmatrix} = \begin{bmatrix} h_0 + \Delta h_\xi & 0 \\ 0 & h_0 + 0 \end{bmatrix} \begin{Bmatrix} f_\xi \\ f_\eta \end{Bmatrix} \tag{5.63}$$

where h_0 is the compliance of the round un-cracked shaft and

$$\Delta h_\xi = \frac{\hat{H}_{22} l^2}{16ER^3} \tag{5.64}$$

is the flexural additional compliance of the crack, considering that the span of the shaft is l.

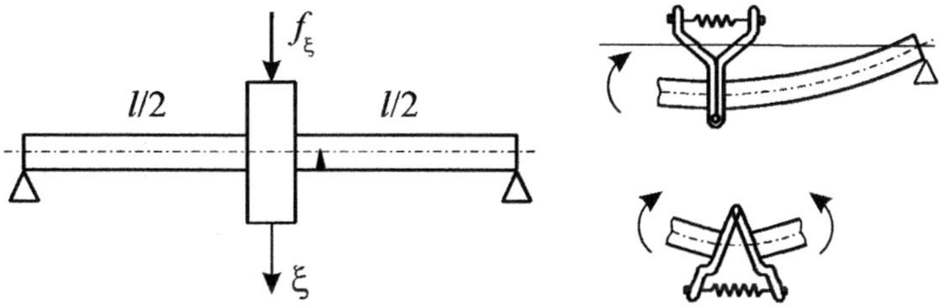

Fig. 5.57. Crack model (hinge model) for the "breathing" crack.

By rotating the shaft with the crack into the compression zone, the crack closes due to the pre-stressing of the weight and the additional compliance Δh_ξ becomes zero; the shaft appears intact again. This kind of behaviour is represented by the *hinge model* shown in figure 5.57. The hinge opens in the tension zone and closes in the compression zone. Thus the crack mechanism is non-linear, or more exactly bi-linear due to the hypotheses assumed. Eq. (5.63) is rewritten in the following form:

$$\begin{Bmatrix} \xi \\ \eta \end{Bmatrix} = \left(\begin{bmatrix} h_0 & 0 \\ 0 & h_0 \end{bmatrix} + \psi(t) \begin{bmatrix} \Delta h_\xi & 0 \\ 0 & 0 \end{bmatrix} \right) \begin{Bmatrix} f_\xi \\ f_\eta \end{Bmatrix} \tag{5.65}$$

where $\psi(t)$ is the breathing steering function and is a rectangular function:

- $\psi(t) = 0$, when the crack is in the compression zone and
- $\psi(t) = 1$, when the crack is in the tension zone.

For very deep cracks, a modification of the breathing steering function is advisable [5.39][5.26][5.41].

To change from the compliance formulation to the more practical stiffness formulation, the compliance matrix in eq. (5.65) must be inverted:

$$\begin{bmatrix} h_0 + \psi(t)\Delta h_\xi & 0 \\ 0 & h_0 \end{bmatrix}^{-1} = \frac{1}{h_0\left(h_0 + \psi(t)\Delta h_\xi\right)} \begin{bmatrix} h_0 & 0 \\ 0 & h_0 + \psi(t)\Delta h_\xi \end{bmatrix} \tag{5.66}$$

By considering that the hinge model is assumed, eq. (5.66) results:

$$\frac{1}{h_0\left(h_0 + \psi(t)\Delta h_\xi\right)} \begin{bmatrix} h_0 & 0 \\ 0 & h_0 + \psi(t)\Delta h_\xi \end{bmatrix} = \begin{cases} \begin{bmatrix} \dfrac{1}{h_0} & 0 \\ 0 & \dfrac{1}{h_0} \end{bmatrix}, \psi(t) = 0 \\ \begin{bmatrix} \dfrac{1}{h_0 + \Delta h_\xi} & 0 \\ 0 & \dfrac{1}{h_0} \end{bmatrix}, \psi(t) = 1 \end{cases} \tag{5.67}$$

If the stiffness of the un-cracked shaft is $k_0 = 1/h_0$ and:

$$\frac{1}{h_0 + \Delta h_\xi} = k_0 - \Delta k_\xi \tag{5.68}$$

where Δk_ξ is the *loss of stiffness* due to the opening of the crack in the hinge model, then the stiffness formulation can be expressed as:

$$\begin{Bmatrix} f_\xi \\ f_\eta \end{Bmatrix} = \left(\begin{bmatrix} k_0 & 0 \\ 0 & k_0 \end{bmatrix} - \psi(t) \begin{bmatrix} \Delta k_\xi & 0 \\ 0 & 0 \end{bmatrix} \right) \begin{Bmatrix} \xi \\ \eta \end{Bmatrix} \tag{5.69}$$

$$\mathbf{F}_{rot} = \left[\mathbf{K}_{0,rot} + \Delta\mathbf{K}_{rot}(\xi_{rot}, t)\right]\xi_{rot} \tag{5.70}$$

Given a rotor resembling a Jeffcott / de Laval rotor with a crack close to the disc, Δk_ξ can be easily determined experimentally by measuring the natural frequencies with open and closed crack, as shown in figure 5.58. If the natural frequency of the shaft with closed crack $\omega_0 = \sqrt{k_0/m}$ is considered as the reference frequency to which the natural frequency ω_ξ with open crack is related, the dimensionless parameter can be defined:

$$\left(\frac{\omega_\xi}{\omega_0}\right)^2 = 1 - \frac{\Delta k_\xi}{k_0} \tag{5.71}$$

The square root ω_ξ / ω_0 will be used later as dimensionless crack-depth parameter.

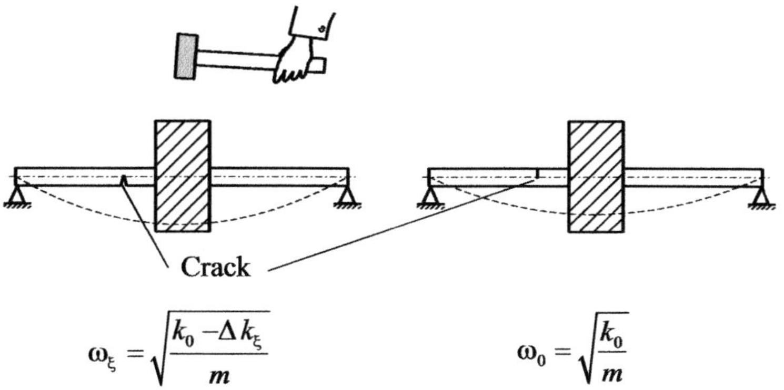

Fig. 5.58. Experimental determination of the stiffness loss Δk_ξ due to the crack.

The stiffness matrix in eq. (5.70) is defined in the rotating reference system, rigidly connected to the rotor. To shift to the inertial reference system, figure 5.59 is considered and the well known transformation matrix for the displacements is obtained:

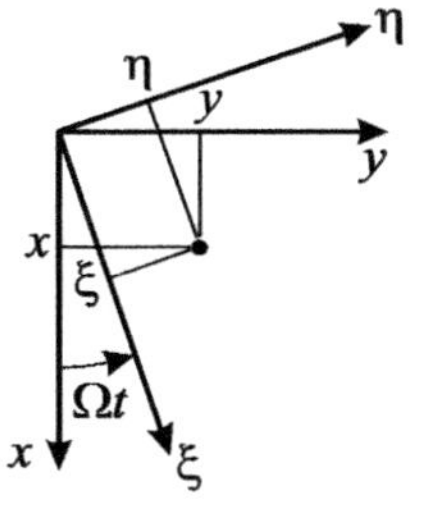

Fig. 5.59. Rotating and non rotating reference systems.

$$\begin{Bmatrix} \xi \\ \eta \end{Bmatrix} = \begin{bmatrix} \cos\Omega t & \sin\Omega t \\ -\sin\Omega t & \cos\Omega t \end{bmatrix} \begin{Bmatrix} x \\ y \end{Bmatrix} \quad \rightarrow \quad \xi_{rot} = [\mathbf{T}(t)]\mathbf{x}_{inertial} \tag{5.72}$$

A similar transformation holds for the forces:

$$\begin{aligned} \mathbf{F}_{rot} &= [\mathbf{T}(t)]\mathbf{F}_{inertial} \\ \mathbf{F}_{inertial} &= [\mathbf{T}(t)]^{-1}\mathbf{F}_{rot} \end{aligned} \tag{5.73}$$

By inserting these transformation matrices, for both the forces and the displacements, into eq. (5.70), the stiffness matrix in the inertial reference system is obtained:

$$[\mathbf{K}(t)] = [\mathbf{K}_0] + [\Delta\mathbf{K}(t)] = [\mathbf{T}]^{-1}[\mathbf{K}_{0,rot} + \Delta\mathbf{K}_{rot}][\mathbf{T}] \tag{5.74}$$

which corresponds explicitly to:

$$[\mathbf{K}_0] + [\Delta\mathbf{K}(t)] = \begin{bmatrix} k_0 & 0 \\ 0 & k_0 \end{bmatrix} - \psi(t)\Delta k_\xi \begin{bmatrix} \cos^2\Omega t & \sin\Omega t\cos\Omega t \\ \sin\Omega t\cos\Omega t & \sin^2\Omega t \end{bmatrix} \tag{5.75}$$

and finally to:

$$[\mathbf{K}_0] + [\Delta\mathbf{K}(t)] = \begin{bmatrix} k_0 & 0 \\ 0 & k_0 \end{bmatrix} - \frac{1}{2}\psi(t)\Delta k_\xi \begin{bmatrix} 1+\cos 2\Omega t & \sin 2\Omega t \\ \sin 2\Omega t & 1-\cos 2\Omega t \end{bmatrix} \tag{5.76}$$

The stiffness matrix $[\mathbf{K}(t)]$ is twofold time dependent: once due to the transformation from rotor-fixed to inertial reference system and further due to the breathing steering function $\psi(t)$, which is equal to 1 or to 0 depending on the crack position in the compression or in the tension zone of the shaft. Actually $\psi(t)$ depends also on the shaft deflection $\mathbf{x}$, thus it is more precise to write:

$$[\mathbf{K}] = [\mathbf{K}(\mathbf{x},t)] = [\mathbf{K}_0] + [\Delta\mathbf{K}(\mathbf{x},t)] \tag{5.77}$$

5.5.1 Equations of Motion and Linearization for a Horizontal Shaft with Breathing Crack

It is rather easy to write the equations of motion of the simple Jeffcott / de Laval rotor shown in figure 5.60, introducing the disc mass m, the damping coefficient c and the stiffness matrix $[\mathbf{K}(\mathbf{x},t)]$ just derived in eq. (5.76). Supports are considered rigid. The external force system on the right hand side (RHS) is composed by the weight $\mathbf{W}$ and by the unbalance $\mathbf{U}$ forces.

$$\begin{bmatrix} m & 0 \\ 0 & m \end{bmatrix} \begin{Bmatrix} \ddot{x} \\ \ddot{y} \end{Bmatrix} + \begin{bmatrix} c & 0 \\ 0 & c \end{bmatrix} \begin{Bmatrix} \dot{x} \\ \dot{y} \end{Bmatrix} + \begin{bmatrix} k_{11} & k_{12} \\ k_{21} & k_{22} \end{bmatrix} \begin{Bmatrix} x \\ y \end{Bmatrix} = \\ = \begin{Bmatrix} mg \\ 0 \end{Bmatrix} + me\Omega^2 \begin{Bmatrix} \cos(\beta + \Omega t) \\ \sin(\beta + \Omega t) \end{Bmatrix} \tag{5.78}$$

$$[\mathbf{M}]\ddot{\mathbf{x}} + [\mathbf{C}]\dot{\mathbf{x}} + [\mathbf{K}(\mathbf{x}, t)]\mathbf{x} = \mathbf{W} + \mathbf{U} \tag{5.79}$$

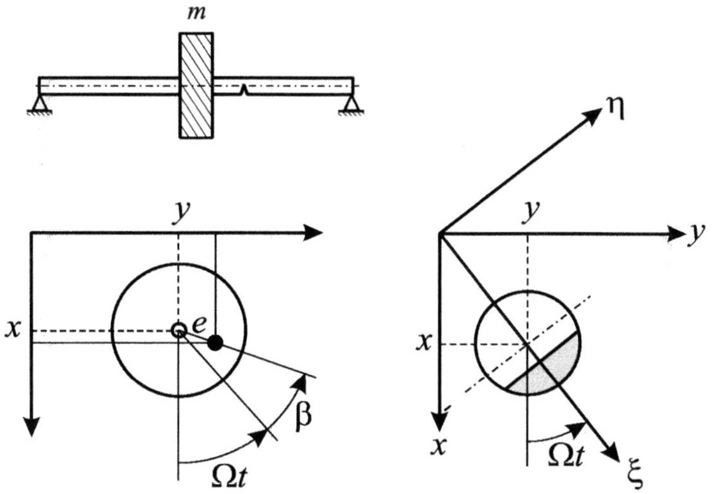

Fig. 5.60. Jeffcott / de Laval rotor with a cracked shaft.

The weight vector on the RHS is of great importance when the equations of motion are linearized. Considering a heavy horizontal rotor, not damaged by any crack, the static deflection $\mathbf{x}_0$ is large compared to the additional deflections caused by the dynamics of the rotating shaft. This is confirmed by a glance considering figure 5.61, in which the relationship between the natural frequency and the static deflection is plotted, bearing in mind that:

$$x_{stat} = \frac{mg}{k_0} \text{ and } \omega_0^2 = \frac{k_0}{m} \quad \rightarrow \quad x_{stat} = \frac{g}{\omega_0^2} \tag{5.80}$$

where ω_0 is the natural frequency (critical speed) of the un-cracked system (see figure 5.58).

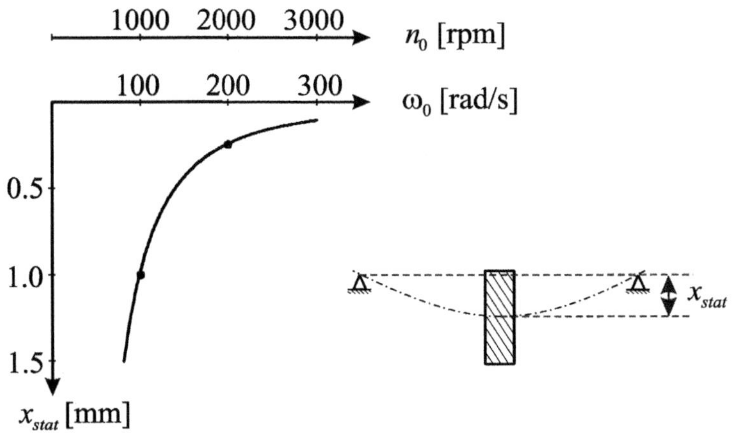

Fig. 5.61. Static deflection and natural frequency ω_0.

A rotating machine with its first critical speed between 600 and 800 rpm, like heavy generators or low pressure turbines, has static deflections of 1 mm or more, as shown in figure 5.61. On the contrary, dynamic deflections of 0.1 mm are already considered as very serious. So it is obvious to split the total deflection in the static and dynamic parts:

$$\mathbf{x}(t) = \mathbf{x}_0 + \Delta\mathbf{x}(t) = \begin{Bmatrix} mg/k_0 \\ 0 \end{Bmatrix} + \Delta\mathbf{x}(t) \tag{5.81}$$

bearing in mind that $\Delta\mathbf{x}(t)$ is small, i.e. $\Delta\mathbf{x}(t) \ll \mathbf{x}_0$. In other words, the deflection due to the weight is predominant and rules the opening and closing of the crack, because the vibration amplitudes $\Delta\mathbf{x}(t)$ are small compared to the static deflection $\mathbf{x}_0$. Figure 5.62 shows qualitatively the *breathing* of the crack under the weight influence when the shaft is slowly turned. More precise discussion about the modelling of breathing is reported in section 5.3.

Please note that, contrarily to the steps in the breathing steering function $\psi(t)$ assumed in the simple model, the deflection of the slowly rotating shaft is softly and continuously changing with Ωt. However, in case of a small crack ($a/R < 0.5$), the crack opens and closes rather abruptly when passing 90° and 270°, where the transition occurs from the tension zone to the compression zone and vice versa. The transition is less abrupt in case of deeper cracks. Mayes suggested the continuous breathing steering function $\psi(t) = (1 + \cos\Omega t)/2$ for deeper cracks, because it comes closer to reality [5.26][5.38][5.39][5.41]. For the basic considerations of this section, the rectangular function of figure 5.62 is sufficient since a crack in early stage ($a/R < 0.5$) is considered. Anyway, the influence of the type of the breathing steering function on the results of this simple model is astonishingly small, as shown in [5.39].

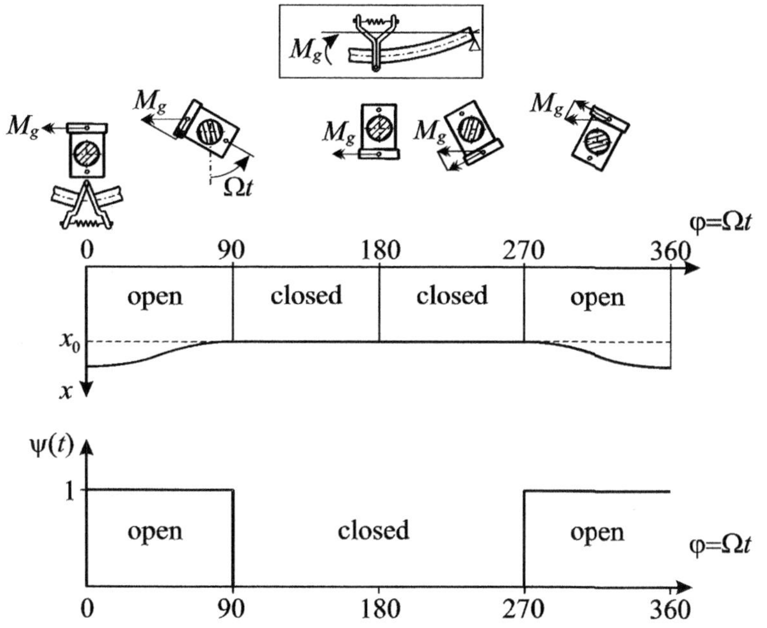

Fig. 5.62. Static deflection and opening and closing of the crack versus angle of rotation; steering function $\psi(t)$.

In order to linearize the equation of motion, it is suitable to highlight the weight dominance in eq. (5.79) by considering eq. (5.77) for the stiffnesses:

$$[\mathbf{M}]\ddot{\mathbf{x}}+[\mathbf{C}]\dot{\mathbf{x}}+\left([\mathbf{K}_0]+[\Delta\mathbf{K}(t)]\right)\mathbf{x}=\mathbf{W}+\mathbf{U} \tag{5.82}$$

and by using the rectangular function as the breathing steering function $\psi(t)$. The deflections are split up into the static $\mathbf{x}_0$ and the dynamic part $\Delta\mathbf{x}(t)$ according to eq. (5.81):

$$[\mathbf{M}]\Delta\ddot{\mathbf{x}}+[\mathbf{C}]\Delta\dot{\mathbf{x}}+\left([\mathbf{K}_0]+[\Delta\mathbf{K}(t)]\right)\left(\mathbf{x}_0+\Delta\mathbf{x}\right)=\mathbf{W}+\mathbf{U} \tag{5.83}$$

Considering that for the static equilibrium:

$$\begin{bmatrix} k_0 & 0 \\ 0 & k_0 \end{bmatrix}\begin{Bmatrix} x_{stat} \\ 0 \end{Bmatrix}=\begin{Bmatrix} mg \\ 0 \end{Bmatrix} \tag{5.84}$$

$$[\mathbf{K}_0]\mathbf{x}_0=\mathbf{W} \tag{5.85}$$

The equations of motion becomes a linear, but periodically time-variant equation for the dynamic vibrations $\Delta\mathbf{x}(t)$:

$$[\mathbf{M}]\Delta\ddot{\mathbf{x}} + [\mathbf{C}]\Delta\dot{\mathbf{x}} + \left([\mathbf{K}_0] + [\Delta\mathbf{K}(t)]\right)\Delta\mathbf{x} = -[\Delta\mathbf{K}(t)]\mathbf{x}_0 + \mathbf{U} \tag{5.86}$$

as the static equilibrium can be separated.

The periodic time-variant stiffness matrix $[\Delta\mathbf{K}(t)]$ appears twice. On the left hand side it determines (together with the damping) whether the system is stable or not. On the right hand side, this matrix times the weight deflections $\mathbf{x}_0$ produces an additional excitation force – the *equivalent crack excitation* – that is periodic in time.

If the stability of the solution of the homogenous equation is guaranteed, this of course must be checked, then a further simplification for the calculation of the forced vibrations due to the right hand side in eq. (5.86) is possible. Due to the stability, the homogenous part $\Delta\mathbf{x}_g(t)$ of the complete solution:

$$\Delta\mathbf{x}(t) = \Delta\mathbf{x}_g(t) + \Delta\mathbf{x}_p(t) \tag{5.87}$$

is fading out ($\Delta\mathbf{x}_g(t) \rightarrow 0$) and also the term $[\Delta\mathbf{K}(t)]\Delta\mathbf{x}(t)$ on the left hand side is very small:

$$[\Delta\mathbf{K}(t)]\Delta\mathbf{x}(t) \rightarrow 0 \tag{5.88}$$

By neglecting this last term for the calculation of the forced vibrations, a time-invariant differential equation is obtained:

$$[\mathbf{M}]\Delta\ddot{\mathbf{x}} + [\mathbf{C}]\Delta\dot{\mathbf{x}} + [\mathbf{K}_0]\Delta\mathbf{x} = -[\Delta\mathbf{K}(t)]\mathbf{x}_0 + \mathbf{U} \tag{5.89}$$

The left hand side of eq. (5.89) is the equation of motion of a round, un-cracked shaft, while the forcing system in the right hand side is composed by the unbalance and by the equivalent crack excitation forces $-[\Delta\mathbf{K}(t)]\mathbf{x}_0$.

5.5.2 Stability

The analysis of the stability of the system is not trivial because the homogeneous part of the equations of motion eq. (5.86):

$$[\mathbf{M}]\Delta\ddot{\mathbf{x}} + [\mathbf{C}]\Delta\dot{\mathbf{x}} + \left([\mathbf{K}_0] + [\Delta\mathbf{K}(t)]\right)\Delta\mathbf{x} = 0 \tag{5.90}$$

has periodically time-variant coefficients due to the presence of the stiffness matrix $[\Delta\mathbf{K}(t)]$. Eq. (5.90) is more conveniently re-written in the state-space as:

$$\begin{bmatrix} [\mathbf{M}] & [\mathbf{C}] \\ 0 & [\mathbf{M}] \end{bmatrix} \begin{Bmatrix} \Delta\ddot{\mathbf{x}} \\ \Delta\dot{\mathbf{x}} \end{Bmatrix} = \begin{bmatrix} 0 & -[\mathbf{K}_0] - [\Delta\mathbf{K}(t)] \\ [\mathbf{M}] & 0 \end{bmatrix} \begin{Bmatrix} \Delta\dot{\mathbf{x}} \\ \Delta\mathbf{x} \end{Bmatrix} \tag{5.91}$$

and finally

$$\dot{\mathbf{z}} = [\mathbf{P}(t)]\mathbf{z} \tag{5.92}$$

where:

$$\mathbf{z} = \begin{Bmatrix} \Delta\dot{x} \\ \Delta\dot{y} \\ \Delta x \\ \Delta y \end{Bmatrix} \tag{5.93}$$

$$[\mathbf{P}(t)] = \begin{bmatrix} [\mathbf{M}] & [\mathbf{C}] \\ 0 & [\mathbf{M}] \end{bmatrix}^{-1} \begin{bmatrix} 0 & -[\mathbf{K}_0]-[\Delta\mathbf{K}(t)] \\ [\mathbf{M}] & 0 \end{bmatrix} =$$

$$= \begin{bmatrix} -\dfrac{c}{m} & 0 & \dfrac{\psi(t)\Delta k_\xi(1+\cos 2\Omega t)-2k_0}{2m} & \dfrac{\psi(t)\Delta k_\xi \sin 2\Omega t}{2m} \\ 0 & -\dfrac{c}{m} & \dfrac{\psi(t)\Delta k_\xi \sin 2\Omega t}{2m} & \dfrac{\psi(t)\Delta k_\xi(1-\cos 2\Omega t)-2k_0}{2m} \\ 1 & 0 & 0 & 0 \\ 0 & 1 & 0 & 0 \end{bmatrix} \tag{5.94}$$

Eq. (5.92) is a first order system with the periodic coefficients $[\mathbf{P}(t)]$, the stability of which can be evaluated using Floquet's theorem [5.42][5.43], especially using the its proof and consequences.

The matrix of coefficients $[\mathbf{P}(t)]$ has minimal period $T = 2\pi/\Omega$ in this case, due to the presence of the breathing steering function $\psi(t)$:

$$[\mathbf{P}(t+T)] = [\mathbf{P}(t)] \tag{5.95}$$

Floquet's theorem states that *a regular system of order* n*, like that of eq. (5.92), where* $[\mathbf{P}(t)]$ *is periodic with minimal period* T*, has at least one non-trivial solution* $\mathbf{z} = \boldsymbol{\chi}(t)$ *such that*:

$$\boldsymbol{\chi}(t+T) = \mu\boldsymbol{\chi}(t) \tag{5.96}$$

where μ *is a constant.*

A fundamental matrix $[\boldsymbol{\Phi}(t)] = [\phi_{ij}(t)]$ of the system is considered for the proof. Note that actually at this stage is not necessary to know explicitly a fundamental matrix. Due to the property of the fundamental matrix, then:

$$[\dot{\boldsymbol{\Phi}}(t)] = [\mathbf{P}(t)][\boldsymbol{\Phi}(t)] \tag{5.97}$$

Because of eq. (5.95), also $[\boldsymbol{\Phi}(t+T)]$ satisfies eq. (5.97). Considering that it is possible to show that $\det[\boldsymbol{\Phi}(t+T)] \neq 0$, then $[\boldsymbol{\Phi}(t+T)]$ is another fundamental matrix. Their columns, i.e. the solutions of the system, are linear combinations of those of $[\boldsymbol{\Phi}(t)]$:

$$\phi_{ij}(t+T) = \sum_{k=1}^{n} \phi_{ik}(t)e_{kj} \tag{5.98}$$

for some constants e_{kj}, so that:

$$[\mathbf{\Phi}(t+T)] = [\mathbf{\Phi}(t)][\mathbf{E}] \tag{5.99}$$

Matrix $[\mathbf{E}] = [e_{kj}]$ is called *transition matrix* and is non-singular, because $\det[\mathbf{\Phi}(t+T)] = \det[\mathbf{\Phi}(t)]\det[\mathbf{E}]$ and therefore $\det[\mathbf{E}] \neq 0$. Given a convenient value t_0 of t, eq. (5.99) can be used to find $[\mathbf{E}]$:

$$[\mathbf{E}] = [\mathbf{\Phi}(t_0)]^{-1}[\mathbf{\Phi}(t_0+T)] \tag{5.100}$$

Let μ and $\mathbf{s}$ respectively be an eigenvalue and its associated eigenvector of $[\mathbf{E}]$:

$$\det\left([\mathbf{E}] - \mu[\mathbf{I}]\right) = 0 \text{ and } \left([\mathbf{E}] - \mu[\mathbf{I}]\right)\mathbf{s} = 0 \tag{5.101}$$

Consider that the solution $\boldsymbol{\chi}(t)$ of eq. (5.92) given by $\mathbf{z} = [\mathbf{\Phi}(t)]\mathbf{s} = \boldsymbol{\chi}(t)$ is really a solution being a linear combination of the columns of $[\mathbf{\Phi}(t)]$, then:

$$\boldsymbol{\chi}(t+T) = [\mathbf{\Phi}(t+T)]\mathbf{s} = [\mathbf{\Phi}(t)]\mathbf{E}\mathbf{s} = [\mathbf{\Phi}(t)]\mu\mathbf{s} = \mu\boldsymbol{\chi}(t) \tag{5.102}$$

A solution of eq. (5.92) that satisfies eq. (5.96) is called a *normal solution.*

The eigenvalues μ of the transition matrix $[\mathbf{E}]$ are called *characteristic numbers* or *multipliers* of eq. (5.92). It is interesting to note that it is possible to proof that the constants μ are independent of the choice of $[\mathbf{\Phi}(t)]$, thus they are really the characteristic numbers of the system. Now, they are used to evaluate also the stability of the system, introducing the concept of the *characteristic* or *Floquet's exponent* of the system in eq. (5.92). Given a characteristic number μ, corresponding to the minimal period T, the characteristic exponent ρ is defined as:

$$\mathrm{e}^{\rho T} = \mu \quad \rightarrow \quad \rho = \frac{1}{T}\ln\mu \tag{5.103}$$

Note that Floquet's exponent is not unique because $\mathrm{e}^{\rho T} = \mathrm{e}^{\left(\rho + \mathrm{i}\frac{k2\pi}{T}\right)T}$ with $k \in \mathbb{N}, \mathrm{i} = \sqrt{-1}$, it can be fixed considering the principal value of the logarithm in eq. (5.103).

The system of eq. (5.92) is stable if the characteristic multipliers have modulus less than one:

$$|\mu_k| < 1 \tag{5.104}$$

In fact a further theorem states that *if the transition matrix* $[\mathbf{E}]$ *of eq. (5.99) has* n *distinct eigenvalues* μ_k, *then the system of eq. (5.92) has* n *linearly independent normal solutions of the form:*

$$\mathbf{z}_k = \mathbf{p}_k(t)\mathrm{e}^{\rho_k t} \tag{5.105}$$

where ρ_k *are the Floquet's exponents corresponding to* μ_k *and* $\mathbf{p}_k(t)$ *are functions of period* T.

Before proving the theorem in eq. (5.105), this last expression explains the stability condition given by eq. (5.104): considering eq. (5.103), if $|\mu_k|<1$ then the homogeneous solution of the equations of motion eq. (5.86) is modulated by an exponential function that is null for $t \to +\infty$.

As regards the proof, Floquet's theorem, see eq. (5.96), states that a solution in the form of $\mathbf{z}_k(t+T) = \mu_k\, \mathbf{z}_k(t) = \mathrm{e}^{\rho_k T}\, \mathbf{z}_k(t)$ corresponds to each μ_k, thus:

$$\mathbf{z}_k(t+T)\mathrm{e}^{-\rho_k T} = \mathbf{z}_k(t) \tag{5.106}$$

and for every t:

$$\mathbf{z}_k(t+T)\mathrm{e}^{-\rho_k(t+T)} = \mathbf{z}_k(t)\mathrm{e}^{-\rho_k t} \tag{5.107}$$

Considering that eq. (5.105) can be written also as:

$$\mathbf{p}_k(t) = \mathbf{z}_k\, \mathrm{e}^{-\rho_k t} \tag{5.108}$$

eq. (5.108) imply that $\mathbf{p}_k(t)$ has period equal to T. The linear independence of the solutions $\mathbf{z}_k(t)$ is guaranteed by their method of construction in Floquet's theorem, i.e. $\mathbf{z}_k = [\mathbf{\Phi}(t)]\mathbf{s}_k$. Since $\mathbf{s}_k$ are the eigenvectors associated to different eigenvalues μ_k, they are linearly independent. Being $[\mathbf{\Phi}(t)]$ non-singular, also the $\mathbf{z}_k(t)$ are linearly independent. The case of eigenvalues μ_k not all distinct is not relevant here and is discussed in [5.43]. Finally, note that the real part of the Floquet's exponents are corresponding to the Lyapunov's exponents, commonly used for the study of (linearized) system stability.

Floquet's theorem has also an interesting mechanical interpretation. Since for eq. (5.96)

$$\mathbf{z}_k(t+T) = \mu_k\, \mathbf{z}_k(t) \tag{5.109}$$

the modulus of the characteristic multipliers $|\mu_k|$ gives us the factor of increase or decrease of the solution after one revolution (cycle) of the rotor. If it is compared with instabilities due to fluid-film bearings, crack instability can be very brutal: a characteristic multipliers with $|\mu_k| = 2$ indicated that after only 10 revolutions of the rotor the amplitude has grown with a factor $2^{10} = 1024$.

Coming back to the problem of eq. (5.90), unfortunately it is impossible to find an analytical form for the transition matrix $[\mathbf{E}]$, due to impossibility to express analytically the fundamental matrix $[\mathbf{\Phi}(t)]$. A numerical method must be used to calculate the characteristic multipliers and a suitable one is presented in

[5.42][5.44]. Using numerical calculations, a survey on the stability of the cracked rotor is presented in figure 5.63 as a function of the rotational speed and the crack depth. Here damping was assumed to be zero, i.e. the worst case for the stability. The (dimensionless) crack depth parameter in this figure is the frequency ratio ω_ξ / ω_0, see figure 5.58. Also the rotational frequency Ω was made dimensionless with the natural frequency of the un-cracked rotor ω_0:

$$\varsigma = \frac{\Omega}{\omega_0} \tag{5.110}$$

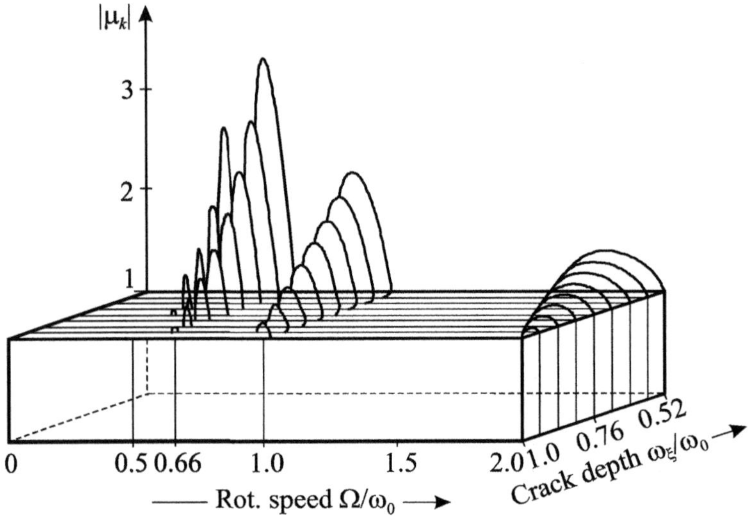

Fig. 5.63. Stability of the cracked Jeffcott/Laval rotor versus rotational speed and crack depth, damping $[\mathbf{C}] = 0$. Unstable zones $|\mu_k| > 1$.

In [5.44][5.45] the borderlines of the unstable zones for zero damping, $c = 0$, were calculated numerically. An analytical approximation of the borderlines of the unstable regions is given by:

$$\frac{2}{n}\left(1 - \frac{1}{4}\frac{\Delta k_\xi}{k_0}\right) < \left(\frac{\Omega}{\omega_0}\right) < \left(\frac{2}{n}\right) \tag{5.111}$$

where $n \in \mathbb{N}$. The influence of fluid-film bearings is discussed in [5.45][5.46].

Instability regions disappear when the dimensionless rotational speed is greater than 2. The unstable zone close to $\varsigma = 2$ becomes very broad when cracks are deeper. The unstable zone close to $\varsigma = 1$ is familiar recalling a rotor with unequal bending stiffnesses ($EI_\xi \neq EI_\eta$). The regions of instability at $\varsigma = 2/3; 2/4\cdots$ are very small. As soon as we have some damping, they disappear [5.39].

Fortunately, some small damping in the system is very helpful to guarantee stability, concerning this compare the width of the unstable zones for $c = 0$ and $c = 0.01$ in figure 5.64. The reader must remember that, in real rotating machines, damping is normally sufficient to avoid the instability when a crack is present.

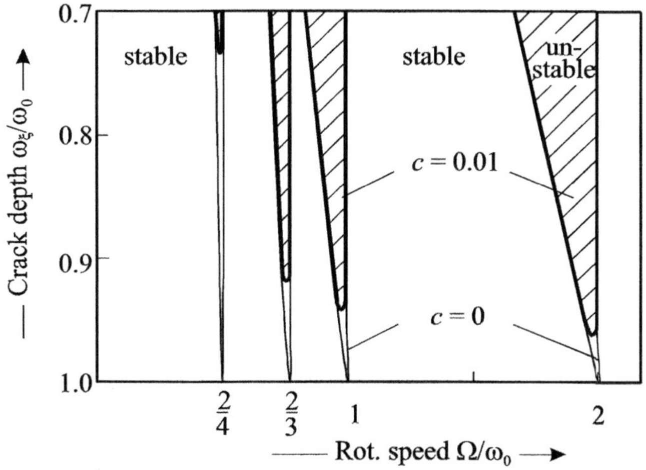

Fig. 5.64. Borderlines of stability with a damping of 1% ($c = 0.01$) and no damping ($c = 0$).

5.5.3 Vibration Forced by the Crack Influence

Once the stability of the system is verified, the forced vibration response may be calculated using the time invariant equations of motion, eq. (5.89):

$$\begin{bmatrix} m & 0 \\ 0 & m \end{bmatrix}\begin{Bmatrix} \Delta\ddot{x} \\ \Delta\ddot{y} \end{Bmatrix} + \begin{bmatrix} c & 0 \\ 0 & c \end{bmatrix}\begin{Bmatrix} \Delta\dot{x} \\ \Delta\dot{y} \end{Bmatrix} + \begin{bmatrix} k_0 & 0 \\ 0 & k_0 \end{bmatrix}\begin{Bmatrix} \Delta x \\ \Delta y \end{Bmatrix} = $$
$$= \frac{1}{2}\psi(t)\Delta k_\xi \begin{bmatrix} 1+\cos 2\Omega t & \sin 2\Omega t \\ \sin 2\Omega t & 1-\cos 2\Omega t \end{bmatrix}\begin{Bmatrix} x_{stat} \\ 0 \end{Bmatrix} + me\Omega^2 \begin{Bmatrix} \cos(\Omega t+\beta) \\ \sin(\Omega t+\beta) \end{Bmatrix} \tag{5.112}$$

This equation can be simplified by introducing a complex notation, as shown in figure 5.65, where Δr is the vector containing both deflections of the shaft Δx and Δy :

$$\Delta r = \Delta x + \mathrm{i}\,\Delta y \quad \text{with } \mathrm{i} = \sqrt{-1} \tag{5.113}$$

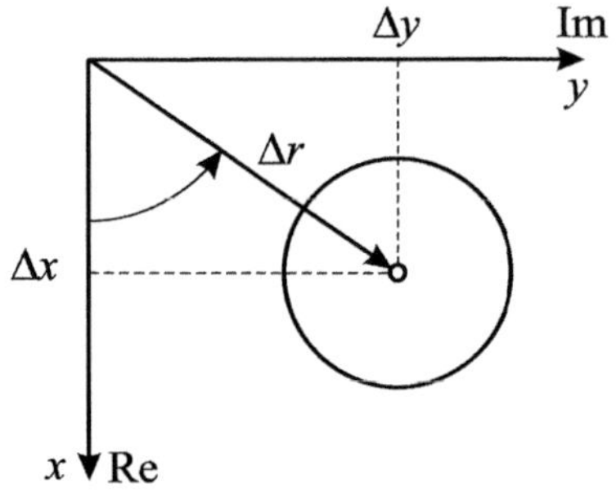

Fig. 5.65. Complex coordinates Δr for the deflections Δx and Δy .

Formally the second equation in eq. (5.112) is multiplied times the imaginary unit i and is added to the first equation. Recalling Euler's formula, the new second order differential equation is obtained:

$$m\Delta\ddot{r} + c\Delta\dot{r} + k_0\Delta r = \frac{1}{2} x_{stat}\psi(t)\Delta k_{\xi}\left(1 + \mathrm{e}^{2\mathrm{i}\Omega t}\right) + me\Omega^2\,\mathrm{e}^{\mathrm{i}(\Omega t+\beta)} \tag{5.114}$$

The first analysis is performed on the pure crack response with no unbalance, $e = 0$, that is proportional to the static deflection x_{stat} and to the stiffness loss due to the crack Δk_{ξ}. The rectangular function of opening and closing of the crack is expanded in Fourier's series:

$$\psi(t) = \frac{1}{2} + \frac{2}{\pi}\left(\cos\Omega t - \frac{2}{3\pi}\cos 3\Omega t + \frac{2}{5\pi}\cos 5\Omega t - \ldots\right) \tag{5.115}$$

Considering that in complex exponential notation:

$$\cos n\Omega t = \frac{1}{2}\left(\mathrm{e}^{\mathrm{i}n\Omega t} + \mathrm{e}^{-\mathrm{i}n\Omega t}\right) \tag{5.116}$$

then the crack force F_c on the right hand side of eq. (5.114) can be expressed in complex Fourier's series as:

$$F_c = \Delta k_{\xi} x_{stat} \sum_{k=-\infty}^{+\infty} b_k\,\mathrm{e}^{\mathrm{i}k\Omega t} \tag{5.117}$$

where b_k are the participation coefficients of the harmonics. In table 5.2 they are presented with an interpretation of their physical meaning.

Table 5.2. Crack excitation force and its harmonics.

k		Excitation	b_k
-4		four times per revolution	0
-3	Backward	three times per revolution	-0.021
-2		twice per revolution	0
-1		once per revolution	0.106
0	Additional static deflection		0.250
+1		once per revolution	0.318
+2	Forward	twice per revolution	0.250
+3		three times per revolution	0.106
+4		four times per revolution	0

Now the response of the rotating shaft due to a single harmonic b_k of the crack excitation can easily be determined by imposing a harmonic solution:

$$\Delta r = \Delta \hat{r}_k \, \mathrm{e}^{\mathrm{i}k\Omega t} \tag{5.118}$$

and replacing it in eq. (5.114), neglecting the unbalance:

$$m\Delta\ddot{r} + c\Delta\dot{r} + k_0 \Delta r = \Delta k_\xi x_{stat} b_k \, \mathrm{e}^{\mathrm{i}k\Omega t} \tag{5.119}$$

the equation is obtained:

$$\left(-\Omega^2 k^2 m + \mathrm{i}\Omega k c + k_0\right)\Delta \hat{r}_k = \Delta k_\xi x_{stat} b_k \tag{5.120}$$

Thus the response is:

$$\Delta \hat{r}_k = \Delta k_\xi x_{stat} \frac{b_k}{\left(k_0 - \Omega^2 k^2 m\right) + \mathrm{i}\Omega k c} \tag{5.121}$$

By considering the superposition of all components k, the complete response of the rotor to the crack excitation is obtained. Since the high order components in the crack force series, eq. (5.117), are very small or exactly equal to zero, the series is truncated after $k = \pm 3$ and results:

$$\Delta r(t) \cong x_{stat} \frac{\Delta k_\xi}{k_0} \sum_{k=-3}^{+3} \frac{b_k \, \mathrm{e}^{\mathrm{i}k\Omega t}}{1 - k^2 \varsigma^2 + 2\mathrm{i}ck\varsigma} \tag{5.122}$$

The contribution $k=0$, $b_0=0.25$ is an *additional static deflection* due to the crack, averaged over one revolution, with the magnitude:

$$0.25 x_{stat} \frac{\Delta k_\xi}{k_0} \tag{5.123}$$

that gives the reference level for the vibrations.

The contribution $k=\pm 1$, $b_{+1}=0.318$ and $b_{-1}=0.106$, describes a forward and a backward circular orbit, once per revolution (1X).

$$\Delta r_{+1}(t) = x_{stat} \frac{\Delta k_\xi}{k_0} \frac{b_{+1}}{1-\left(\frac{\Omega}{\omega_0}\right)^2 + 2\mathrm{i}c\left(\frac{\Omega}{\omega_0}\right)} \mathrm{e}^{\mathrm{i}\Omega t} = \Delta\hat{r}_{+1}\,\mathrm{e}^{\mathrm{i}\Omega t} \tag{5.124}$$

$$\Delta r_{-1}(t) = x_{stat} \frac{\Delta k_\xi}{k_0} \frac{b_{-1}}{1-\left(\frac{\Omega}{\omega_0}\right)^2 - 2\mathrm{i}c\left(\frac{\Omega}{\omega_0}\right)} \mathrm{e}^{-\mathrm{i}\Omega t} = \Delta\hat{r}_{-1}\,\mathrm{e}^{-\mathrm{i}\Omega t} \tag{5.125}$$

The two components superimpose in an elliptic orbit once per revolution in a forward sense, figure 5.66. The resonance (i.e. the critical speed) belonging to these components is at $\Omega=\omega_0$ (For deeper cracks at $\Omega=\omega_0^*$, see following section 5.5.5).

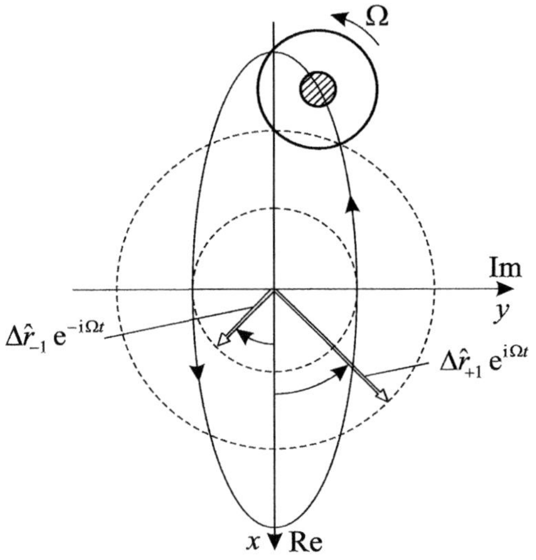

Fig. 5.66. Elliptic orbit superimposed from the forward (Ω) and the backward ($-\Omega$) circular vibration.

The contribution $k = \pm 2$, $b_{-2} = 0$ and $b_{+2} = 0.25$, describes a forward circular orbit twice per revolution (2X) of the rotor:

$$\Delta r_{\pm 2}(t) = 0 + x_{stat}\frac{\Delta k_{\xi}}{k_0}\frac{b_{+2}}{1-\left(\frac{2\Omega}{\omega_0}\right)^2 - 2ic\left(\frac{2\Omega}{\omega_0}\right)}e^{i2\Omega t} \tag{5.126}$$

Here the resonance occurs at $\Omega = \omega_0/2$.

Finally the contribution $k = \pm 3$, $b_{-3} = 0.021$ and $b_{+3} = 0.106$, results in a practically forward elliptical orbit, three times per revolution of the rotor (3X). The resonance is at $\Omega = \omega_0/3$.

The magnification functions for $k = -1, +1, +2, +3$ are shown in figure 5.67 (the magnification for $k = -3$ is very small and is omitted).

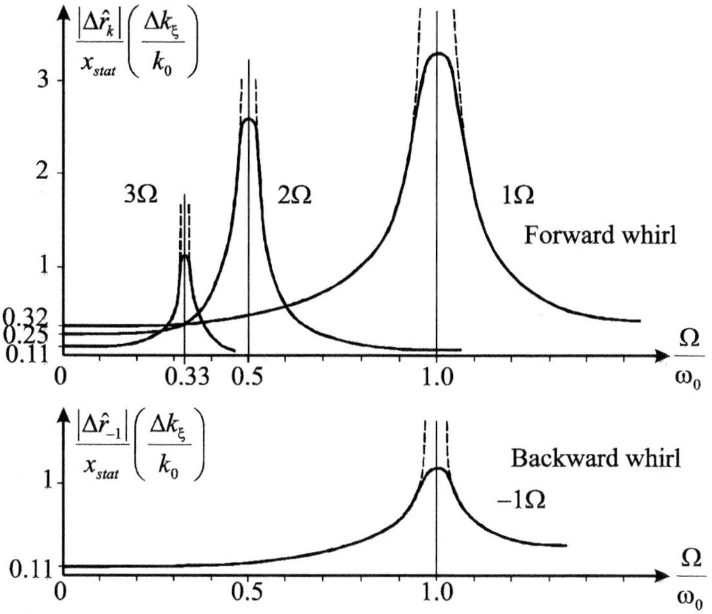

Fig. 5.67. Magnification functions of the forward Ω, 2Ω and 3Ω circular vibrations and the backward $-\Omega$ circular vibration; $c = 0$ and $c = 0.05$.

However it is more instructive to consider figure 5.68, where the circular and elliptic orbits of the 1X, 2X and 3X vibrations are shown as function of the dimensionless rotational speed $\varsigma = \Omega/\omega_0$ for the case of 5% damping.

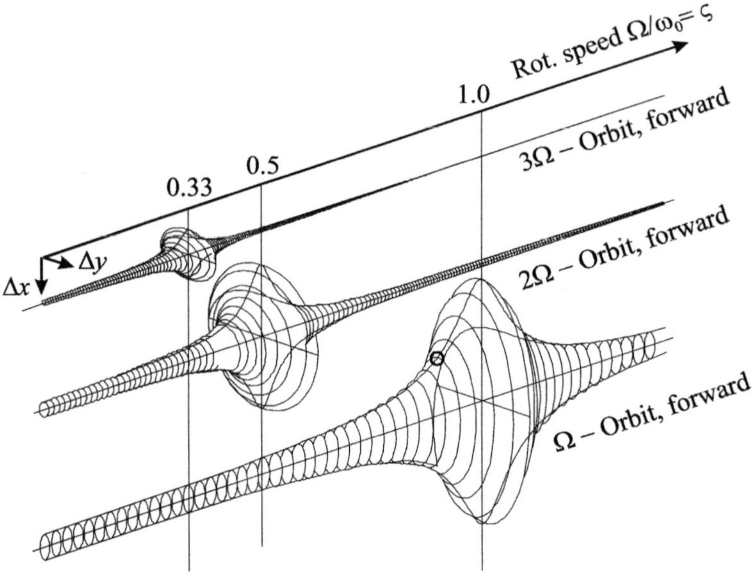

Fig. 5.68. Elliptical forward orbits (Ω and 3Ω) and the circular 2Ω (forward) orbit versus dimensionless rotational speed; $c = 0.05$. They are superimposed in figure 5.69.

The curve touching the left arm of the cross (O) at $\varsigma = 1$ in the 1X orbit is presenting the phase angle resulting from equations (5.124)(5.125). The clarity of the contribution of the component is lost as soon as the three contributions are superimposed to the complete orbit of the system figure 5.69.

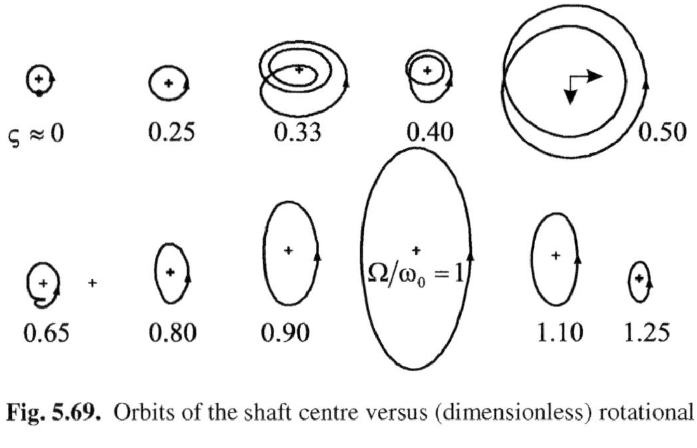

Fig. 5.69. Orbits of the shaft centre versus (dimensionless) rotational speed ς, $c = 0.05$.

At $\varsigma = 1/3$ three loops per one revolution of the shaft are present, as this is the resonance of the 3X excitation. With $\varsigma = 1/2$ there are two loops within one revolution of the shaft and so on. A comprehensive way to represent the cracked rotor behaviour is the two-sided spectral order analysis, an example is shown in figure 5.70.

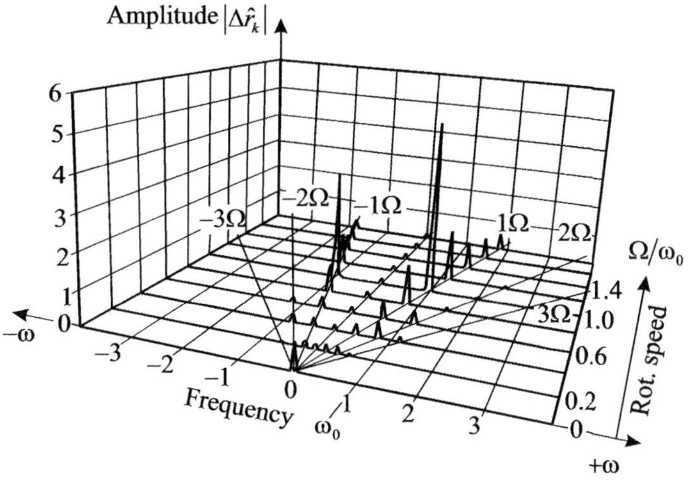

Fig. 5.70. Two sided spectral order analysis of the Jeffcott/Laval rotor with a crack. Dimensionless amplitude versus dimensionless rotational speed and frequency.

The analysis using this simple model shows that the analysis of the spectral components of the vibration amplitudes, combined with those of the phases (not shown here), is a very helpful instrument for early crack detection, as shown in chapter 2.

5.5.4 Crack and Unbalance Response

In practice there is always an overlap of the crack and the unbalance response. Thus both solutions of eq. (5.114) are superposed:

$$\Delta r(t) = \Delta r_U + \Delta r_{crack} = \\ = e\,\mathrm{e}^{\mathrm{i}\beta} \frac{\varsigma^2}{1-\varsigma^2-2\mathrm{i}c\varsigma} \mathrm{e}^{\mathrm{i}\Omega t} + x_{stat} \frac{\Delta k_\xi}{k_0} \sum_{k=-3}^{3} \frac{b_k}{1-(k\varsigma)^2-2\mathrm{i}ck\varsigma} \mathrm{e}^{\mathrm{i}k\Omega t} \tag{5.127}$$

Now the synchronous response has two contributions: one caused by the crack and one by the unbalance, as it is evident by the comparison of eq. (5.114) and eqs. (5.124)(5.125). The resultant of this superposition depends on the angular position β between eccentricity and crack, see figure 5.60. These two contributions can intensify or compensate each other, since the exciting 1X forces are equal to:

$$\left(me\Omega^2\,\mathrm{e}^{\mathrm{i}\beta} + x_{stat}\Delta k_\xi b_{+1}\right)\mathrm{e}^{\mathrm{i}\Omega t} \tag{5.128}$$

This fact indicates that, if the unbalance can be set so that β becomes 180°, the 1X response due to the crack for the rotational speed Ω considered can be completely suppressed. This is obtained by balancing the rotor so that $\beta = 180°$ and:

$$me\Omega^2 = x_{stat}\Delta k_{\xi} b_{+1} \tag{5.129}$$

By balancing in a way that the equation:

$$me\omega_0^2 = x_{stat}\Delta k_{\xi} b_{+1} \tag{5.130}$$

is fulfilled, even the resonance at $\Omega = \omega_0$ is suppressed. However it is impossible to exert an influence using the balancing on the $-\Omega$ response or on the 2Ω and 3Ω contributions of the crack and on the additional static deflection due to b_0.

5.5.5 Deeper Cracks

Eq. (5.112) has been used in the sections 5.5.3 and 5.5.4 for the calculation of crack and unbalance response. The left hand side of eq. (5.112) is actually that of the un-cracked shaft, so its natural frequency $\omega_0 = \sqrt{k_0/m}$ is considered as the reference frequency. This is acceptable only for small cracks and all results presented in these sections are valid only in this case.

When the crack is deeper, it is well known (see chapter 3 for experimental evidences) that the critical speeds that appears in the frequency response diagrams are a little bit lower than that of the un-cracked shaft ω_0. This fact can be introduced into the previous results in the following way.

Considering only the static part, $k = 0$, $b_0 = 0.25$, the solution of eq. (5.122) gives the additional static deflection:

$$\Delta r_{stat,add} = 0.25 x_{stat} \frac{\Delta k_{\xi}}{k_0} \tag{5.131}$$

averaged over one revolution of the rotor.

The total static deflection – including both the static deflection of the un-cracked shaft $x_{stat} = mg/k_0$ and the additional averaged contribution of the crack of eq. (5.131) – is:

$$x_{stat,tot} = x_{stat} + \Delta r_{stat,add} = \frac{mg}{k_0}\left(1 + \frac{1}{4}\frac{\Delta k_{\xi}}{k_0}\right) = \frac{mg}{k_0^*} \tag{5.132}$$

Stiffness k_0^* can be taken as the reduced stiffness of the rotating shaft that now includes also the weakening due to the crack. Considering that $\omega_0^* = \sqrt{k_0^*/m}$, the lowered natural frequency is:

$$\omega_0^* = \omega_0\left(1 - \frac{1}{8}\frac{\Delta k_{\xi}}{k_0}\right) \tag{5.133}$$

For deeper cracks, the lowered, crack depth dependent, natural frequency ω_0^* has to substitute ω_0 in the equations (5.120) to (5.126) of section 5.5.3, to calculate the resonance behaviour.

5.5.6 *Final Remarks*

Although the dynamics of an idealized rotating shaft with a crack is theoretically transparent, in practice early crack detection is still not easy. Hardly a crack was detected in a turbo-set running for month with constant rotational speed having a crack depth of less than 25 percent [5.47][5.48]. At that depth there is already a slight change in the critical speeds of some 2 or 3 percent. By considering the symptoms analyzed in chapter 2 – and of course the 2X resonance at half critical speed – a crack can be detected. The fluid-film damping does not contribute to the early crack detection and the 3X resonance is usually completely suppressed.

However permanent monitoring enables to trend analysis (i.e. today's vibration minus vibration a fortnight before), that is very helpful for successful early crack detection.

5.6 Modelling the Cracked Rotor Dynamical Behaviour

Once the simple Jeffcott / de Laval model has been presented, it is time to consider how to model the dynamical behaviour of a real cracked rotor of a horizontal axis industrial machine. The rotor system, i.e. the shaft-line, the bearings and the supporting structure, is modelled by means of a standard rotor-dynamic FE model, as shown for instance in [5.49].

By indicating with $[\mathbf{M}]$ the mass matrix, with $[\mathbf{C}]$ the damping matrix, with $[\mathbf{Gyr}]$ the gyroscopic matrix and with $[\mathbf{K}]$ the stiffness matrix, the equation of motion of the system is:

$$[\mathbf{M}]\ddot{\mathbf{x}}+([\mathbf{C}]+[\mathbf{Gyr}]\Omega)\dot{\mathbf{x}}+[\mathbf{K}]\mathbf{x}=\sum_n \mathbf{F}_{e_n}\,\mathrm{e}^{\mathrm{i}n\Omega t}+\mathbf{W} \tag{5.134}$$

where $\mathbf{W}$ is the weight force vector and $\mathbf{F}_{e_n}$ the vectors that contains all the exciting forces acting on the rotor (e.g. unbalances, bow etc.) up to the n-th harmonic component. If the breathing crack is included in the model, i.e. when the equivalent cracked beam having a reduced cross section and a suitable length is inserted in the rotor-dynamic FEM of the rotor, the stiffness matrix will have variable values in the 8×8 elements that correspond to the cracked beam element, which may vary between a maximum stiffness (corresponding to closed crack) and a minimum stiffness (corresponding to open crack), instead of constant values. Stiffness variation is defined by crack breathing that is function of the stresses acting in correspondence of the crack. Stresses are due to static loads (bending moment due to weight and bearing alignment condition) and to dynamic loads (bending moment due to the actual dynamic deformation shape). The dynamic deformation shape is function of the external force distribution, like residual unbalances, bows and any other exciting force, including also the equivalent crack forces, and of the corresponding exciting frequency. This makes the problem non-linear: crack breathing depends on vibration, and vibration depends on crack forces that are function of breathing condition.

In this general approach the non-linearity of the problem is given by the fact that the breathing is defined not only by the angular position Ωt of the shaft with respect to the static load, but also by the vibration $\mathbf{x}$ (this happen also for the simplified Jeffcott / de Laval model, see section 5.5):

$$[\mathbf{K}] = [\mathbf{K}(\Omega t, \mathbf{x})] \tag{5.135}$$

Eq. (5.134) becomes:

$$[\mathbf{M}]\ddot{\mathbf{x}} + ([\mathbf{C}] + [\mathbf{Gyr}]\Omega)\dot{\mathbf{x}} + [\mathbf{K}(\Omega t, \mathbf{x})]\mathbf{x} = \sum_n \mathbf{F}_{e_n} e^{\mathrm{i} n\Omega t} + \mathbf{W} \tag{5.136}$$

Now, the force vectors $\mathbf{F}_{e_n}$ can include also the resultant moments of the thermal stress distribution acting on the cracked section, for simulating the thermal sensitivity of the cracked shaft, which has two aspects.

One aspect is the effect of thermal stresses on the breathing mechanism, which can be accounted for as described in section 5.3.

The other aspect is the thermal bow generated by the axially symmetrical thermal stresses that are not compensated in the open parts of the crack, where stresses cannot be transmitted, thus generating a breathing dependent resultant bending moment that is null only for completely closed crack. Also this effect can be simulated using eq. (5.136), including the above said bending moments in vector $\mathbf{F}_{e_n}$.

Eq. (5.136) can be integrated in the time domain, dividing one revolution of the shaft in several angular steps, calculating step by step the crack breathing condition. Transient behaviour, instabilities and steady state forced vibrations are all taken into account in the solution of eq. (5.136), as shown in section 7.3.

When the stresses due to dynamic loads are small with respect to those due to static loads, and can therefore be neglected, condition which is called *weight governed breathing*, then the breathing becomes function of the angular position Ωt of the shaft only. Therefore the stiffness variation $\mathbf{K}(\Omega t)$ becomes periodical and independent of rotating speed and of exciting forces.

The global stiffness matrix can be written according to eq. (5.137):

$$[\mathbf{K}(\Omega t)] = [\mathbf{K}_m] + \sum_j [\Delta\mathbf{K}_j] e^{\mathrm{i} j\Omega t}, \; j = 1,2,3 \tag{5.137}$$

The Fourier's expansion of the periodic stiffness matrix is truncated at the third harmonic component. Therefore, the equation of motion of the rotor can be expressed with eq. (5.138).

$$\begin{aligned} &[\mathbf{M}]\ddot{\mathbf{x}} + ([\mathbf{C}] + [\mathbf{Gyr}]\Omega)\dot{\mathbf{x}} + \left([\mathbf{K}_m] + \sum_j [\Delta\mathbf{K}_j] e^{\mathrm{i} j\Omega t} \right)\mathbf{x} = \\ &= \sum_n \mathbf{F}_{e_n} e^{\mathrm{i} n\Omega t} + \mathbf{W}, \; j = 1,2,3 \end{aligned} \tag{5.138}$$

that is linear with variable parameters. The rotor displacements $\mathbf{x}$ can be split in their static and dynamic components $\mathbf{x}_s$ and $\mathbf{x}_d$:

$$\mathbf{x} = \mathbf{x}_s + \mathbf{x}_d \tag{5.139}$$

where

$$\mathbf{x}_s = [\mathbf{K}_m]^{-1} \left(\mathbf{W} + \mathbf{F}_0 \right) \tag{5.140}$$

and $\mathbf{F}_0$ is a static force vector defined below in eq. (5.147) due to the presence of the crack.

Then, eq. (5.138) can be rewritten in the following form:

$$\begin{aligned} &[\mathbf{M}]\ddot{\mathbf{x}}_d + ([\mathbf{C}] + [\mathbf{Gyr}]\Omega)\dot{\mathbf{x}}_d + [\mathbf{K}_m]\mathbf{x}_d = \\ &= \sum_n \mathbf{F}_{e_n} \mathrm{e}^{\mathrm{i}\, n\Omega t} - \sum_j [\Delta\mathbf{K}_j]\mathrm{e}^{\mathrm{i}\, j\Omega t} \left(\mathbf{x}_s + \mathbf{x}_d \right),\ j = 1,2,3 \end{aligned} \tag{5.141}$$

If only the steady state solution due to the exciting forces is of interest, being the exciting forces periodical, then it is possible to shift from time domain to the frequency domain.

Let $\mathbf{F}_j$ indicate the j-th component of the forces due to the crack and $\mathbf{F}_{t_j}$ that of the total forces:

$$\mathbf{F}_{t_j} \mathrm{e}^{\mathrm{i}\, j\Omega t} = \mathbf{F}_{e_j} \mathrm{e}^{\mathrm{i}\, j\Omega t} + \mathbf{F}_j \mathrm{e}^{\mathrm{i}\, j\Omega t},\ j = 1,2,3 \tag{5.142}$$

The vibrations $\mathbf{x}_d$ in steady state conditions are also periodical and can be expanded in a Fourier's series:

$$\mathbf{x}_d = \sum_j \mathbf{x}_j \mathrm{e}^{\mathrm{i}\, j\Omega t},\ j = 1,2,3 \tag{5.143}$$

where $\mathbf{x}_1$, $\mathbf{x}_2$ and $\mathbf{x}_3$ are 1X, 2X and 3X complex harmonic vibration components, respectively.

By applying the harmonic balance approach to eq. (5.141), the following equations are obtained:

$$\left(-(j\Omega)^2[\mathbf{M}] + \mathrm{i}\, j\Omega([\mathbf{C}] + [\mathbf{Gyr}]\Omega) + [\mathbf{K}_m] \right) \mathbf{X}_j = \mathbf{F}_{t_j},\ j = 1,2,3 \tag{5.144}$$

Then, the equivalent crack forces (and moments) $\mathbf{F}_j$ given by eqs. (5.147)-(5.150) applied at the nodes of the equivalent cracked beam element are calculated by substituting eqs. (5.145) and (5.146) into the last term of eq. (5.141).

$$\mathrm{Re}\left(\mathbf{x}_j \mathrm{e}^{\mathrm{i}\, j\Omega t} \right) = \frac{1}{2} \mathbf{x}_j \mathrm{e}^{\mathrm{i}\, j\Omega t} + \frac{1}{2} \mathbf{x}_j^* \mathrm{e}^{-\mathrm{i}\, j\Omega t},\ j = 1,2,3 \tag{5.145}$$

$$[\Delta\mathbf{K}_j]\mathrm{e}^{\mathrm{i}\, j\Omega t} = \frac{1}{2}[\Delta\mathbf{K}_j]\mathrm{e}^{\mathrm{i}\, j\Omega t} + \frac{1}{2}[\Delta\mathbf{K}_j]^* \mathrm{e}^{-\mathrm{i}\, j\Omega t},\ j = 1,2,3 \tag{5.146}$$

$$\mathbf{F}_0 = \frac{[\Delta \mathbf{K}_1]}{4}\mathbf{x}_1^* + \frac{[\Delta \mathbf{K}_1]^*}{4}\mathbf{x}_1 + \frac{[\Delta \mathbf{K}_2]}{4}\mathbf{x}_2^* + \frac{[\Delta \mathbf{K}_2]^*}{4}\mathbf{x}_2 + \frac{[\Delta \mathbf{K}_3]}{4}\mathbf{x}_3^* + \frac{[\Delta \mathbf{K}_3]^*}{4}\mathbf{x}_3 \quad (5.147)$$

$$\mathbf{F}_1\, \mathrm{e}^{\mathrm{i}\Omega t} = \left([\Delta \mathbf{K}_1]\mathbf{x}_s + \frac{[\Delta \mathbf{K}_2]}{2}\mathbf{x}_1^* + \frac{[\Delta \mathbf{K}_3]}{2}\mathbf{x}_2^* + \frac{[\Delta \mathbf{K}_1]^*}{2}\mathbf{x}_2 + \frac{[\Delta \mathbf{K}_2]^*}{2}\mathbf{x}_3 \right)\mathrm{e}^{\mathrm{i}\Omega t} \quad (5.148)$$

$$\mathbf{F}_2\, \mathrm{e}^{\mathrm{i}2\Omega t} = \left([\Delta \mathbf{K}_2]\mathbf{x}_s + \frac{[\Delta \mathbf{K}_3]}{2}\mathbf{x}_1^* + \frac{[\Delta \mathbf{K}_1]}{2}\mathbf{x}_1 + \frac{[\Delta \mathbf{K}_1]^*}{2}\mathbf{x}_3 \right)\mathrm{e}^{\mathrm{i}2\Omega t} \quad (5.149)$$

$$\mathbf{F}_3\, \mathrm{e}^{\mathrm{i}3\Omega t} = \left([\Delta \mathbf{K}_3]\mathbf{x}_s + \frac{[\Delta \mathbf{K}_1]}{2}\mathbf{x}_2 + \frac{[\Delta \mathbf{K}_2]}{2}\mathbf{x}_1 \right)\mathrm{e}^{\mathrm{i}3\Omega t} \quad (5.150)$$

where $\mathbf{x}_j^*$ and $[\Delta \mathbf{K}_j]^*$ are conjugate complex quantities of $\mathbf{x}_j$ and $[\Delta \mathbf{K}_j]$.

The main components of the above force vectors are the moments according to the vertical and horizontal coordinates rather than the forces. The static and dynamic response of the cracked rotor can be evaluated using eq. (5.141) that is solved with an iterative procedure, shown in the flowchart of figure 5.71, for the three harmonic components, by considering the force vectors obtained with eqs. (5.147)-(5.150).

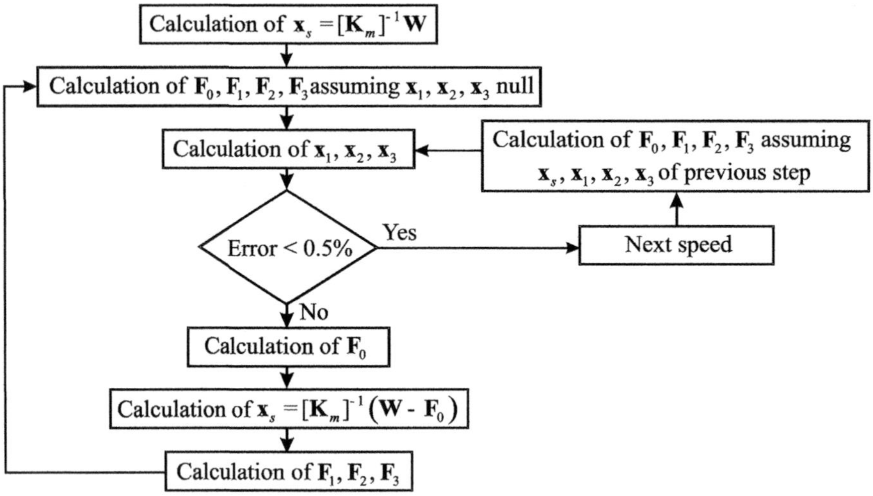

Fig. 5.71. Flow chart of the iterative procedure used to evaluate the dynamic response of the cracked rotor.

In this linear approach, only the static bending moment is considered, because the bending moment due to the inertia force distribution associated to the rotor vibrations usually gives a small contribution in respect to the static bending moment (often only near the rotor critical speeds it can become significant). This assumption is generally acceptable for heavy horizontal rotating machines.

If this condition is not fulfilled and a non-linear approach is desired, then an iterative procedure, which considers also the rotor vibrations, can be applied. For each rotating speed, at the first iteration step, this approach considers the vibrations obtained at the last step of the previous speed, so that a refined evaluation of the bending moment due also to the actual vibrations can be carried out.

As an example, the methods for transverse crack modelling and cracked rotor dynamic behaviour calculation described above have been used to simulate the dynamic response of a cracked shaft of a turbo generator shaft of a power plant. Figure 5.72 shows the finite element model of the unit and the position of the crack. Figure 5.73 shows the crack shape and depth and the amplitudes of the 1X, 2X and 3X vibrations evaluated, in vertical direction at one of the bearings of the cracked shaft, using the FLEX method, based on the equivalent beam, for the stiffness matrix calculation. Then the dynamical behaviour has been calculated according to eq. (5.141) and to the flow chart of figure 5.71.

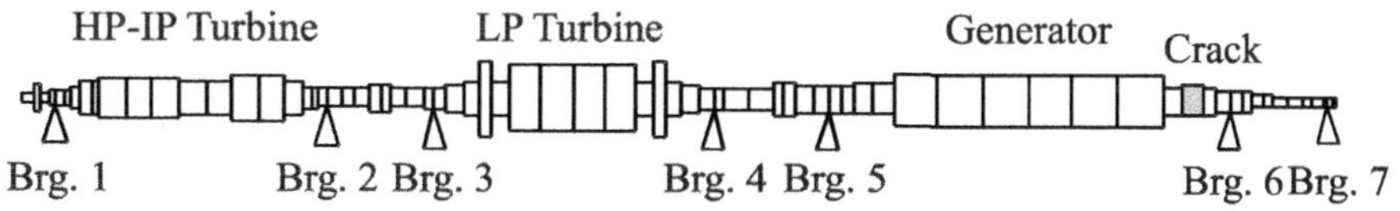

Fig. 5.72. Finite element model of a turbo generator.

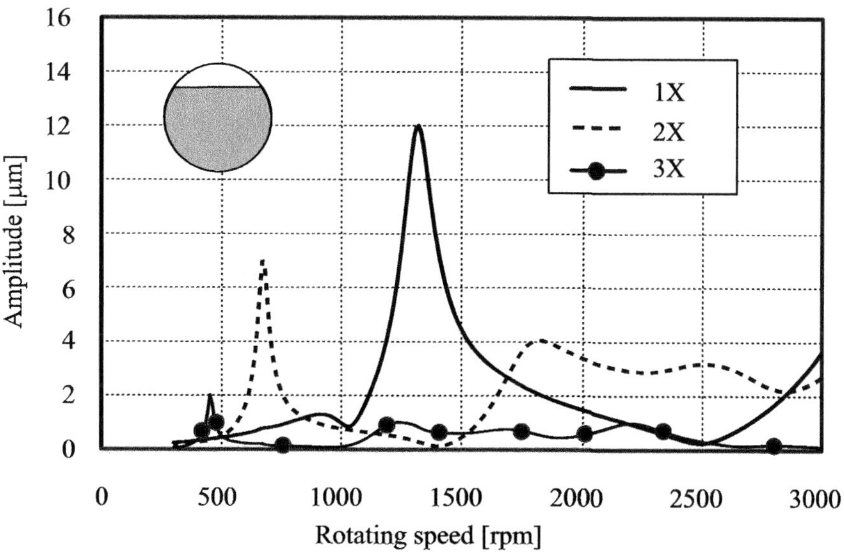

Fig. 5.73. Vibration amplitudes of the 1X, 2X and 3X components in bearing #6 of the turbo generator, as a function of the rotating speed, 25% crack relative depth.

In the following chapters many applications of the described procedure will be shown for calculating the vibrations excited by cracked rotors in different conditions.

References

[5.1] Wauer J (1990) On the dynamics of cracked rotors: a literature survey. Applied Mechanics Review, 43(1), 13-17.

[5.2] Gasch R (1993) A survey of the dynamic behaviour of a simple rotating shaft with a transverse crack. Journal of Sound and Vibration, 160(2), 313-332.

[5.3] Dimarogonas AD (1996) Vibration of cracked structures: a state of the art review. Engineering Fracture Mechanics, 55(5), 831-857.

[5.4] Sabnavis G, Kirk RG, Kasarda M, Quinn D (2004) Cracked shafts detection and diagnostics: a literature review. The Shock and Vibration Digest, 36(4), 287-296.

[5.5] Bachschmid N, Pennacchi P (2008) Special Issue: Crack effects in rotordynamics. Mechanical Systems and Signal Processing, 22(4), 761-762.

[5.6] Inglis CE (1913) Stresses in a plate due to the presence of cracks and sharp corners. Transactions of the Institution of Naval Architects, 55, 219-230.

[5.7] Griffith AA (1920) The phenomena of rupture and flow in solids. Philosophical Transactions of the Royal Society, A221, 163-198.

[5.8] Irwin GR (1948) Fracture Dynamics. In: Fracturing of Metals. American Society for Metals, Cleveland, Ohio, 147-166.

[5.9] Irwin GR (1957) Analysis of stresses and strains near the end of a crack traversing a plate. Journal of Applied Mechanics, 24, 361-364.

[5.10] Williams ML (1957) On the stress distribution at the base of a stationary crack. Journal of Applied Mechanics, 24, 109-114.

[5.11] Zienkiewicz (1987) The finite element method. 4th edition, McGraw-Hill Co.

[5.12] Barsoum RS (1976) On the Use of isoparametric finite elements in linear fracture mechanics. International Journal for Numeric Methods in Engineering, 10, 25-37.

[5.13] Henshell RD, Shaw KG (1975) Crack tip elements are unnecessary. International Journal for Numeric Methods in Engineering, 9, 495-507.

[5.14] Manu C (1983) Quarter-point elements for curved cracks fronts. Computer & Structures, 17, 227-231.

[5.15] Pang HLJ (1993) Linear elastic fracture mechanics benchmarks: 2D finite element test cases. Engineering Fracture Mechanics, 44(5), 741-751.

[5.16] Broek D (1989) The practical use of fracture mechanics. Kluwer Academic Publisher, 271-274.

[5.17] Hartranft RJ, Sih GC (1977) Stress singularity for a crack with an arbitrarily curved front. Engineering Fracture Mechanics. 9(3), 705-718.

[5.18] Parks DM (1974) A stiffness derivative finite element technique for determination of elastic crack tip stress intensity factor. International Journal of Fracture, 10, 487-502.

[5.19] Suresh (1998), Fatigue of materials, Cambridge University Press.

[5.20] Buch A. (1998), Fatigue data handbook, John Wiley and Sons, USA.

[5.21] Paris PC, Erdogan F (1960) A critical analysis of crack propagation laws. Journal of Basic Engineering, 85, 528-534.

[5.22] Elber W (1970) Fatigue crack closure under cyclic tension. Engineering Fracture Mechanics, 2, 37-45.

[5.23] Dentsoras AJ, Dimarogonas AD (1989) Fatigue Crack Propagation in Resonating Structures. Engineering Fracture Mechanics, 34(3), 721-728.

[5.24] Shih Y, Chen J (1997) Analysis of Fatigue Crack Growth on a Cracked Shaft. International Journal of Fatigue, 19(6), 477-485.

[5.25] Gasch R (1976) Dynamic behaviour of a simple rotor with a cross-sectional crack. IMechE Conf. Vibration in Rotating Machinery, paper C 178/76, Sept. 1976, Cambridge, UK, 123-128.

[5.26] Mayes IW, Davies WGR (1976) The vibrational behaviour of a rotating shaft system containing a transverse crack. IMechE Conf. Vibration in Rotating Machinery, paper C168/76, Sept. 1976, Cambridge, UK, 53-64.

[5.27] Darpe AK, Gupta K, Chawla A (2004) Coupled bending longitudinal and torsional vibrations of a cracked rotor. Journal of Sound and Vibration, 269(1-2), 33-60.

[5.28] Papadopoulos CA (2004) Some comments on the calculation of the local flexibility of cracked shafts. Journal of Sound and Vibration, 278(4-5), 1205-1211.

[5.29] Wu X, Friswell MI, Sawicki JT (2005) Finite element analysis of coupled lateral and torsional vibrations of a rotor with multiple cracks. Proceedings of ASME Turbo Expo 2005 Gas Turbine Technology: Focus for the Future, Jun 2005, vol 4, 841 850.

[5.30] Varé C, Andrieux S (2001) Cracked beam section model applied to turbine rotors. ISCORMA Int. Symposium on Stability Control of Rotating Machinery, Aug. 2001, South Lake Tahoe, California, USA.

[5.31] Papadopoulos CA (2008) The strain energy release approach for modeling cracks in rotors: A state of the art review. Mechanical Systems and Signal Processing, 22(4), 763-789.

[5.32] Dimarogonas AD, Papadopoulos CA (1983) Vibration of cracked shafts in bending. Journal of Sound and Vibration, 91(4), 583-593.

[5.33] Chasalevris AC, Papadopoulos CA (2006) Identification of multiple cracks under bending Mechanical Systems and Signal Processing, 20(7), 1631-1673.

[5.34] Andrieux S, Varé C (2002) A 3D cracked beam model with unilateral contact. Application to rotors. European Journal of Mechanics A/Solids, 21, 793-810.

[5.35] Code_Aster@, (2006) Numerical simulation software for structural analysis, http://www.code-aster.org.

[5.36] Yamamoto T, Ishida Y (2001) Linear and nonlinear rotordynamics. John Wiley & Sons, Inc. New York NY, USA.

[5.37] Muszyńska A (2005) Rotordynamics. CRC Press, Boca Raton FL, USA.

[5.38] Theis W (1990) Längs- und Torsionsschwingungen bei quer angerissenen Wellen (Longitudinal and torsional vibrations of rotating shafts with a transverse crack). VDI-Verlag series 11 no 131, Düsseldorf.

[5.39] Gasch R, Person M, Weitz B (1988) Dynamic behaviour of the Laval rotor with a cracked hollow shaft – a comparison of crack models. IMechE Conf. Vibration in Rotating Machinery, paper C314/88, Sept. 1988, Edinburgh, UK.

[5.40] Schmalhorst B (1989) Experimentelle und theoretische Untersuchung zum Schwingungsverhalten angerissener Rotoren (Experimental and theoretical approach to the vibrational behaviour of cracked rotors). VDI-Forschungs-berichte series 11 no 117, Düsseldorf.

[5.41] Mayes IW, Davies WGR (1984) Analysis of the response of a multi-rotor-bearing system containing a transverse crack in a rotor. Journal of Vibration, Acoustics, Stress and Reliability in Design, 106, 139-145.

[5.42] Friedmann PP (1990) Numerical methods for the treatment of periodic systems with applications to structural dynamics and helicopter rotor dynamics. Computer & Structures, 35(4), 329-347.

[5.43] Jordan DW, Smith P (1999) Nonlinear Ordinary Equation- An Introduction to Dynamical System. Third Edition. Oxford University Press, Oxford, UK.

[5.44] Meng G, Gasch R (2000) Stability and stability degree of a cracked flexible rotor supported on journal bearings. ASME Journal of Vibration and Acoustics, 122, 116-125.

[5.45] Meng G, Gasch R (1992) Stabilität eines gleitgelagerten, einfachen Rotors mit Riß. (Stability of a cracked rotor in fluidfilm bearings). SIRM-Tagung "Schwingungen in rotierenden Maschinen", Febr. 1992, Wien.

[5.46] Hiß F (1996) Nichtlineare Dynamik und Zustandsbeobachtung gleitgelagerter elastischer Rotoren mit angerissenem Wellenquerschnitt (Nonlinear dynamics and monitoring of flexible rotors in fluidfilm bearings having a transverse crack). Dissertation TU Berlin.

[5.47] Sanderson AFP (1992) The vibration behaviour of a large steam turbine generator during crack propagation through the generator rotor. IMechE Conf. Vibration in Rotating Machinery, paper C 432/102, Sept. 1992, Bath, UK.

[5.48] Allianz-Berichte (1987) Schwingungsüberwachung von Turbosätzen – ein Weg zur Erkenntnis von Wellenrissen (Vibration monitoring in turbosets – A way to early crack detection). Bericht 24.

[5.49] Lalanne M, Ferraris G (1998) Rotordynamics Prediction in Engineering. John Wiley & Sons Inc, Chichester, England.

Chapter 6
Results Obtained Using Simulations

6.1 Simulations Compared to Experimental Results

In this section some comparisons between experimental results (obtained by means of test-rigs and real machines) and simulations are shown. The calculated results are obtained with the approach described in sections 5.3.1 and 5.4.3. The good agreement between the results confirm the validity of the described approach. Other simulated results will be presented also in chapter 7 of this book, in order to highlight different peculiar behaviours.

6.1.1 Results Obtained on EDF EURoPE Test-Rig

The test-rig and the model of the shaft are described in detail in section 8.5.2. The results for one of the cracks that have been introduced in the shaft are shown here.

In general it should be reminded that the different harmonic components can present different degrees of accuracy in the comparison:

- the 1X component can be masked by a permanent bow that is generated by crack propagation or by some thermal effects that generate 1X components. It is therefore difficult to isolate the effect of the crack from total vibration;
- the 3X component is small, at least in the bearings, and can be also masked by some surface irregularity.

Therefore the 2X component is the most suitable one for making comparisons between simulated and experimental values.

The selected case is related to a 41% deep crack: the vibration amplitudes of the shaft have been measured in correspondence of the cracked section of the shaft (see figure 6.1) during a run-down transient from 1500 rpm to 100 rpm, using the EURoPE test-rig in EDF laboratories and are shown in figure 2.20 of section 2.4.3.

Figure 6.2 shows the comparison between experimental data and simulated results using the FLEX model. As already specified, the 1X calculated component is due to crack only, whilst the measured one in figure 2.20 is due also to the unknown unbalance and bow distribution of the shaft specimen, which has been generated during the crack propagation using the fatigue cracking machine. Therefore the 1X components cannot be directly compared.

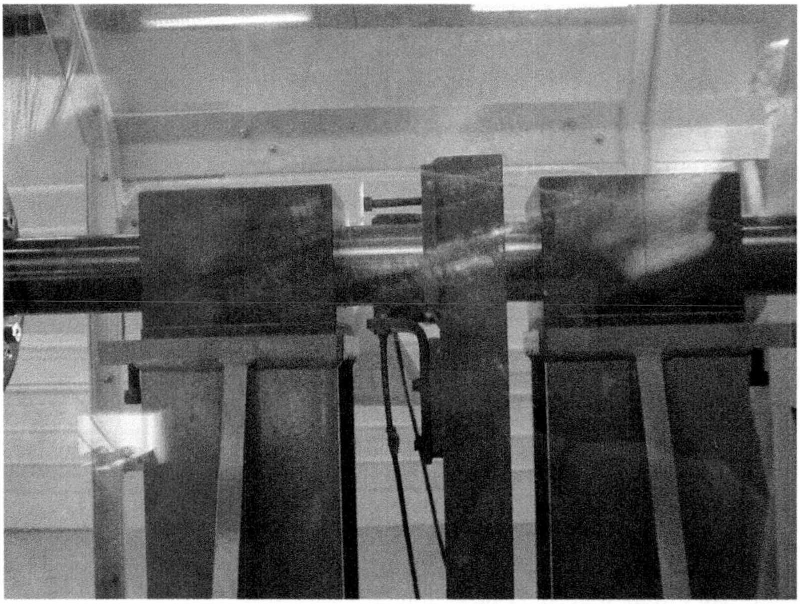

Fig. 6.1. Close-up of the cracked section: note the proximity probe just below the shaft.

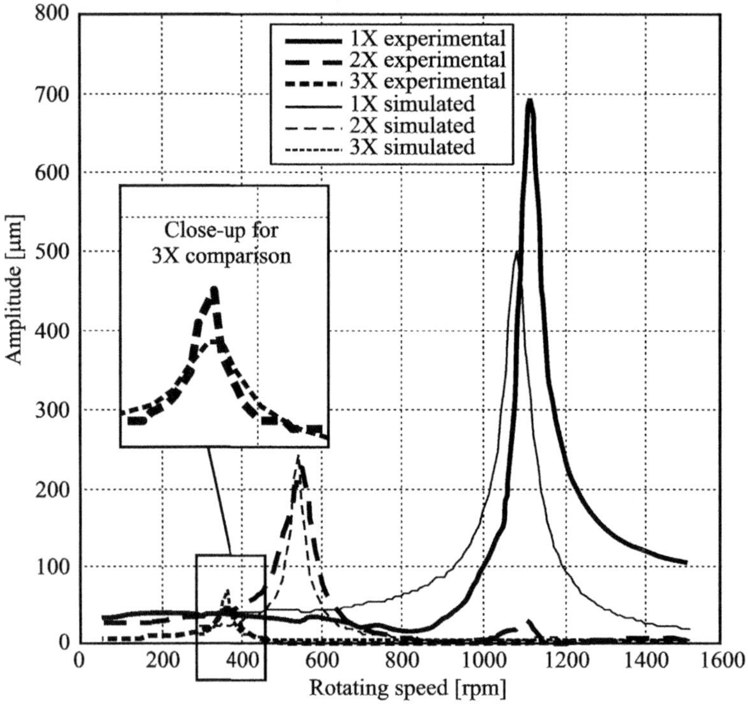

Fig. 6.2. Vibration components excited by a 41% deep crack with rectilinear tip calculated in correspondence of the cracked element.

The agreement between experimental and simulated results is very good for the 2X and the 3X component. This confirms the validity of the proposed model.

Other numerical results are compared to experimental results obtained on the EURoPE test-rig in section 8.5.2. In those cases, the simulated responses are obtained by means of equivalent bending moments, which have been obtained using the model based identification procedure described in section 8.4. These equivalent bending moments, although obtained by means the modelling procedure described in chapter 5, are an approximation of the equivalent crack forces of eqs. (5.148)–(5.150).

6.1.2 Results Obtained on PdM Test-Rig

Also the MODIAROT test-rig of Dept. of Mechanical Engineering of Politecnico di Milano, shown in figure 6.3, has been equipped with a cracked shaft: the crack maximum depth was about 38% of the diameter. The model of the shaft is shown in figure 6.4.

Fig. 6.3. 2-shafts and 4-oil film bearings MODIAROT test-rig.

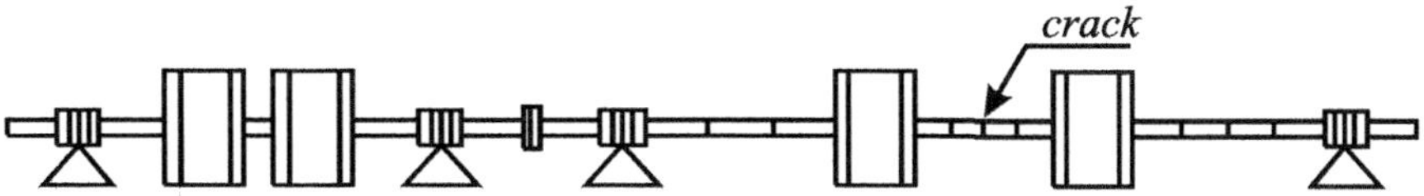

Fig. 6.4. Model of the test-rig shaft.

In this case, 2X and 3X components have been measured in correspondence of the cracked shaft section and compared to the simulated results.

The comparisons, illustrated in figure 6.5 and figure 6.6, show a good agreement. All the presented results confirm the validity of the described approach for modelling cracks and cracked shaft dynamic behaviour.

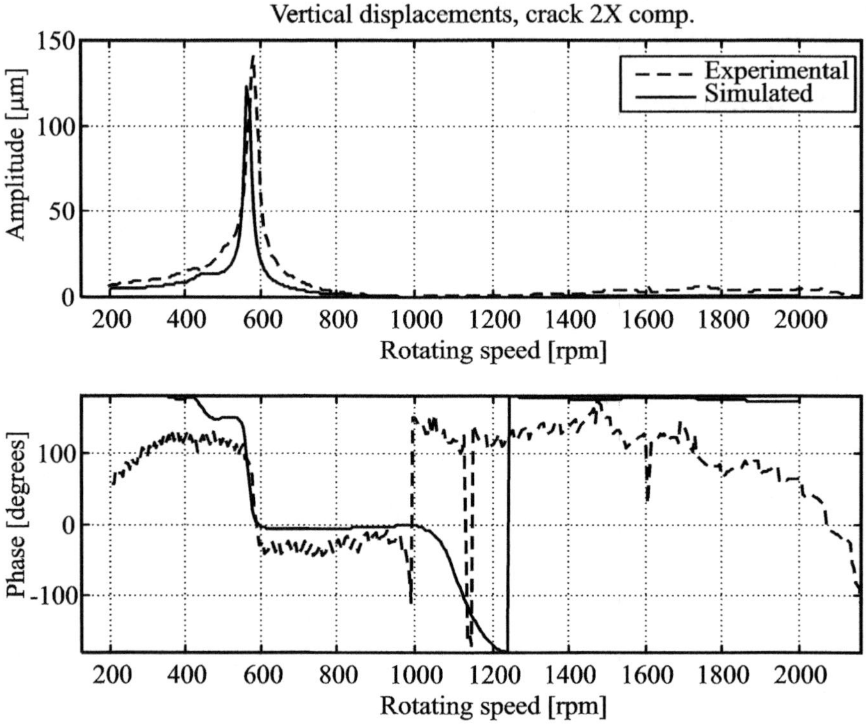

Fig. 6.5. 2X component Bode plot: simulated and measured values in correspondence of the cracked section.

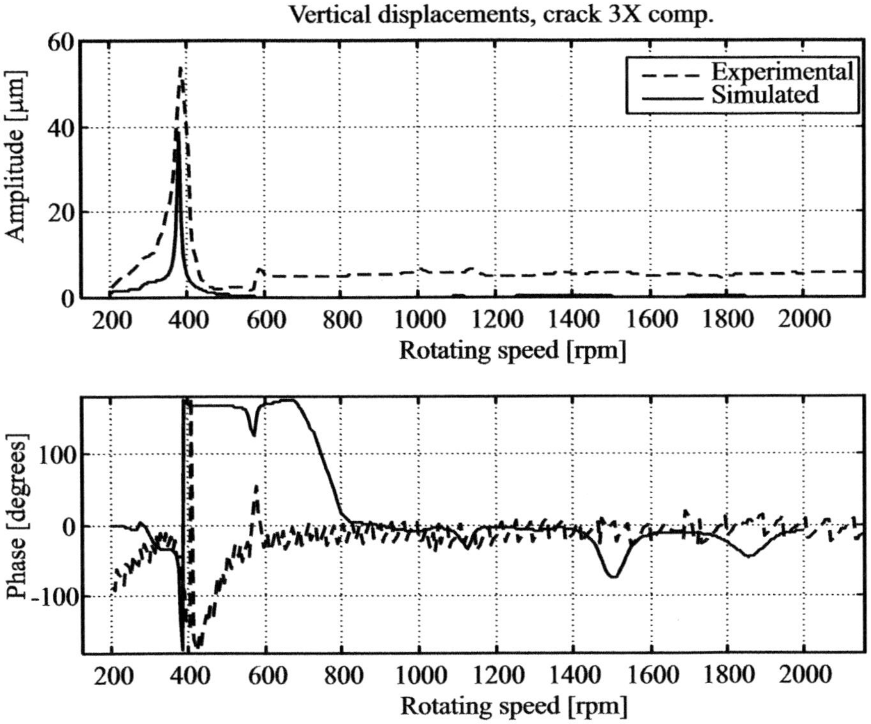

Fig. 6.6. 3X component Bode plot: simulated and measured values in correspondence of the cracked section.

6.1.3 Results Obtained on Real Machines

Some comparisons had been made also on real machines, but in this case the crack model was an early version of that described in sections 5.3.1 and 5.4.3. This "non-tuned" crack model is less accurate than the late model, which had been tuned using 3D non-linear finite elements. The characteristics of this early model were:

1. the breathing mechanism was calculated exactly as in the final model described in section 5.3.1, taking into account the thermal stresses;
2. a bending moment due to thermal stresses had been included in the model. In fact since tensile thermal stresses are not transmitted across the cracked area, the thermal stress distribution loses its polar symmetry and generates a resultant bending moment that causes a bow;
3. the equivalent length of the cracked element was assumed simply twice the depth of the crack. This is the main difference with the late model, in which the equivalent length has been accurately tuned with 3D calculations for each different depth;

4. the dynamical behaviour has been calculated similarly to the method described in section 5.4.3, but with less accuracy in the iterative procedure, considering as dynamic loads both the thermal bow and the equivalent crack forces.

The results are here presented to discuss the reasons of the poor agreement, which are not only due to the less accurate model, and to contribute to a better understanding of cracked rotor behaviour.

The first case analyzed is relative to the cracked generator described in section 2.4.1. The vibration amplitudes measured during the run-down transient are shown in figure 6.7 (that is part of figure 2.12) and can be compared to the simulated results shown in figure 6.8. The simulated results have been obtained, using the early model, considering a rectilinear tip crack with 45% depth and also a cooling thermal effect, with a temperature gradient on the outer surface of the shaft of −0.8°C/min, which tends to open the crack.

The simulated 1X results show smaller amplitudes than the measured values. This could be attributed to an actual sharper thermal transient (that generates a higher bow), to the contribution of unknown unbalance distribution, which had been disregarded in the model, and to rotor model inaccuracies: in fact the linear model exhibits much higher damping than the actual system.

The 2X simulation results should be compared to the "additional" 2X components, generated by the crack only, which could not be separated from original 2X component due to the axial asymmetry of this one pole-pair generator.

Anyway, also the simulated 2X components are smaller than the measured ones. The reason is due again to higher model damping as well as probably to a fully open crack, forced by a thermal cooling transient sharper than the transient considered in the simulation.

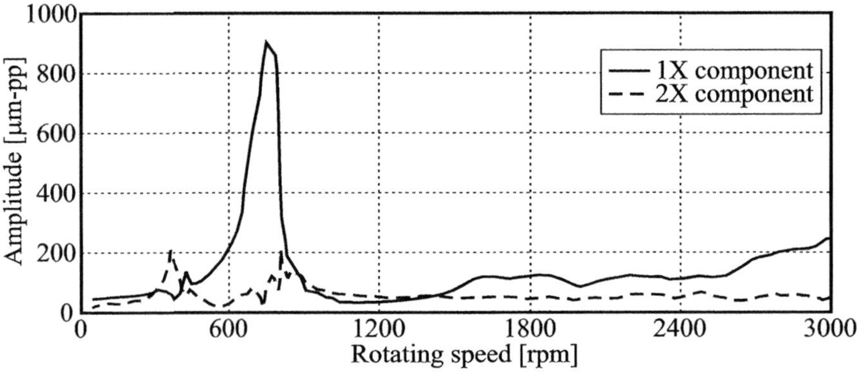

Fig. 6.7. Cracked rotor behaviour measured in correspondence of a bearing during the dramatic run-down transient.

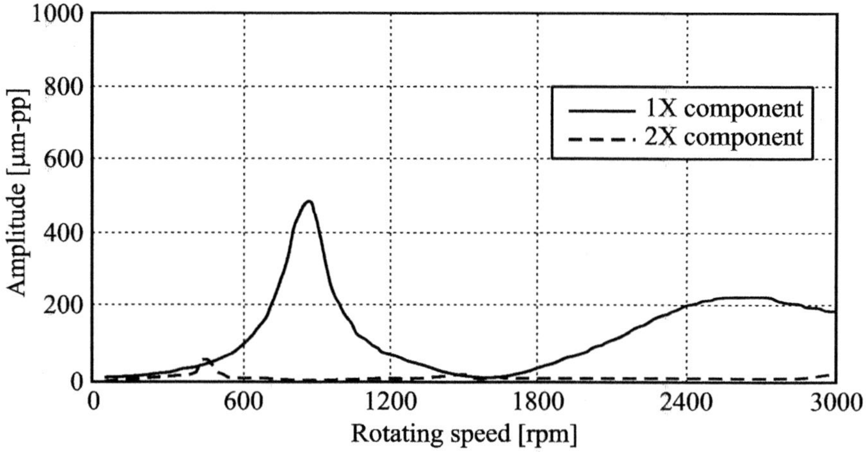

Fig. 6.8. Simulated behaviour in the same bearing (from [6.1]).

The last case analyzed is the comparison between measured results of the cracked steam turbine shaft shown in figure 2.6 and simulated results. Simulated results take into account the effects of thermal stresses on opening and closing the crack, totally or partially, and of the generation of a thermal bow. The simulated results have been obtained using the early non-tuned model, assuming a roughly estimated thermal stress distribution around the crack.

Figure 6.9 shows 1X vertical displacements of the shaft measured in correspondence of the crack in the speed range 0 – 2500 rpm. Figure 6.10 shows instead the 2X component. Both components are related to a heating transient,

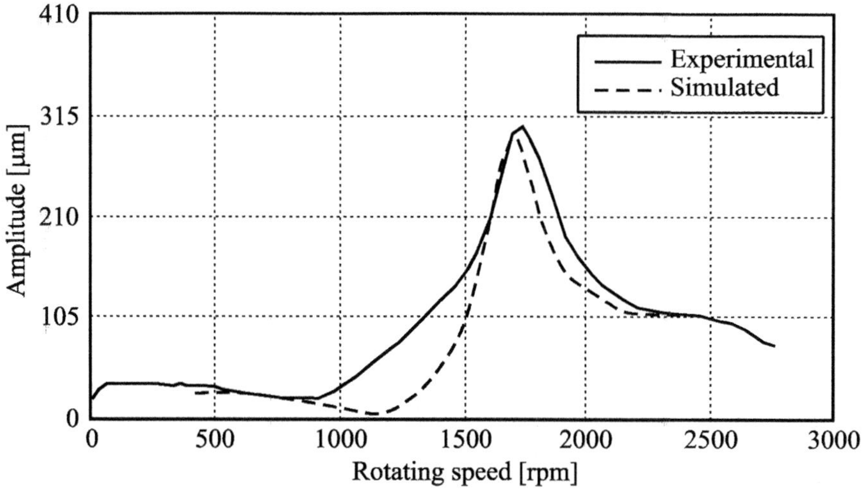

Fig. 6.9. 1X behaviour during heating thermal transient.

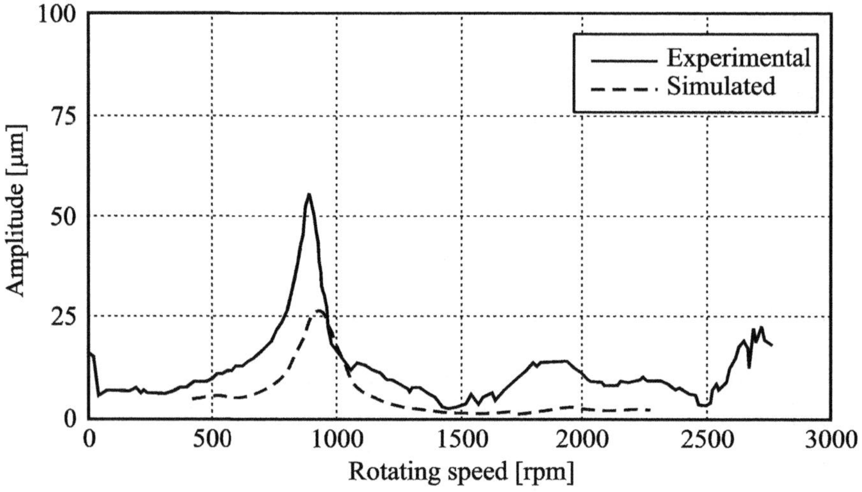

Fig. 6.10. 2X behaviour during heating thermal transient.

which tends to close the crack. The resonance peak amplitude at the shaft 1X critical speed due to the unbalance is clearly represented. The 2X resonance peak amplitude is underestimated by the model.

Considering instead the cooling transient, the crack opens, unbalance is compensated by the local bow in the crack and by the 1X crack effect. The 1X resonance peak disappears (figure 6.11) and the 2X resonance at ½ critical speed is strongly excited (figure 6.12) by the nearly open crack.

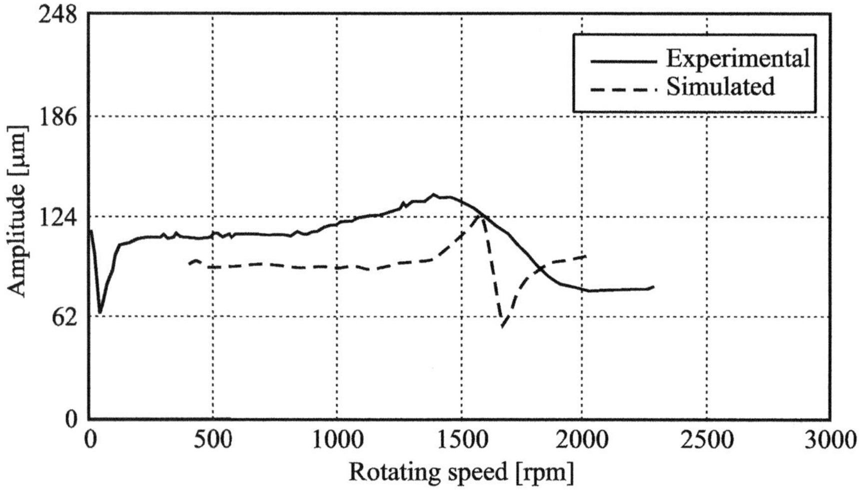

Fig. 6.11. 1X behaviour during cooling thermal transient.

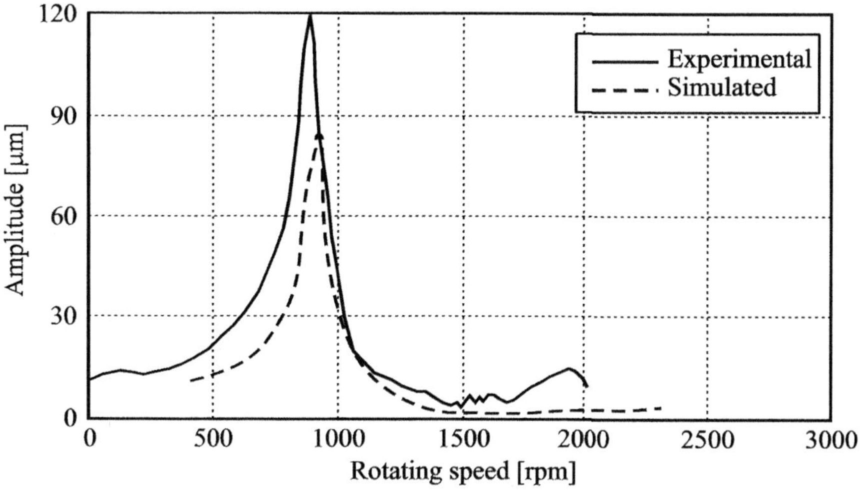

Fig. 6.12. 2X behaviour during cooling thermal transient.

These results show that, even using a poor crack model, the thermal effects have been well reproduced.

6.2 Sensitivity of Crack Induced Vibrations to Different Parameters

Once the model of the crack and the model of the cracked shaft have been verified, the sensitivity of crack induced vibrations to different crack configurations and to different loading conditions can be investigated by means of simulations. The effects of crack position and depth are investigated in section 6.2.1, the effects of crack shape in section 6.2.2 and the effect of shear forces in section 6.2.3.

The study of other effects, which require modifications in modelling, are presented in chapter 7.

6.2.1 Sensitivity of Crack Excited Vibrations to Crack Depth and Position

Using the described models the vibrations that occur in correspondence of the bearings of the cracked shaft, where they are measured in real machines, can be calculated and their severity allows to predict the possibility of detecting the presence of the crack. Therefore it is interesting to investigate how deep a crack has to be in order to generate symptoms that can be easily measured. This depends obviously also on the position of the crack along the shaft.

As shown in section 5.3, the breathing mechanism is the result of stress and strain distribution around the cracked area, which is due to static loads – like the weight –, to the bearing reaction forces and so on, and to dynamical loads – like the unbalance and distribution of the inertia force induced by the vibration –. When the static loads overcome the dynamical ones, which is normal in horizontal axis industrial machinery, the breathing is governed by the angular position of the shaft with respect to the stationary load direction and the crack opens and closes again completely once each revolution.

The FLEX simplified model, described in section 5.4.3 proved to be very accurate and will be used in this section. Once the breathing mechanism and the second moments of area have been defined for the different angular positions of the cracked section, the stiffness matrix of the equivalent cracked beam element of suitable length l_c can be calculated for each different angular position, as shown in section 5.4.3. The angular position dependent stiffness matrix $[\mathbf{K}_c(\Omega t)]$ as defined by eq. (5.46) of section 5.4.3 is once per revolution periodical, has a constant term (the mean stiffness $[\mathbf{K}_m]$) and several harmonic components $[\Delta\mathbf{K}_i]$, of which only the first three components are significant.

In the present analysis, only 4 degrees of freedom per node are used for evaluating the lateral vibrations only and the possible coupling effects of torsion and bending are neglected. All the cracks considered in this section have been assumed with rectilinear tip (see figure 6.13).

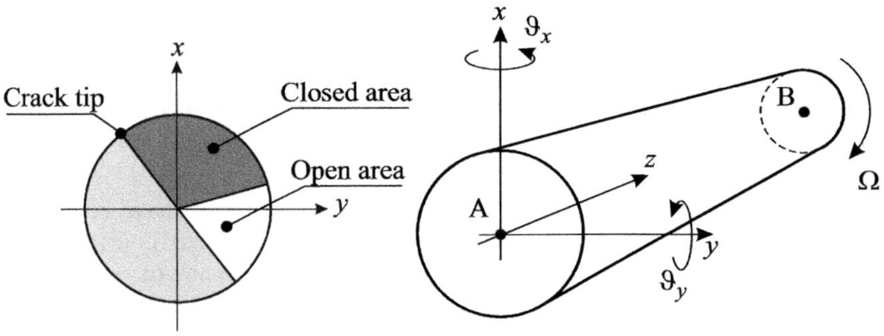

Fig. 6.13. Crack with rectilinear tip and reference frame.

Note that the stiffness coefficients in eq. (5.46) are all proportional to the corresponding second moments of area, as shown in eq. (5.47), that are functions of the depth and of the angular position Ωt, because they depend on the one per revolution periodical breathing of the crack.

The length l_c, tuned by means of 3D finite element calculations, depends only on the diameter D of the shaft and on the crack depth a. The value of the dimensionless ratio l_c/a as function of the dimensionless crack depth $p = a/D$ has been defined in section 5.4.3.

When the equivalent cracked beam, having a reduced cross section and a suitable length, is inserted in the standard beam rotor-dynamic model of the rotor, the global stiffness matrix (including bearing and supporting structure stiffness) is one per revolution periodical, due to the presence of the breathing crack, and can be expanded in Fourier's series.

6.2.1.1 Crack Effects on Rotating Shaft Lateral Vibrations

The static and dynamic response of the cracked rotor can be evaluated using eq. (5.141) of section 5.6 that is solved with iterative procedure by considering the force vectors obtained with eqs. (5.147)-(5.150), for the first three harmonic components.

The crack effect on the bending stiffness is much higher than that one on the shear stiffness as will be shown in section 6.2.3, where shear effects are evaluated. Thus, it results that the *equivalent force vector* $\mathbf{F}_j$ defined by eqs. (5.148)-(5.150) is composed mainly by *bending moments* (given by bending stiffness $[\Delta\mathbf{K}_j]$ of the cracked beam element, multiplied by the angular deflections $\mathbf{x}_j$ of the end nodes of the cracked element, or by its curvature).

In other words the *equivalent crack forces*, which excite in the un-cracked shaft with mean stiffness $[\mathbf{K}_m]$ the vibrations that are caused by the crack in the cracked shaft, are proportional to the different harmonic components $[\Delta\mathbf{K}_j]$ of the crack dependent bending stiffness variation, multiplied by the static $\mathbf{x}_s$ and dynamic $\mathbf{x}_j$ *curvatures* of the cracked beam element. The static curvature $\mathbf{x}_s$ of the cracked element of length l_c is proportional to the static bending moment $\mathbf{M}_b$ and inversely proportional to the second moment of area of the shaft section (which means to fourth power of the shaft diameter), multiplied by l_c which is proportional to the shaft diameter D. Therefore $\mathbf{x}_s$ is proportional to the ratio of bending moment $\mathbf{M}_b$ to the third power of shaft diameter D:

$$\mathbf{x}_s \propto \frac{\mathbf{M}_b}{D^3} \tag{6.1}$$

Also the dynamic curvatures $\mathbf{x}_j$ caused by the crack are proportional to the static bending moment, divided by the third power of the diameter, as can be seen developing the iterative procedure. Since the vibrations $\mathbf{x}_j$ are caused not only by the crack but also by other exciting causes (like unbalance, bow and so on), the equivalent crack forces may be modified by the presence of vibrations excited by other causes, but this modification is generally small and can be neglected in a first approximation.

Therefore it results that the equivalent crack forces are proportional to the static curvature of the cracked beam:

$$\mathbf{F}_j \propto \mathbf{x}_s \tag{6.2}$$

It is possible to say that the equivalent crack forces of eqs. (5.148)-(5.150) impose additional deflections (see figure 6.14) to the shaft that in reality are generated by the periodical additional bending flexibility introduced by the breathing crack in the cracked element. The equivalent rotating crack force vector is composed by pairs of bending moments, applied to the end nodes of the cracked beam element, which generate the additional bending deflections of the cracked element. These crack forces are proportional to the static non-rotating bending moment, divided by the third power of shaft diameter.

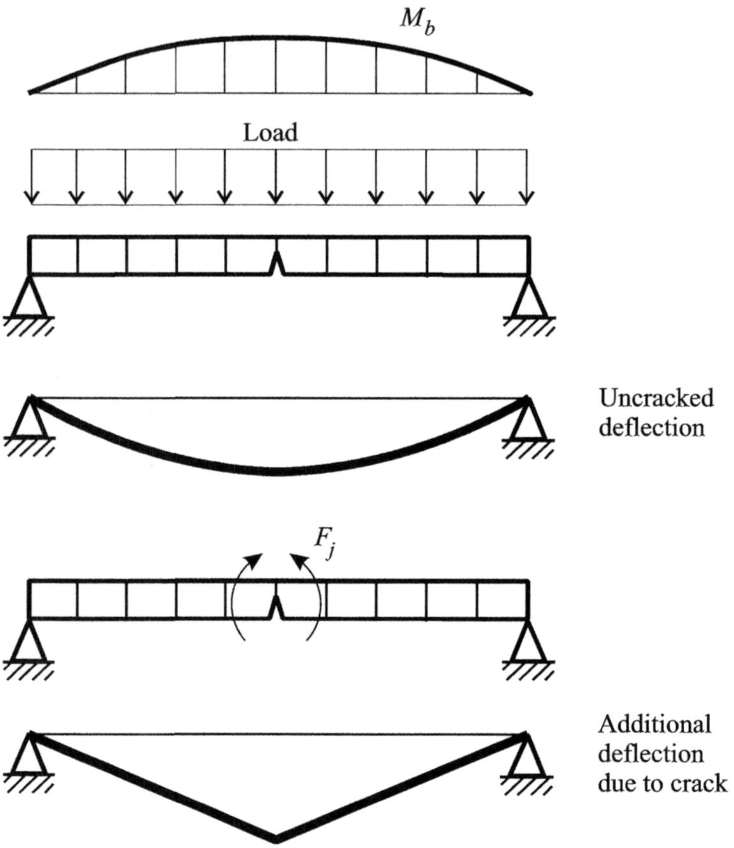

Fig. 6.14. Additional deflections due to the crack.

When the crack is located in a position along the shaft where the static bending moment is small (for instance close to bearings) the excitation of vibrations is weak. When instead it is located where the static bending moment is large (at midspan between bearings) the excitation is much stronger, for a given constant shaft diameter. When also the shaft diameter variations are considered, then maximum excitation is found when the crack is located in positions where the ratio of bending moment to third power of diameter is maximum, as will be shown also in following numerical simulation.

Moreover, the excitation of each harmonic component is proportional to the corresponding stiffness variation $[\Delta\mathbf{K}_j]$ (see eqs. (5.148)-(5.150)), which is in turn proportional to the variation of the second moments of area ΔJ_j over the equivalent length l_c, which are both *function of the crack depth only*, as can be deduced from the structure of $[\mathbf{K}_c(\Omega t)]$ in eq. (5.46):

$$[\Delta\mathbf{K}_j] \propto \frac{\Delta J_j}{l_c} \tag{6.3}$$

Summarizing, the crack forces that excite the vibrations due to crack, depend on *crack depth* only and are proportional to the ratio of *static bending moment* to the *third power of shaft diameter.*

6.2.1.2 *Evaluation of Vibration Components Excitation as Function of Crack Depth*

The effect of the full expressions of the equivalent crack forces eqs. (5.148)-(5.150) is evaluated by calculating the first three harmonic vibration components for different crack depths. The model of a simple symmetrical shaft affected by a crack with rectilinear tip located at mid-span between the two bearings is used for analysing the dependence of vibration excitation on crack depth. The vibration amplitudes for different crack depths are calculated at mid-span of the shaft at very low rotating speed, in order to avoid any dynamic effect. The amplitudes excited are divided by the mean static deflection in the same position, in order to get the dimensionless trend of vibration excitation as function of the depth.

The simple model is shown in figure 6.14. The shaft is loaded by its own weight only. The diagram over the shaft shows the resulting bending moment and the diagram under the shaft represents the static deflection shape, from which also the beam curvature in the different positions can be evaluated.

Figure 6.15 shows the first three harmonic component amplitudes as a function of the crack depth; they are divided by the mean static deflection to obtain dimensionless quantities.

These dimensionless ratios are independent of the used shaft model and give an idea on how the vibration components increase as the crack propagates deeper.

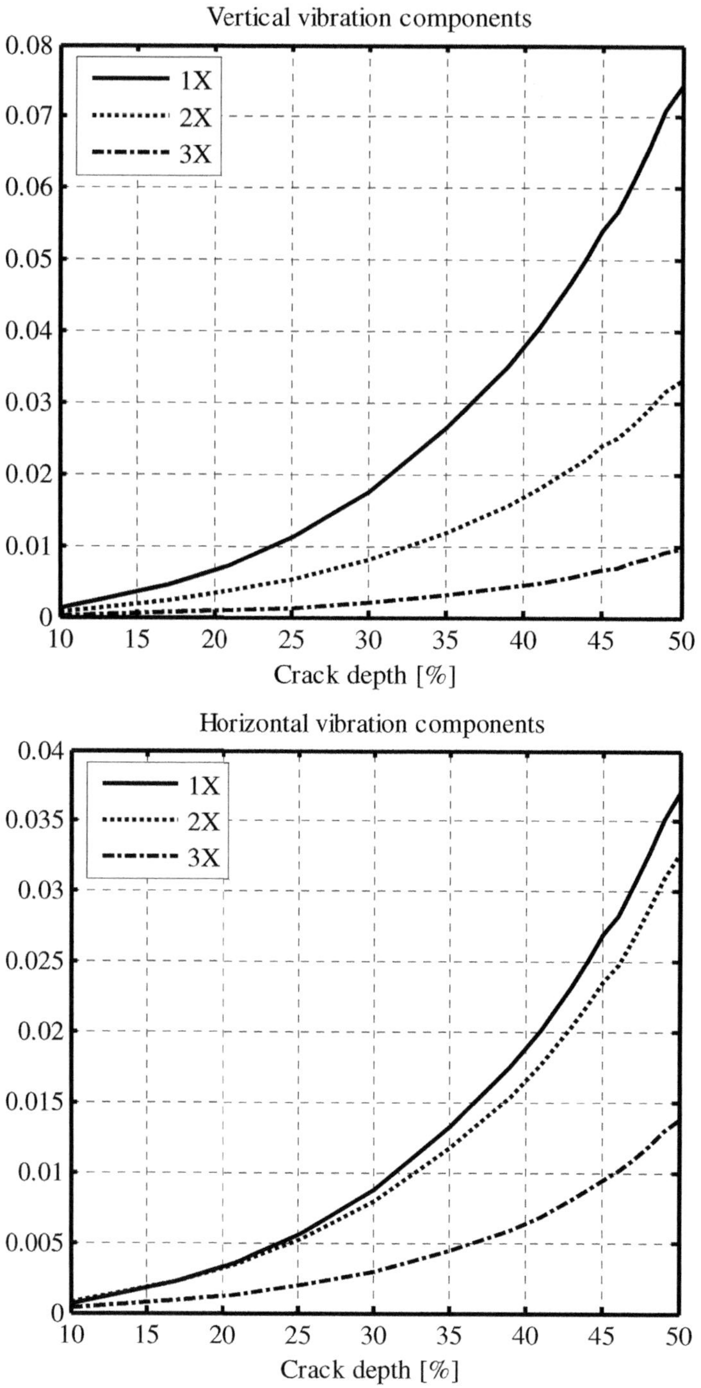

Fig. 6.15. Crack induced dimensionless vibration components x_n/x_s versus crack depth.

The 1X excitation is higher (roughly double) in vertical direction than in horizontal direction, 2X is almost equal in both directions, 3X is slightly higher in horizontal direction than in vertical direction, for any depth. So the trend of vibration excitation of the different harmonic components as function of the depth can be forecasted.

The response of the rotating shaft to the vibration excitation depends obviously on the rotating speed dependent dynamical behaviour of the shaft and on the position where the equivalent crack forces are applied. This will be shown with the aid of a model of a turbo generator unit.

6.2.1.3 Dynamic Behavior of a Cracked Shaft-Line

a) Natural frequencies (critical speeds)

Due to the reduction of the bending and torsion stiffness of the cracked shaft with respect to the un-cracked shaft, also natural frequencies or critical speed values will be reduced. In open crack conditions two different values of bending natural frequencies will be found due to different cracked shaft stiffness according to two orthogonal principal directions. In closed crack condition the original natural frequency of the un-cracked shaft will be found. When the shaft is rotating, due to the breathing and the periodical shaft stiffness, the natural frequencies cannot be directly calculated. The critical speed of a rotating cracked shaft can then be better evaluated by computing the frequency response curves of the cracked shaft and checking the speed where the resonance peak is maximum. A numerical example will be given in the following using the model of a turbo generator unit.

The change of critical speed values depends strongly on the depth and on the position of the crack.

The influence of the crack on the critical speed value is maximum when the crack is located where the corresponding mode shape has maximum curvature and can vanish when the crack is located where the shape has zero curvature. In the first case the reduction of bending stiffness may change consistently the mode shape (and the associated frequency), in the second case the mode shape and frequency remain unaffected by the presence of the crack. In a symmetrical shaft a crack at mid-span changes consistently the first bending critical speed but leaves unchanged the second critical speed.

Similar results are obtained also for the torsion natural frequencies.

b) Frequency response of the rotating shaft

The intensity of excitation of the different harmonic components of the vibrations depends on the depth of the crack (according to figure 6.15) and is proportional to the static bending moment $\mathbf{M}_b$ and inversely proportional to third power of the shaft diameter D, see eq. (6.1), which are both function of the position of the crack. The resulting vibrations depend obviously not only on the intensity of excitation, but also on the position of the excitation: using the concepts of modal analysis we can forecast the severity of the resulting vibrations. This depends on the intensity of the generalized force component of the exciting equivalent bending moments, which are function of the position of the cracked element with respect to the different vibration modes (eigenvectors) and on the dynamic amplification factor (Q factor), which is function of the ratio between frequency of excitation and the corresponding eigenfrequency and of the associated damping factor.

Each harmonic component of the vibration $\mathbf{x}_j(n\Omega)$ according to the j-th degree of freedom, can be expressed by the sum of its modal components, as shown in (6.4).

$$\mathbf{x}_j(n\Omega) = \sum_{r=1}^{m} \frac{\psi_{rj} \sum_{k=k_1}^{k_2} \psi_{rk} \ \mathbf{F}_{nk}(n\Omega)}{m_r \left[\left(\omega_r^2 - n^2\Omega^2 \right) + \mathrm{i}\, h_r \, 2\omega_r n\Omega \right]} \tag{6.4}$$

where n indicates the order of the harmonic component ($n = 1,2,3,\dots$), r indicates the order of the eigenvalue and m is its highest value considered for the analysis ($r = 1,\dots,m$), ω_r is the r-th eigenfrequency, h_r is the associated damping factor, m_r is the modal mass, ψ_{rj} is the displacement according to the j-th degree of freedom in the r-th eigenmode, ψ_{rk} is the displacement of the cracked element nodes in the r-th eigenmode according to the k-th degree of freedom, $\mathbf{F}_{nk}(n\Omega)$ is the component according to the k-th degree of freedom of the crack force vector $\mathbf{F}_n$.

By modelling the shaft with 4 degrees of freedom per node only, which are x, y, ϑ_x and ϑ_y with reference to figure 5.37, and assuming that the crack has developed in beam element c, in between nodes c and $c+1$, k_1 and k_2 assume the following values:

$$k_1 = 4(c-1)+1; \quad k_2 = k_1 + 7 \tag{6.5}$$

The generalized crack force component shown in (6.6) represents the work done (or the energy introduced) by the crack forces in the r-th mode of vibration.

$$\sum_{k=k_1}^{k_2} \psi_{rk} \ \mathbf{F}_{nk}(n\Omega) \tag{6.6}$$

Bearing in mind that crack forces are mainly bending moments, the introduced energy becomes high when the modal displacements ψ_{rk} are composed by angular deflections with high relative values between the two nodes (c and $c+1$) of the cracked element, in other words when the r-th vibration mode presents high curvature in correspondence of the crack.

On the contrary the introduced energy becomes negligible small when the curvature is zero.

The ratio (6.7) is proportional to the dynamic amplification factor (Q factor) which is obviously highest in resonance (for $\omega_r = n\Omega$) provided that the damping factor is not very high.

$$\frac{1}{m_r\left[\left(\omega_r^2 - n^2\Omega^2\right) + \mathrm{i}\, h_r 2\omega_r n\Omega\right]} \tag{6.7}$$

Obviously in the sum of eq. (6.4), only the terms with high generalized crack force components give important contributions that are emphasized by high values of the dynamic amplification factors. Other components give contributions that can be generally neglected.

A crack located in a position where the considered eigenmode presents a high curvature, excites strongly this mode. If moreover the eigenfrequency of this mode is close to the 1X, or 2X or 3X exciting frequency, then this mode is more strongly excited due to its closeness to a resonance.

If the exciting forces are large, but their generalized component is small in the position of the crack with respect to the vibration mode corresponding to the eigenfrequency that is the closest to the $n\Omega$ exciting frequency, the resulting vibrations are small. If also the dynamic amplification factor is small (e.g. because the mode has a high damping factor or because the exciting frequency is far away from resonance), then the vibrations can be hardly measured and the crack can be hardly detected.

Another related problem can be expressed as: are these resulting vibrations large enough to be measured in correspondence of the measuring stations, which are generally the bearings? This depends again on the amplitude of the excited eigenvectors in correspondence of the measuring station.

Only taking into account all these factors, the measurable vibrations can be predicted.

Claiming that a crack of depth $x\%$ can be detected by vibration measurements, without defining all the above specified factors is nonsense.

Some results related to above considerations will be shown with the aid of the model of a turbo generator unit.

6.2.1.4 Description of a Typical Turbo Generator Unit

Figure 6.16 shows the FE model of a 320 MW a turbo generator unit composed by a HP-IP steam turbine, a LP steam turbine and a generator. The unit is equipped with 7 oil-film bearings, which are modeled by their linearized stiffness and damping coefficients. The supporting structure is modeled by means of simple pedestals.

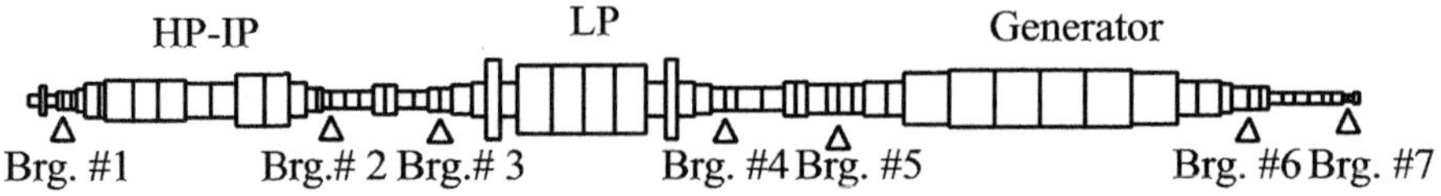

Fig. 6.16. Finite element model of a turbo generator unit.

Let us now examine the static bending moment distribution along the shaft-line with reference to this typical turbo generator unit. This analysis, along with the evaluation of the shaft diameters in the different positions along the shaft-line allows understanding where and when crack excitation can be either severe or weak.

6.2.1.5 Evaluation of Static Bending Moments

The actual static bending moment, which is responsible for generating the crack forces and the related vibrations, depends on the distribution of masses along the shaft-line and on the bearing alignment conditions. Generally the first alignment in cold conditions is made by imposing null forces and moments in correspondence of the coupling flanges of the different shafts. With respect to this situation some change may be introduced with the aim of increasing loads on bearings which, being lightly loaded, could operate close to its instability threshold. This occurs e.g. in the 320 MW turbo generator of figure 6.16 for the last bearing of the generator (bearing #7). At the operating speed, this alignment condition is slightly modified by the oil film thickness that builds up in the different bearings and displaces the journals. In some machines also the thermal expansion of the supporting structure may modify the initial alignment conditions. In order to simulate accurately the effect of a crack the actual static bending moment distribution in operating conditions should be considered. Figure 6.17 shows the comparison between the distributions of the static bending moments in the vertical and in the horizontal plane at rest and at the normal operating speed of 3000 rpm. In the vertical direction the two curves are superposed and differences are negligible. The differences are recognizable (due to different scales) only in horizontal direction. At rest, the horizontal bending moments are null, when the alignment in this direction is according to a straight line (as it should theoretically be). At the operating speed, they are due to the oil film which builds up in the bearings and displaces the journals. It can be seen that the bending moments in horizontal direction are negligible with respect to the bending moments in vertical direction in almost all positions along the shaft-line except in positions close to the bearings where the vertical bending moments become very small.

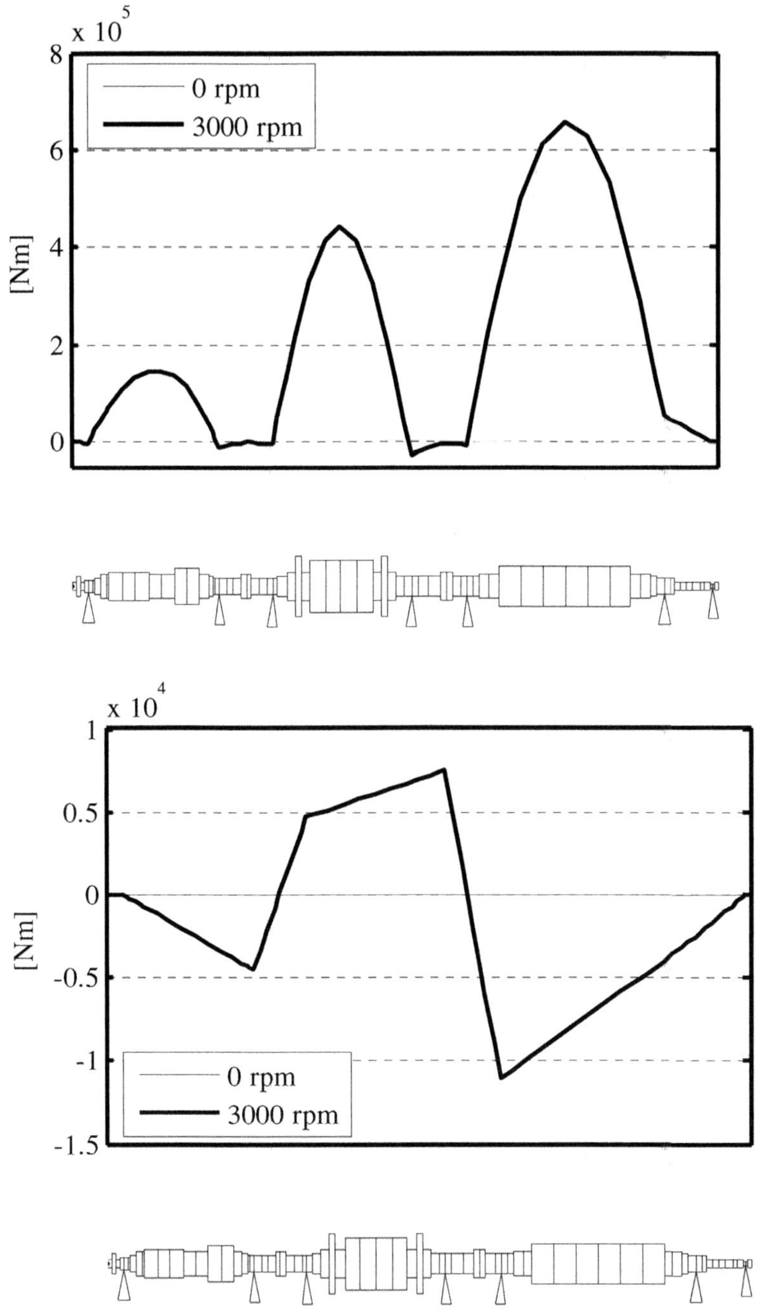

Fig. 6.17. Static bending moments distributions along the shaft-line at rest and at normal operating condition: top in vertical plane and bottom in horizontal plane.

Bearing in mind that the crack forces are proportional to $\mathbf{x}_s$ that is proportional in each node to the static bending moment, divided by the third power of shaft diameter, see eq. (6.1), it results that cracks generate maximum excitation when they are located close to the mid-span of shafts with diameters which are not changing abruptly (like in the HP-IP steam turbine) or when they are located in positions where the bending moment is still rather high, and the shaft diameter is rather low (as it occurs in the LP steam turbine close to the two symmetrical last blade stages, indicated by the two disks in the FE model).

6.2.1.6 Un-cracked Shaft-Line Dynamical Behaviour

The natural frequencies (eigenfrequencies) and the associated mode shapes (eigenvectors) have been calculated at the operating speed of 3000 rpm. The most significant mode shapes and associated frequencies and damping factors are shown in figure 6.18. Some highly damped modes have been disregarded. The first six modes correspond to classical modes of single shafts:

Modes I, II and III correspond respectively to the first bending modes of the generator, of the LP turbine and of the HP-IP turbine shafts. Modes IV, V, and VI are respectively a different first mode of the HP-IP turbine shaft, the second mode of the LP turbine and the second mode of the generator shafts. Mode VII is another generator shaft second mode. Mode X is displayed to show that in the higher frequency range (close to 2X excitation) mode shapes present high amplitudes in parts of the shaft-line in between adjacent bearings, where couplings are present or the exciter of the generator is located. All these mode shapes should be taken into consideration when the generalized crack forces for the different mode shapes are estimated. Being the crack forces equivalent to bending moments applied locally, their generalized forces are strong (which means that the energy introduced by the equivalent bending moments in that particular vibration mode is large), when the crack is located in a position where the mode shape presents high curvature.

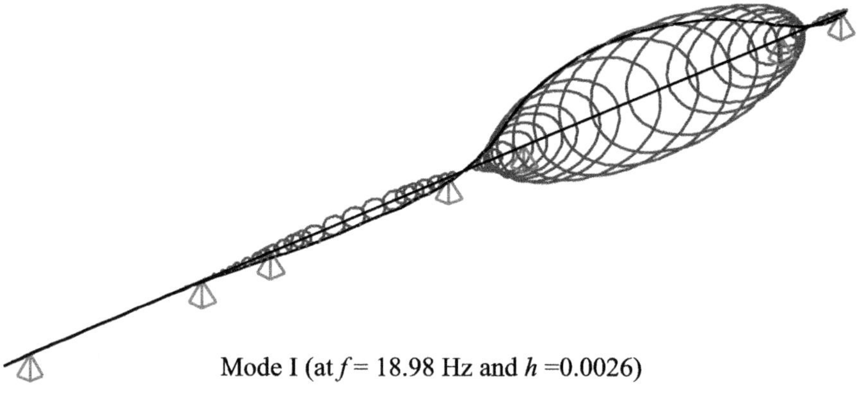

Mode I (at f = 18.98 Hz and h =0.0026)

Fig. 6.18. Most significant mode shapes and damping factors h of the turbo generator unit.

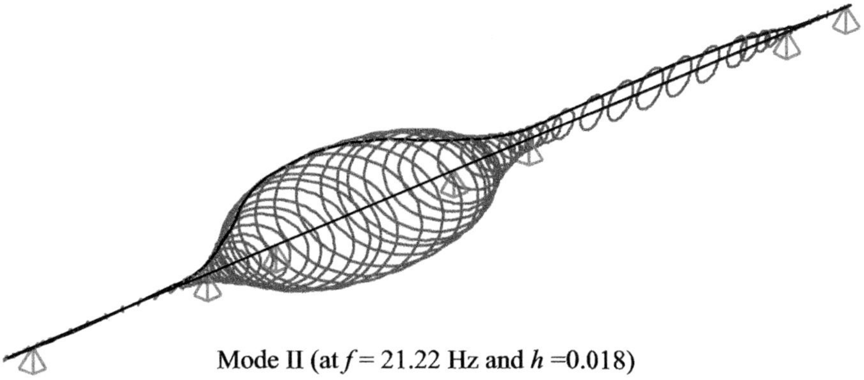

Mode II (at f = 21.22 Hz and h =0.018)

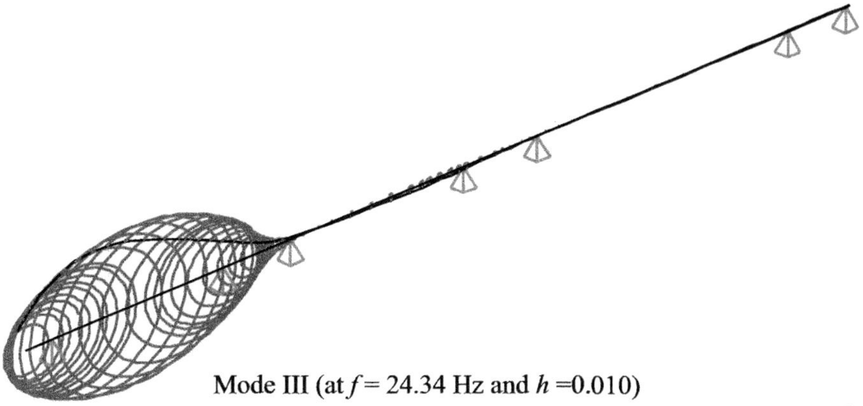

Mode III (at f = 24.34 Hz and h =0.010)

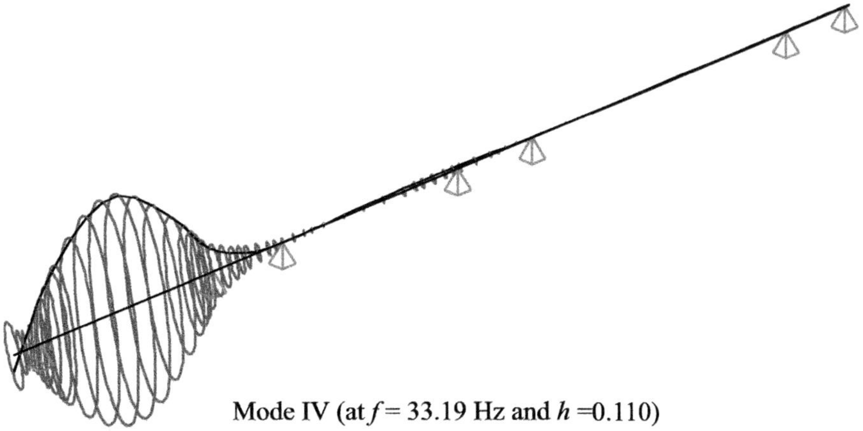

Mode IV (at f = 33.19 Hz and h =0.110)

Fig. 6.18. (*continued*)

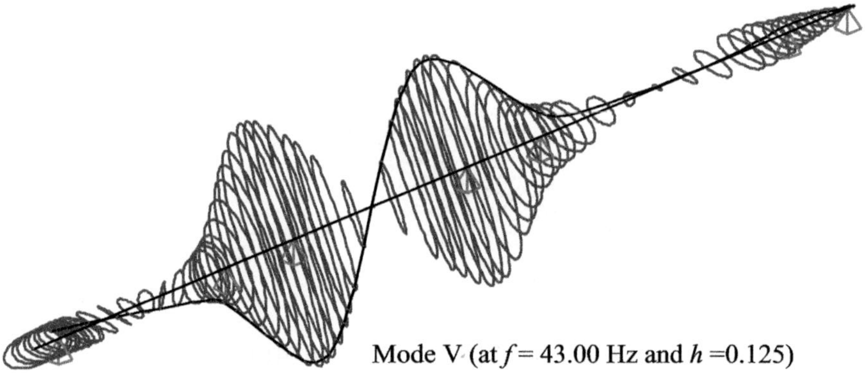

Mode V (at f = 43.00 Hz and h =0.125)

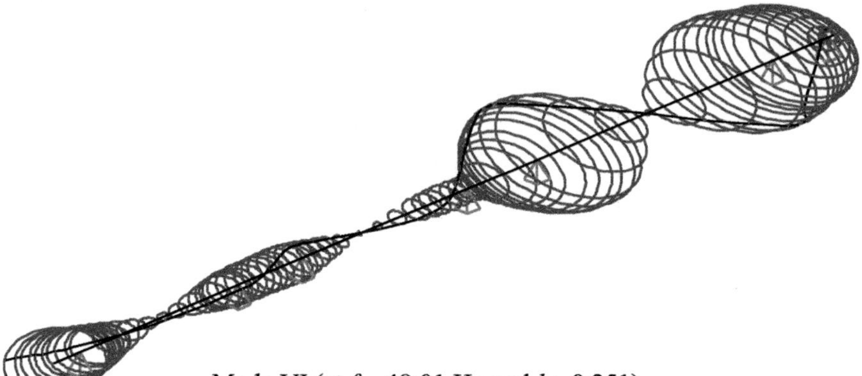

Mode VI (at f = 48.01 Hz and h =0.251)

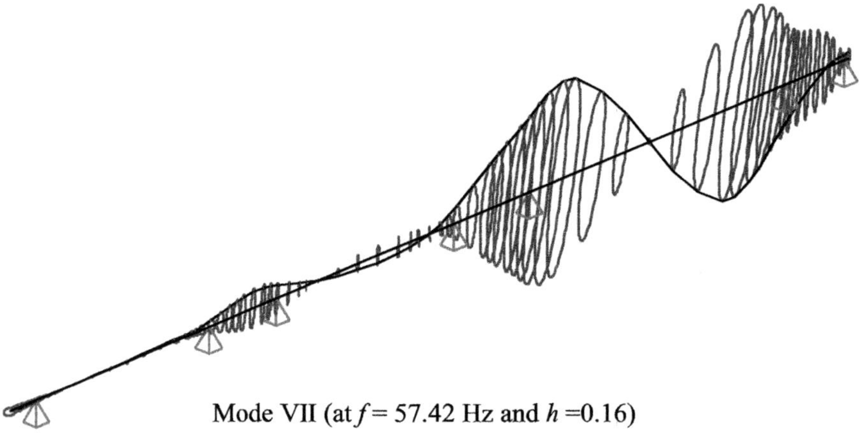

Mode VII (at f = 57.42 Hz and h =0.16)

Fig. 6.18. (*continued*)

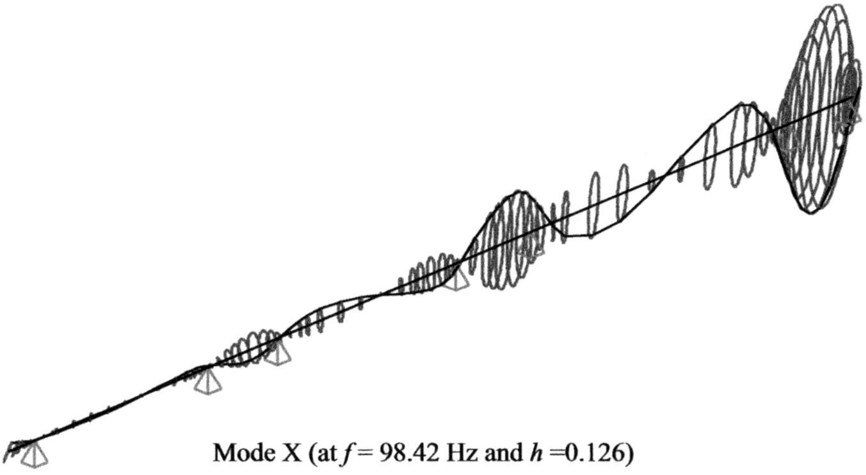

Fig. 6.18. (*continued*)

Cracks at mid-span of single shafts excite strongly their first bending modes when the shaft is rotating at the corresponding frequency (critical speed), but excite very weakly their second bending modes. These are more strongly excited when the crack is located closer to the main bearings (see modes V and VI of figure 6.18). The closest mode to the running speed frequency (mode VI) is the second mode of the generator that could be strongly excited by a crack located on the shaft close to its main bearings, considering only the 1X component of the exciting force. Its 2X component at running speed would excite the generator shaft vibrations of mode X more strongly if the crack would have developed in positions close to bearing #6.

The presence of a transverse crack will affect only slightly the natural frequencies and the associated mode shapes.

This will be shown in the following example: a crack with two different depths (25% and 50% of the diameter) has been applied in the mid-span of the LP turbine rotor, in order to maximize its influence on the first bending critical speed. The mode shape is that of mode II in figure 6.18. Figure 6.19 shows the frequency response curves calculated in bearing #3 of the unit for the un-cracked shaft, excited by a given unbalance, and for the cracked shaft excited by crack only. Resonance peaks at first bending critical speed are found at 1460 rpm for the un-cracked shaft, 1450 rpm for the shaft affected by the small crack (25%) with a frequency shift of 0.6% and at 1414 rpm for the shaft affected by the large crack (50%) with a frequency shift of 3.15%. Note that the natural frequency has moved from the eigenfrequency of 21.22 Hz calculated at rated speed of 3000 rpm (see figure 6.18) to 24.33 Hz (1460 rpm) due to the higher oil film stiffness at lower rotating speeds.

The same cracks in different positions of the same shaft will produce smaller frequency shifts, which can also vanish completely. This confirms the well known opinion that small cracks may not sufficiently affect the bending critical speeds of shafts, so that frequency shifts cannot be considered a consistent symptom of their presence. Bending modes are generally well damped: resonance peaks are very smooth and this fact increases the difficulty of measuring critical frequency shifts in real machines.

Differences of mode shapes would be probably higher, but they have not be taken into considerations due to the difficulties, or the practical impossibility, of measuring the mode shapes in real machines during operating conditions.

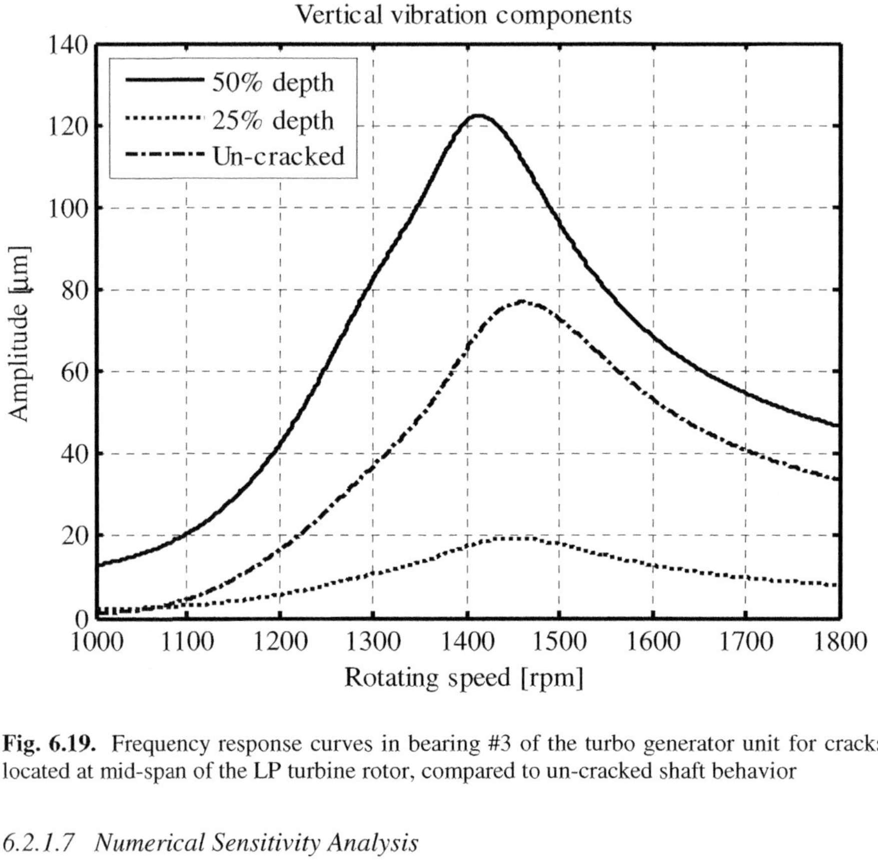

Fig. 6.19. Frequency response curves in bearing #3 of the turbo generator unit for cracks located at mid-span of the LP turbine rotor, compared to un-cracked shaft behavior

6.2.1.7 Numerical Sensitivity Analysis

A numerical analysis has been performed by calculating the vibration behavior of the unit of figure 6.16 affected by transverse cracks only of two different depths in different positions along the shaft: a rather small crack with depth of 25% of the diameter and a rather deep one of 50% of the diameter. The cracks have been located successively into all the elements of the shaft-line. The waterfall diagrams show the deflections along the shaft-line as a function of the different crack locations. For clarity reasons all the deflections between the bearings are not displayed, in order to highlight the vibrations in correspondence of the bearings only.

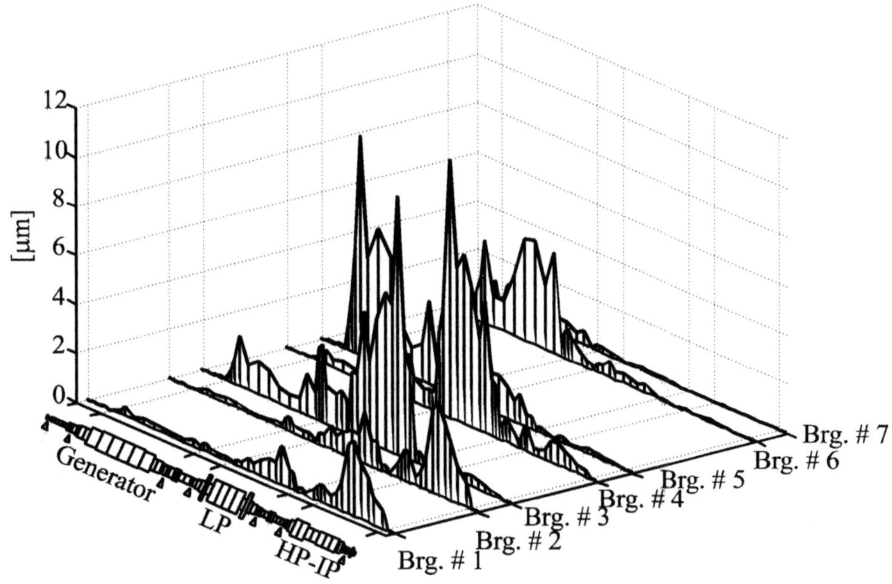

Fig. 6.20. Waterfall plot of 1X vertical vibration amplitudes in correspondence of the bearings (at operating speed of 3000 rpm) as a function of the position of a 25% deep crack.

Figure 6.20 and figure 6.21 show the waterfall plots of respectively the 1X and 2X vibrations in the bearings in vertical direction as a function of the crack position for a crack with depth of 25% of the diameter. 1X vibration excitation is larger in vertical than in horizontal direction, which is not shown for brevity. 2X vibration excitation is instead almost equal in horizontal and vertical direction. Both results were predicted in figure 6.15.

The maximum 1X vibration amplitude in bearings #1 and #2 is less than 4 µm when the crack is located between the HP and the IP sections of the turbine, position in which the vibration excitation is maximum as shown in section 6.2.1.5.

In bearings #3 and #4 the maximum amplitude is about 10 µm when the crack is located in positions of the LP turbine close to one or the other elements modeling the last stages, where the vibration excitation is maximum. High dynamic response is expected for cracks in these positions, bearing in mind the LP turbine mode shape at 43 Hz (mode V of figure 6.18 that is rather close to 50 Hz) and the high curvatures.

On the generator, the maximum amplitude is about 9 μm in bearing #5 when the crack is located on the generator in a position rather close to bearing #6, where the shaft diameter has a strong reduction. In this position the excitation is strong due to a good ratio of bending moment to diameter, as also in the symmetrical position close to bearing #5, but the response due to excitation close to bearing #6 is higher in bearing #5 than the response in bearing #6 due to the same excitation. The response in bearing #6 due to excitation in bearing #5 is lower. This can be explained considering the effect of the dynamical behavior of the unit at the frequency of 50 Hz, which is between the natural frequencies at 48.01 and 57.42 Hz where the modes VI and VII of figure 6.18 show high curvatures in the shaft close to bearing #6 (in mode VI) and high vibration amplitudes in bearing #5 (in mode VII).

Figure 6.21 shows that the maximum 2X vibration amplitudes in bearings #1 and #2 are less than 2 μm when the crack is between the HP and the IP sections of the turbine (maximum excitation), less than 3 μm in bearings #3 and #4 when the crack is in the LP turbine close to one of the last stage elements (again in positions of maximum excitation). In both cases the response is not very high as can be deduced from mode shape X at the frequency of 98.42 Hz, which shows low vibration levels in both turbines. In bearing #5 a maximum of 4.6 μm is found when the crack is on the generator, in a position close to bearing #6. The crack in the same position excites also 3.5 μm in bearings #6 and #7. These effects are due to the dynamical behavior of the group at the frequency of 100 Hz, when the excited vibration mode is close to the mode shape X. High curvature of the mode shape in correspondence of the crack position and high vibration amplitudes in correspondence of bearing #5 justify the calculated behavior.

The 2X excitation is roughly half the 1X excitation, as predicted in figure 6.15, but dynamic effects are different at 50 and 100 Hz. The 2X component, which is the most reliable symptom related to a developing crack, is rather small for the 25% deep crack and could hardly be detected by measurements taken in the bearings at rated speed. This holds for any position of the crack along the shaft-line, with the exception of the few above specified positions.

It can further be noted that cracks propagate generally in positions where bending stresses are high and are magnified by stress intensity factors: this occurs exactly in the positions where high excitation has been found in the numerical simulation.

This fact is confirmed also by some of the experimental results of real machinery presented in Section 2.1, where rather high vibration levels have been found. In fact these are related to cracks that have developed in positions where high bending stresses and consequently high excitation would probably be found.

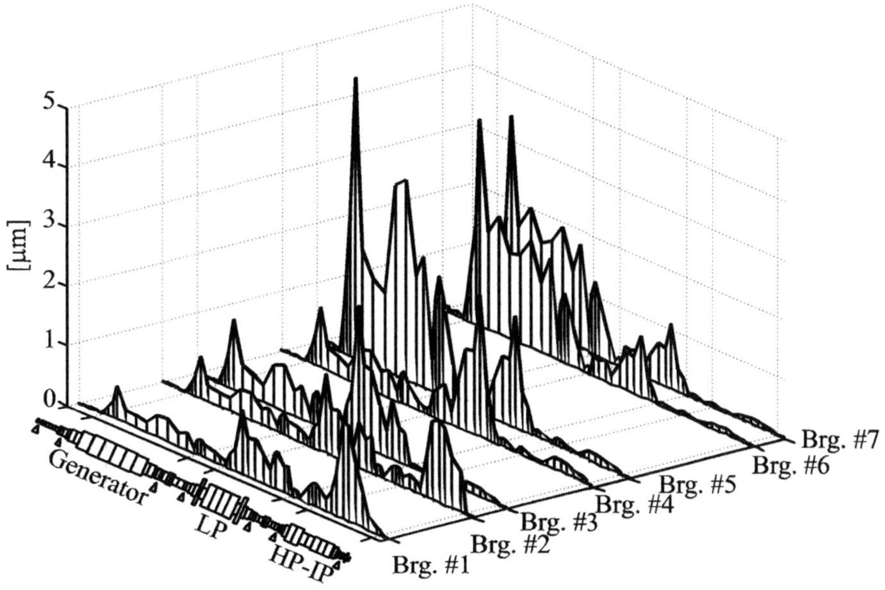

Fig. 6.21. Waterfall plot of 2X vertical vibration amplitudes in correspondence of the bearings (at operating speed of 3000 rpm) as a function of the position of a 25% deep crack.

Figure 6.22 and figure 6.23 show the 1X and 2X components for crack depth of 50%. The expected increase of 1X and 2X excitations from 25% deep crack to the 50% deep crack, as calculated in figure 6.15 for slow rotating conditions, is roughly 7.5 times the values of the 25% crack for the 1X component and 6.6 times for the 2X component. At rated speed, the influence of the dynamic curvature of the cracked element on the equivalent crack forces modifies the equivalent crack force values. This modification increases obviously as the crack depth increases generating higher vibrations. Therefore the calculated increase of vibrations is somewhat different from expected one. The maximum 1X and 2X amplitudes reach now respectively 63 μm and 21 μm, therefore the 1X vibrations are magnified 6.3 times and the 2X component 4.5 times, which is less than the expected values. Similar ratios are found in all other positions.

Rather deep cracks located in any position of the shaft-line, with few exceptions, generate not very large but significant 2X symptoms. Anyway, there remain still some positions in which a crack could develop and grow up to a dangerous depth, without generating measurable symptoms at rated speed. Fortunately these positions correspond to sections of the shaft-line in which the bending moment is very small (as between bearings #2 and #3) that makes the crack force excitation small but also the crack propagation difficult due to the low values of the stresses.

From figure 6.22 and figure 6.23 it can be seen that the severity of the crack excited vibrations in the bearings, at normal operating speed, depend not only on excitation severity but also on the dynamic behaviour of the shaft-line in correspondence of the crack position and in correspondence of the bearings where the vibrations can be measured.

Much more information can be gathered if a run-down speed transient is analyzed, in which resonances can amplify the 2X symptoms.

As an example in figure 6.24 the frequency response curves for the 3 harmonic components that could be measured in bearing #3 of the LP turbine, excited by a 25% crack located at mid-span, are represented. At rated speed of 3000 rpm amplitudes of 6-7 μm for the 1X component and 1-2 μm for the 2X component are found, which is in accordance with figure 6.20 and figure 6.21. These components are obviously magnified in correspondence of critical speeds. At mid-span of the LP turbine, the bending moment is maximum (strong crack force excitation) and the generalized crack force components are relevant in the first mode (at first bending critical speed) and are small in the second mode (at its second bending critical speed) that is not excited at all. Regarding the 2X component, resonances are present at ½ of the values of first and third critical speeds. Finally also a peak at ⅓ of the first bending critical speed is clearly recognizable in the 3X component. These symptoms help strongly to identify the presence of a crack.

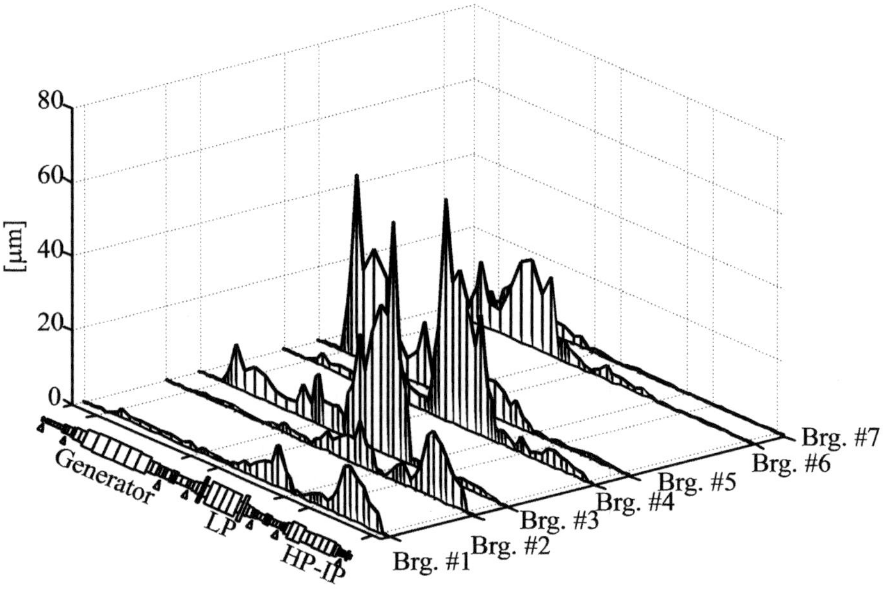

Fig. 6.22. Waterfall plot of 1X vertical vibration amplitudes in correspondence of the bearings (at operating speed of 3000 rpm) as a function of the position of a 50% deep crack.

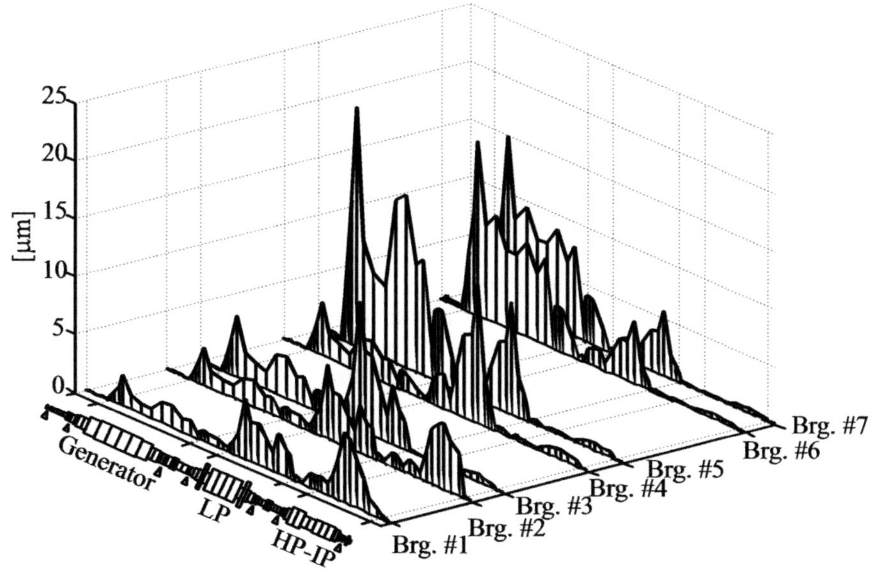

Fig. 6.23. Waterfall plot of 2X vertical vibration amplitudes in correspondence of the bearings (at operating speed of 3000 rpm) as a function of the position of a 50% deep crack.

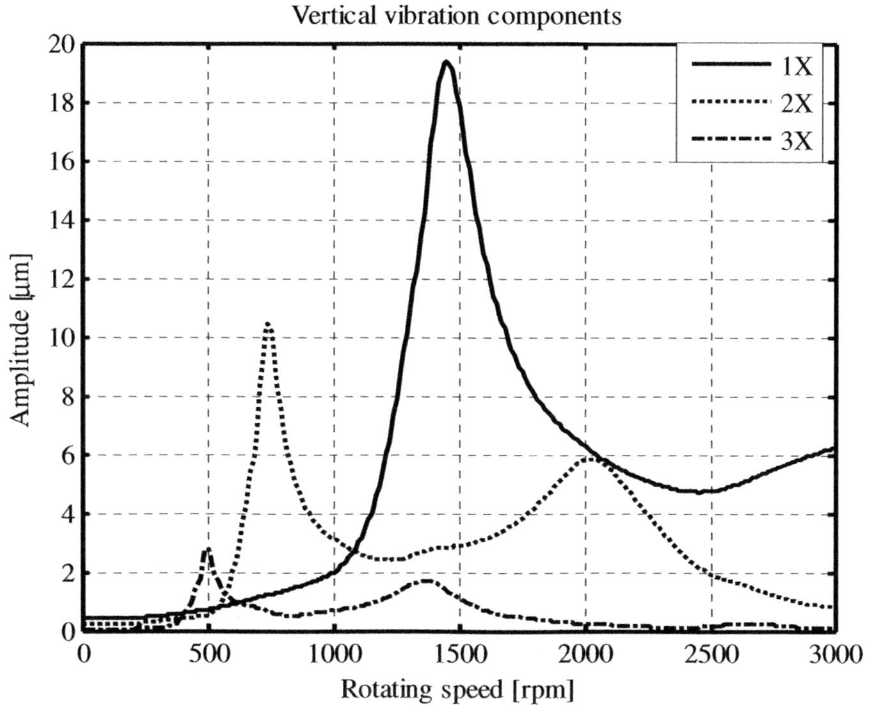

Fig. 6.24. Vertical frequency response curves in bearing #3 for the 25% crack located at mid-span of the LP turbine.

It should be always reminded that the excitation of 1X vibration component can be due to different causes (crack included) whereas the 2X and 3X vibration excitation are due to the stiffness asymmetry associated to the crack, if no other causes of asymmetry are present. In one-pole pair generators (figure 3.6) the original residual asymmetry generates 2X vibrations, to which the vibrations due to the crack are superposed. The 2X vibration, and the 3X vibration when measurable, are the most reliable symptoms of the presence of a transverse crack in the rotating shaft.

Finally a set of calculations has been made for showing the 2X vibration amplitudes that could be measured in the bearings of the unit during a run-down transient. Only the case of the small crack, i.e. depth 25% of the diameter, and the vibrations in one bearing are shown as an example when the crack is located successively in all different beam elements of the unit model. The resulting waterfall diagram of the vibration amplitudes in vertical direction obtained in bearing #4 is shown in figure 6.25. The maximum vibration amplitudes are obtained in correspondence of the first natural frequencies of LP turbine and of generator and are about 10 μm. As can be seen from figure 6.25, bearing #4 is sensitive to cracks which develop in the main bodies of LP turbine and generator but is insensitive to cracks that develop in HP-IP turbine or in the exciter of the generator. Their symptoms could be measured in other bearings.

From this analysis it results once again that the most reliable symptoms of a developing crack are the 2X components that are excited in resonance corresponding to the first natural frequencies of the different shafts during a run-down transient of the unit.

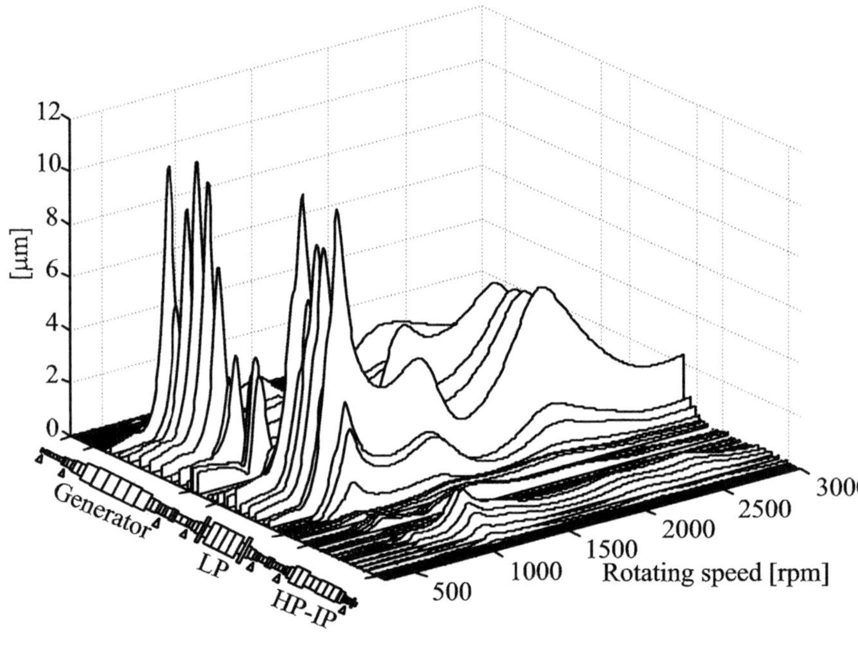

Fig. 6.25. 2X vibration amplitudes in vertical direction in bearing #4 during a run-down transient as function of different crack positions (crack depth 25%).

6.2.2 Effect of Crack Shape

Almost all papers about cracks in shafts deal with cracks having a rectilinear tip. Real cracks may have different shapes, which also change during the propagation of the transverse crack. A crack with a small depth can have a convex elliptical shape, as shown e.g. in figure 2.6 of section 2.3.1, where the relative depth was measured about 25% of the diameter. When the crack propagates deeper, it can develop a convex elliptical shape, even if it starts from a small concave elliptical shape, as shown e.g. in figure 2.15 of section 2.4.1. Otherwise it can also develop a concave elliptical shape, as shown in figure 4.5 of section 4.2.1, passing probably during its propagation through a shape with roughly rectilinear tip. Therefore it is interesting to analyse the effect of different crack shapes on the vibrations.

The results presented in this section will show that the differences between elliptical and rectilinear tips with same relative maximum depths are rather small. These results are obtained by comparing the vibration amplitudes evaluated in correspondence of bearing #6 of the turbo generator unit shown in figure 6.16 affected by a crack in a given position, shown in figure 6.26. The results found for cracks with rectilinear tip can be considered valid, although with some restrictions and less accuracy, also for cracks with different shapes.

Other possibilities of crack morphology are taken into account in this section:

- sometimes permanent stresses that arise during crack propagation hold the crack open during the full rotation of the shaft. The open crack excites a peculiar vibration behaviour in the rotating shaft;
- sometimes a crack can start on one side of the shaft, can then stop propagating on this side and the shaft start cracking on the opposite side.

It could occur also that cracks propagate faster in circumferential direction than in radial direction. When the circumferential extension exceeds half the circumference, then some change in excitation occurs. Some numerical results are discussed in section 7.1 for torsional and lateral excitation.

Finally some cases of occurrence of multiple cracks in the same shaft section have been reported in literature. Therefore a triple crack with elliptical shapes is also analyzed in this section, for calculating the breathing behaviour and the consequent deflections due to given loads. The excitation of the vibrations are then proportional to these static deflections.

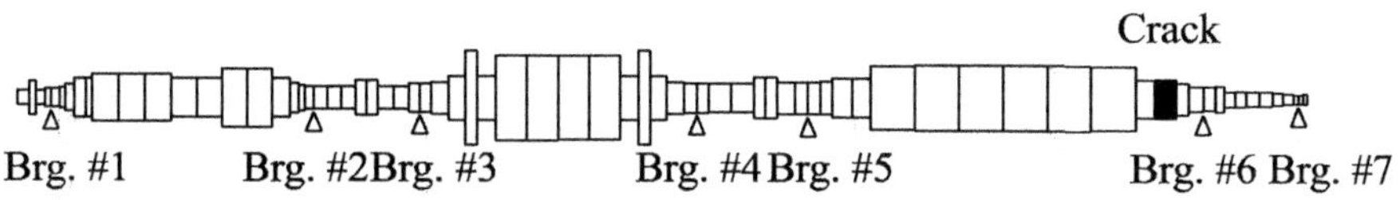

Fig. 6.26. FEM model of the turbo generator unit indicating the position of the crack.

6.2.2.1 Rectilinear Tip Crack Compared to Convex Elliptical Crack

The effects caused by a crack with rectilinear tip have been compared to those of an elliptical crack. The comparison has been made for equal cracked area, i.e. the elliptical crack is slightly deeper than the rectilinear crack. Breathing behaviour has been calculated for both cracks and finally the vibrations have been calculated during a run-down transient of the machine, considering the two cracks in the same position.

Figure 6.27 shows the results of the rectilinear crack that can be compared with the results of the elliptical crack shown in figure 6.28.

The 1X component is somewhat greater for the rectilinear crack, the 3X component is slightly greater for the elliptical crack, but the 2X component is equal for both shapes.

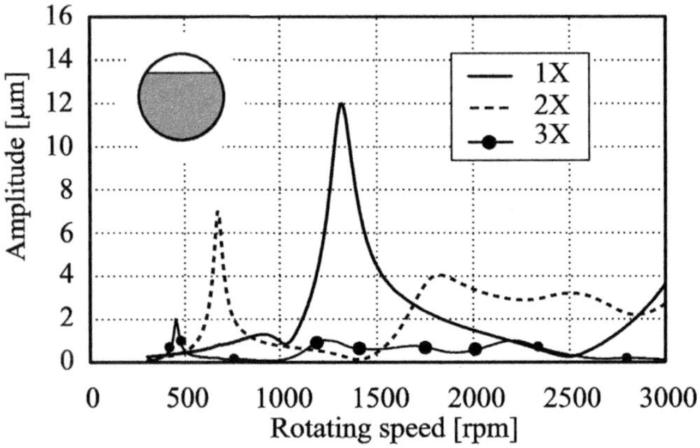

Fig. 6.27. Vibration components measured in bearing #6 of a cracked turbo generator: crack with rectilinear tip.

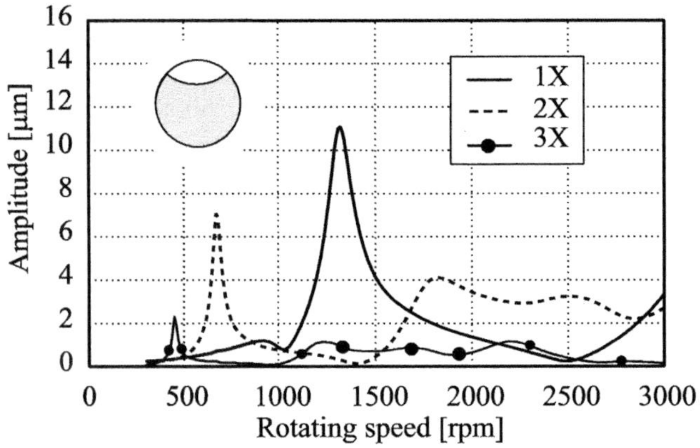

Fig. 6.28. Vibration components measured in bearing #6 of a cracked turbo generator: crack with elliptical tip.

These simulations suggest that the effects caused by elliptical cracks on shaft vibrations can be considered equivalent to those caused by rectilinear cracks, providing that the same cracked area is assumed. The most significant symptom, the 2X component, is equal for both crack shapes and the differences in the other harmonic components are vanishing small.

6.2.2.2 Open Crack

Sometimes it occurs that permanent stresses that arise during crack propagation hold the crack open during the full rotation of the shaft. In this case only 2X vibrations are excited.

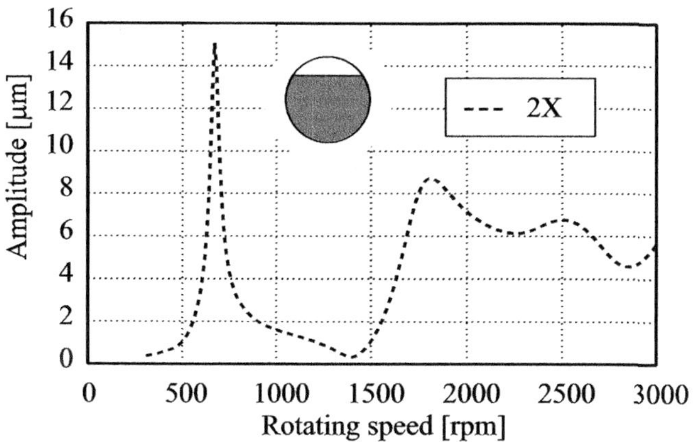

Fig. 6.29. Fully open crack effects.

This is shown in figure 6.29 where the simulated vibrations of the shaft affected by a fully open crack are represented during a run-down of the machine. The crack shape and surface is equal to those of the rectilinear crack shown in figure 6.27. Comparing the vibrations caused by the open crack to those caused by the breathing crack, it results that 1X and 3X components have disappeared and that the 2X vibrations are more than twice. This could be expected because a fully open crack generates only a 2X stiffness variation that is much higher than the 2X variation of the breathing crack with the same shape, which is a crack that closes completely during a consistent angle of rotation. While the crack is closed, there is no stiffness variation.

6.2.2.3 Double Cracks

Sometimes a crack can start on one side of the shaft and then stop propagating on this side and start instead propagating on the opposite side. These cracks are so called *double cracks*. In this case the simulation is performed considering a main crack of the same shape and area of that of figure 6.28, while the second one is diametrically opposite and smaller.

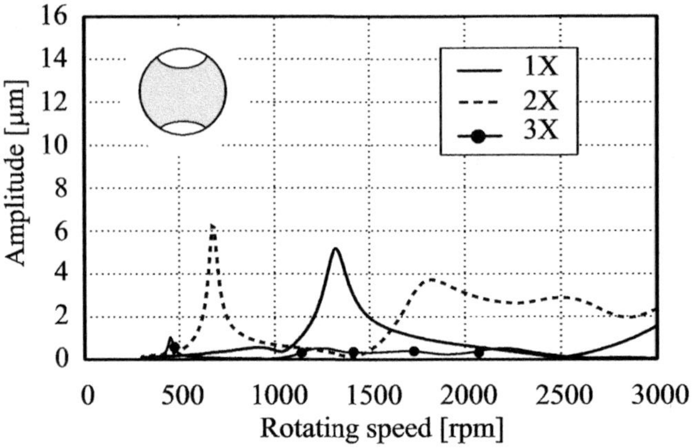

Fig. 6.30. Double crack effects.

Again, breathing behaviour and vibrations have been calculated for the same shaft with the cracks in the same position. Figure 6.30 shows the results obtained in the same bearing #6 of the shaft-train as previous figure 6.27-figure 6.29.

As could be expected the 1X vibration component is much smaller, the 2X is only a little bit smaller and the 3X is again much smaller than the components excited by a single crack. The smaller crack on the opposite side compensates in some way the effect of the main crack. This makes the detection of this kind of crack very difficult by means of lateral vibration measurements. Other symptoms (like shift in torsion natural frequencies or torsion vibration excitation) might be more efficient.

6.2.2.4 Triple Elliptical Cracks

The effect of multiple cracks in different shaft sections has also been analyzed by some authors (see e.g. [6.2]). The occurrence of a triple crack on the same shaft section has been instead reported in literature [6.3]. These cracks have developed from 3 equally spaced keyholes in the shaft of a low pressure steam turbine, similarly to figure 1.3. This case has then been accurately analyzed.

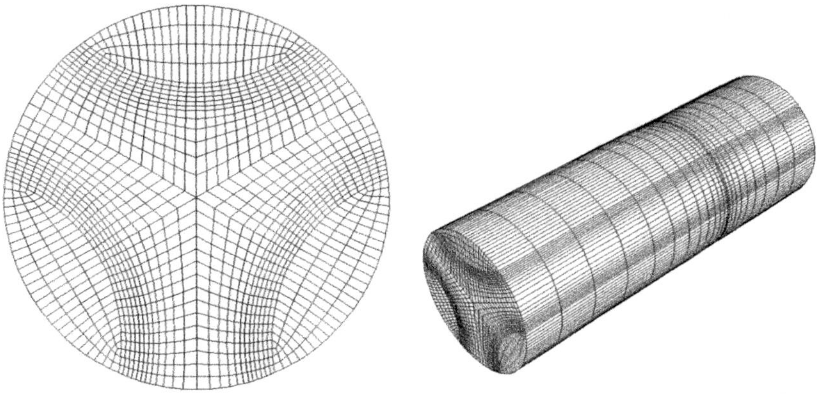

Fig. 6.31. 3D model of the cracked shaft specimen.

The 3D model used is shown in figure 6.31. The breathing behaviour of the triple crack, as results of the 3D non-linear model, is shown in figure 6.32 and has been compared to the FLEX model results, for 3 different angular positions of the shaft: 0°, 30° and 60°. Excellent agreement between the two models has been found.

The behaviour is symmetric for the angular positions in between 60° and 120°, and is duplicated for the angular positions between 120° and 240° as well as between 240° and 360°. As expected, the breathing behaviour is periodical with period equal to $\frac{1}{3}$ of the shaft rotation.

These results have been used for calculating the stiffness variation and the resulting deflections of the specimen in its different angular positions, loaded by a given bending moment. The static deflections of the triple crack as results of the 3D non-linear model compared to the FLEX model results, in different angular positions, are represented in figure 6.33.

The only significant component of the lateral deflections is the 3X component, which is slightly underestimated by FLEX model. The vertical 3X component is 28% greater than the horizontal component.

The same behaviour will be found when calculating the vibrations of a rotating shaft affected by this kind of triple crack. The 3X vibrations are the only available symptom for detecting this kind of multiple crack. Another available symptom is obviously the reduction of the torsional stiffness of the shaft, which could be identified by means of dynamic torsional excitation and measurement of torsional resonant frequency, as also proposed in [6.3].

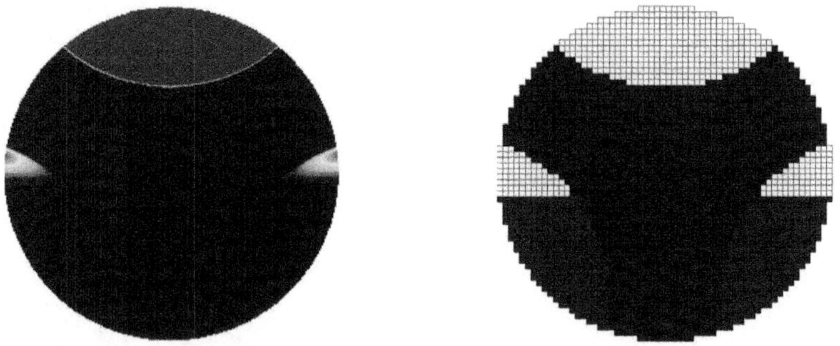

0° angular position of the shaft with respect to the vertical load

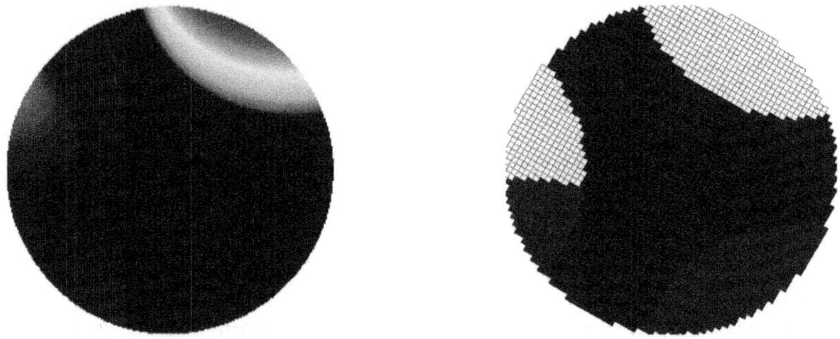

30° angular position of the shaft with respect to the vertical load

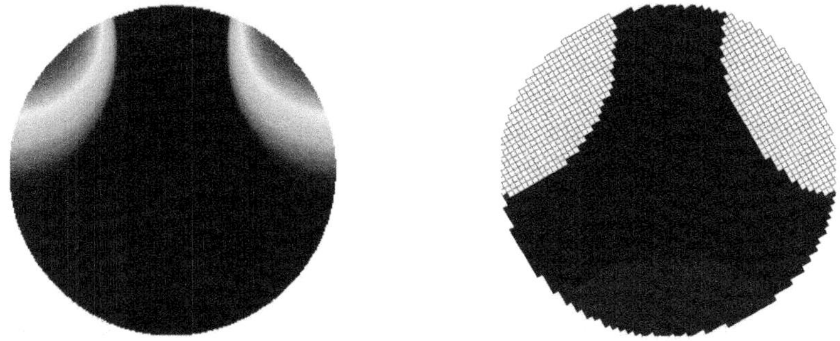

60° angular position of the shaft with respect to the vertical load

Fig. 6.32. Breathing behaviour of the triple crack as results of the 3D non-linear model compared to FLEX model results, in different angular positions.

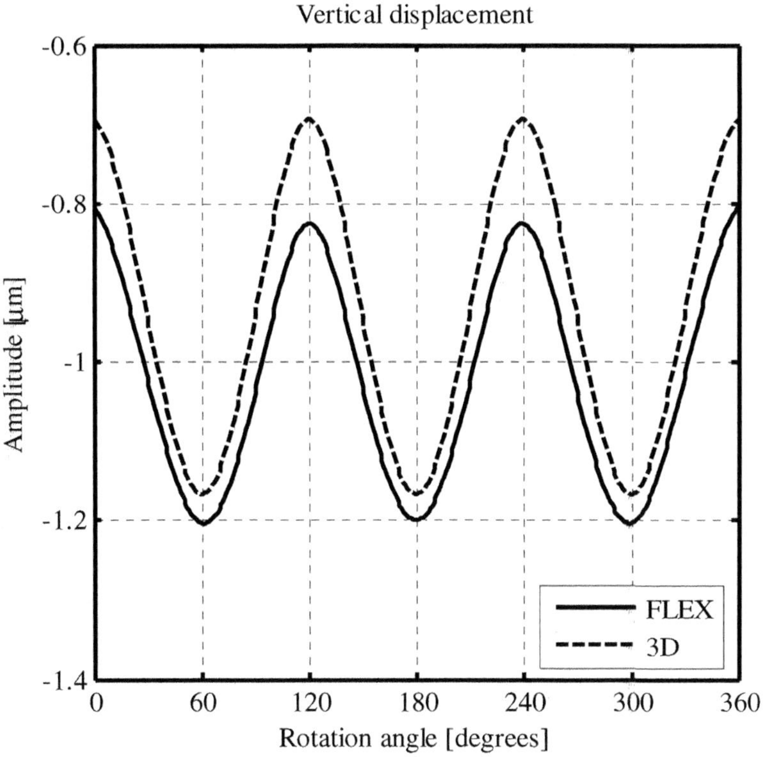

Fig. 6.33. Static deflections of the triple crack as results of the 3D non-linear model compared to FLEX model results, in different angular positions. Top vertical deflection, bottom horizontal deflection.

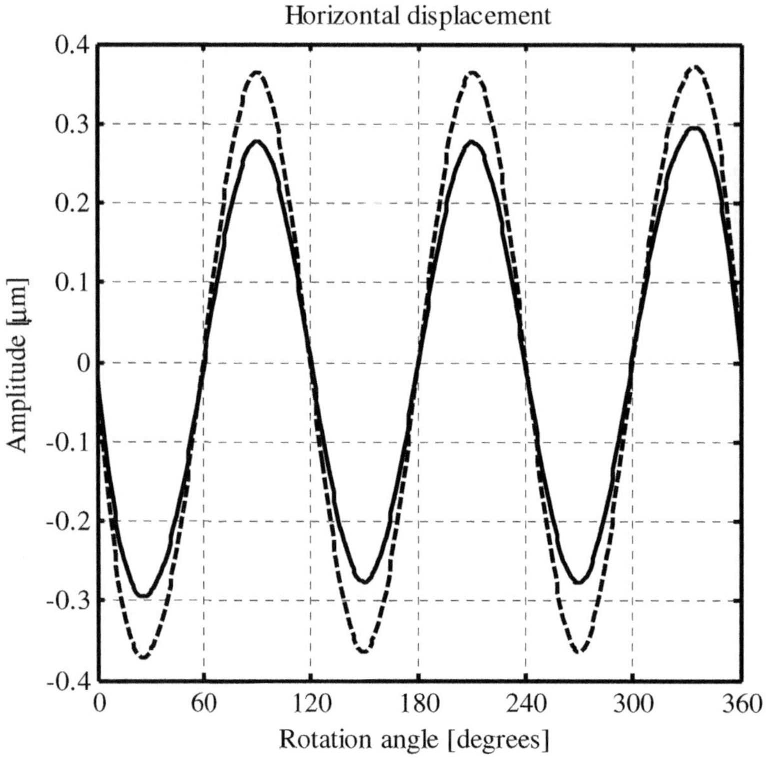

Fig. 6.33. *(continued)*

6.2.3 Effects of Shear Forces on Cracked Shaft Deflections and Vibrations

In this section the contribution to deflections of cracked beams of the additional shear flexibility introduced by the crack is investigated. The analysis is performed by considering a test beam with a diameter of 70 mm with transverse cracks with rectilinear tip and different depths, clamped at one end and loaded by shear forces and bending moments at the other end, as shown in figure 6.34. The deflections according to all the 6 degrees of freedom have been evaluated for the different angular positions of the loads with respect to the crack. For computational convenience a clamped beam and rotating loads have been considered instead of a rotating beam and fixed load as is the situation in real rotating cracked shafts.

The non-linear 3D model with a transverse crack with depth of 30%, 40%, 50%, 60% and 70% of the diameter has been meshed and analyzed. In order to emphasize the effect of the crack and to highlight the differences of its static behaviour attributed to the additional shear flexibility, the deflections of the un-cracked specimen have been subtracted from the deflections of the cracked specimens. These differences are compared in the same diagrams for different values of the shear force. This way only the effect of the crack is shown.

The model used for the analysis is the same of section 5.4.3.

The contact model in the cracked surface is obviously non linear. The value of the friction coefficient that has been introduced in order to account for micro-slip conditions in the cracked area, due to shear forces, has been assumed in this case equal to 0.4.

6.2.3.1 Definition of Loads

The considered load cases are as follows: shear forces and bending moments applied to the extension length of the shaft (see also figure 5.22 and figure 5.41) have been selected in order to have always in correspondence of the cracked section the same bending moment and increasing shear forces. The shear force in load case T_2 is double with respect to that one of load case T_1, the shear force in case T_3 is three times the load of case T_1. Shear and bending loads are rotating with respect to the crack, in order to simulate the behaviour of a rotating shaft loaded with stationary bending and shear loads.

- Load case C_1: bending only moment $M_b = 600\,\text{Nm}$.
- Load case T_1: bending moment $M_b = 600\,\text{Nm}$ and shear force $T = 4285\,\text{N}$.
- Load case T_2: bending moment $M_b = 600\,\text{Nm}$ and shear force $T = 8570\,\text{N}$.
- Load case T_3: bending moment $M_b = 600\,\text{Nm}$ and shear force $T = 12855\,\text{N}$.

Figure 6.34 shows the cracked specimen along with the loads and the deflections according to the 6 degrees of freedom of the extremity node of the specimen and α is the rotation angle of the loads.

In the initial position bending moment and shear force are directed towards the positive x axis: therefore the crack is closed. When the loads are rotated then the crack gradually opens. Axial and torsion deflections are due to so-called *coupling effects*, because axial and torsion loads are not applied to the beam.

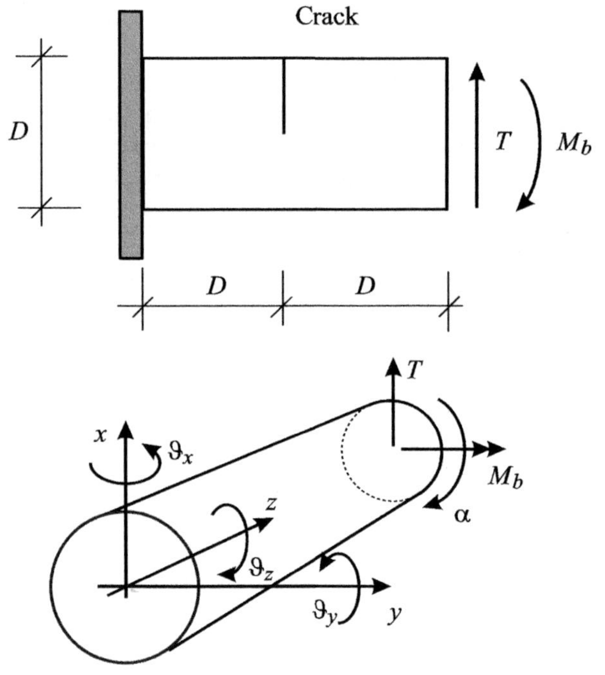

Fig. 6.34. Cracked specimen with loads and associated deflections.

6.2.3.3 Results

Figure 6.35 shows the deflections due to a crack with a depth of 50% of the diameter, according to the 6 degrees of freedom specified in figure 6.34, for the 4 different loading cases. All displacements are given in μm and angular deflections in μrad. The effect of increasing shear loads can be seen in the horizontal deflections that increase as the shear force increases. Anyway, the deflections according to the other degrees of freedom are rather unaffected by the shear force, except for the torsion deflection. This behaviour can be explained by the following considerations.

The shear force is applied to a resisting section of the crack that, considering an open or partially open crack or a crack in which micro-slipping occurs, is eccentric with respect to the beam axis and with respect to the shear force direction, producing eccentric torsion due to a *coupling effect*. Torsion on the cracked beam produce torsional deflections and also horizontal deflections due again to a *coupling effect*.

It should be noted that this effect on horizontal deflections is related to the situation of fixed beam and rotating load; it will be shown that when the results are converted to rotating beam and fixed load the effect of shear forces on the lateral deflections is reduced and spread over both horizontal and vertical directions. Angular deflections and axial displacements are completely unaffected by the shear forces. The main effect is the excitation of torsion deflection ϑ_z.

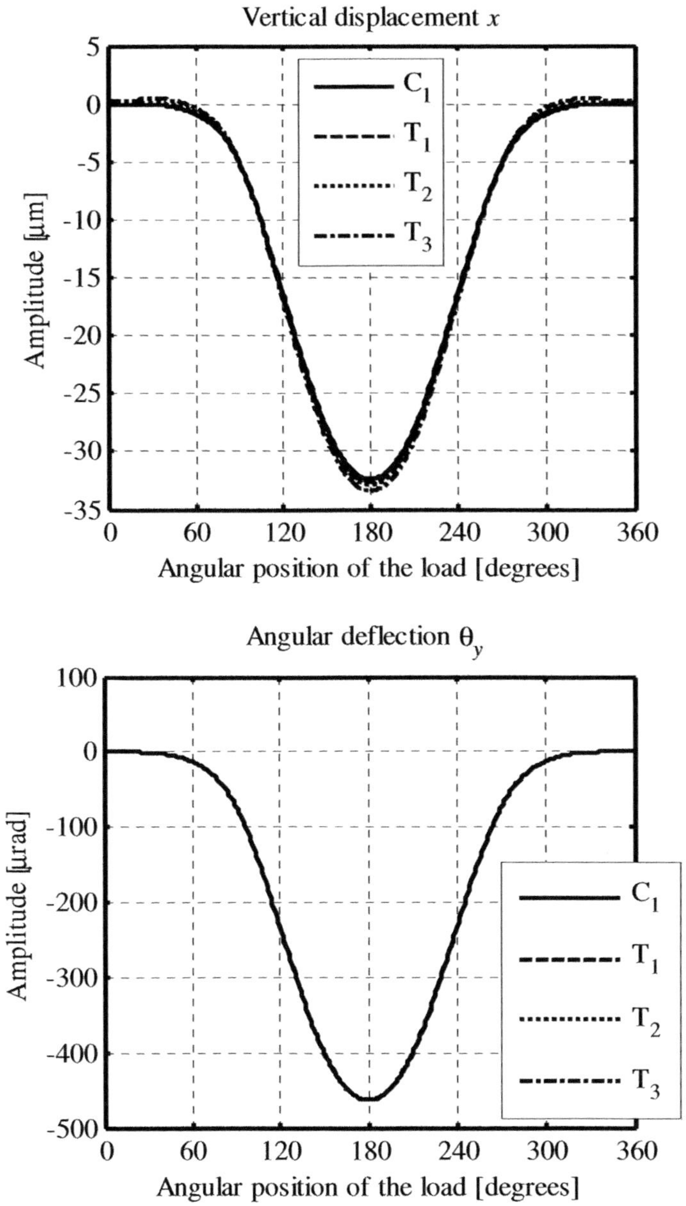

Fig. 6.35. Effect of increasing shear forces on the deflections of a beam with 50% crack.

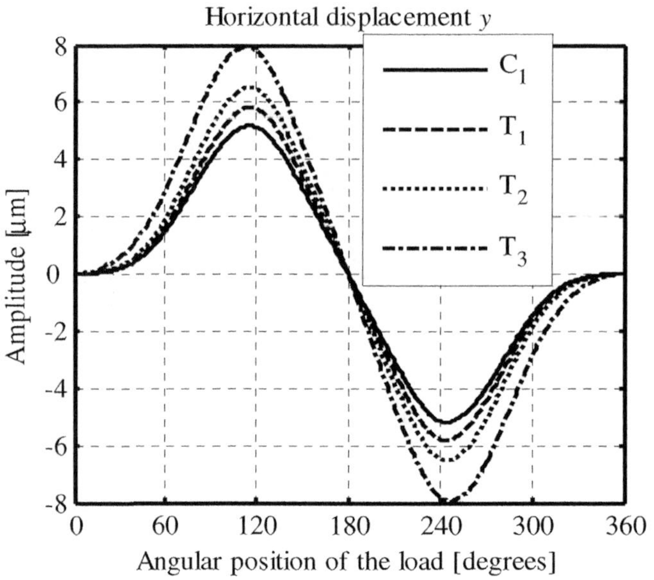

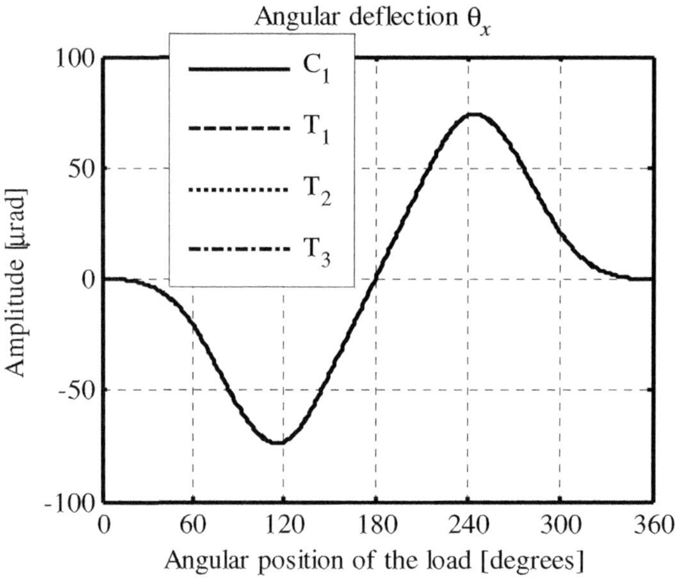

Fig. 6.35. (*continued*)

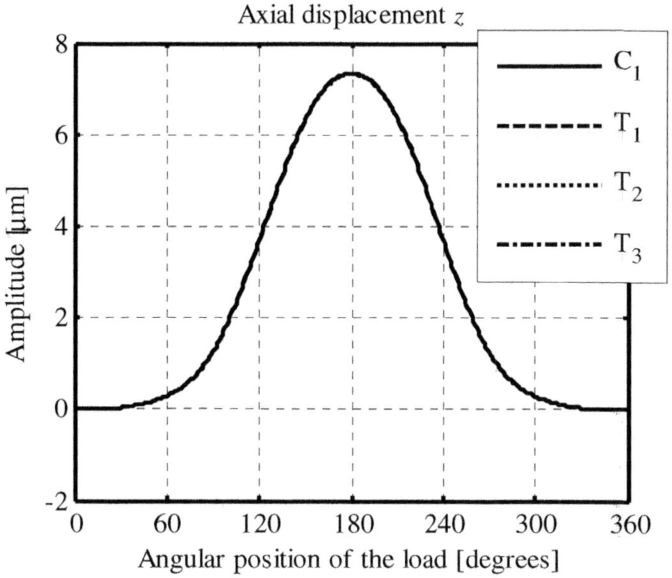

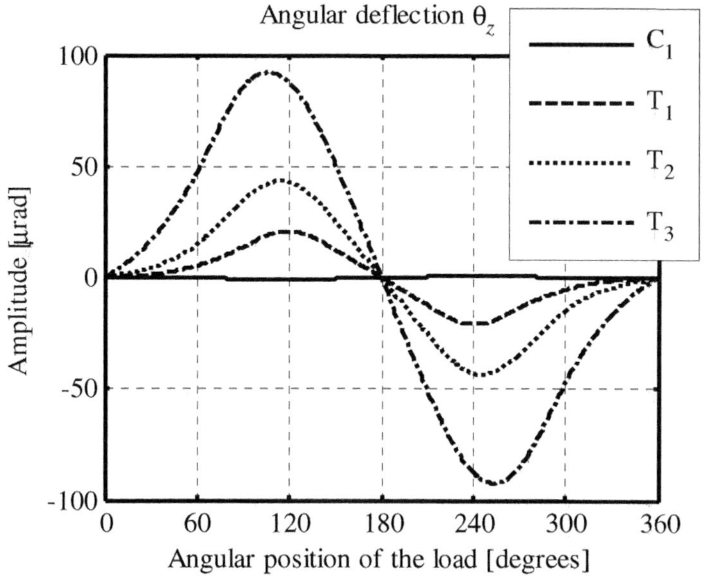

Fig. 6.35. (*continued*)

The same behaviour has been found for all different crack depths analyzed. Obviously the deflections are magnified by high crack depths and reduced for small crack depths. The excitation of torsion deflections has an important consequence considering the application to rotordynamics: the excitation of torsion vibrations in rotating shafts is due to the same coupling mechanism that causes statically the torsion deflection mechanism.

Therefore torsion vibration excitation is directly proportional to the shear forces in correspondence of the crack position, in a similar way as the lateral vibrations are proportional to the static bending moment in correspondence of the crack.

For two extreme crack depths of 30% (minimum depth) and 70% (maximum depth), the results obtained for fixed beam and rotating load have been transformed to represent fixed load and rotating beam. Figure 6.36 shows some results for the 30% crack: angular deflections are not shown for brevity. As already pointed out the axial displacement (as well as the angular deflections ϑ_x and ϑ_y) are completely unaffected by the shear load and horizontal and vertical displacements are only slightly affected by the value of the shear force.

The effect of friction forces can be seen mainly in the torsion deflection. Without friction forces the deflection will have a pure sinusoidal law according to the component of shear parallel to the crack tip which is sinusoidal with the angle of rotation of the crack. This shear component is responsible for the torsion generation due to the eccentricity of the cracked resisting section with respect to the direction of shear load.

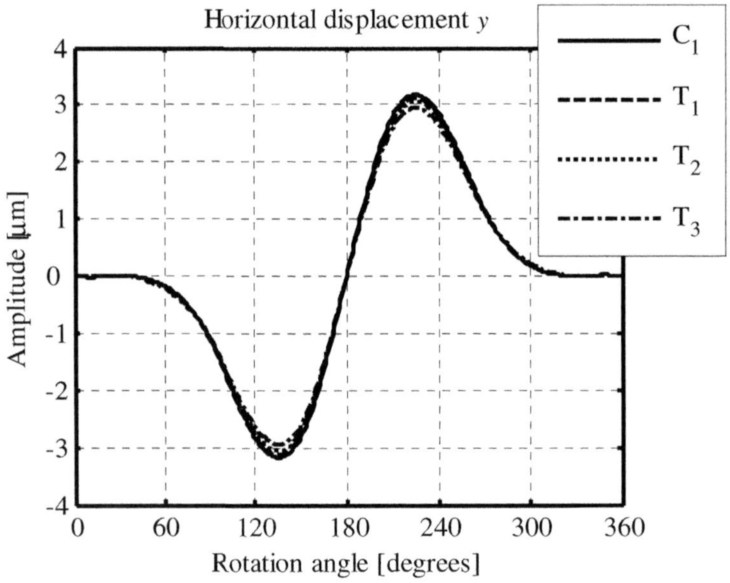

Fig. 6.36. Deflections obtained with 30% deep rotating crack and different shear loads.

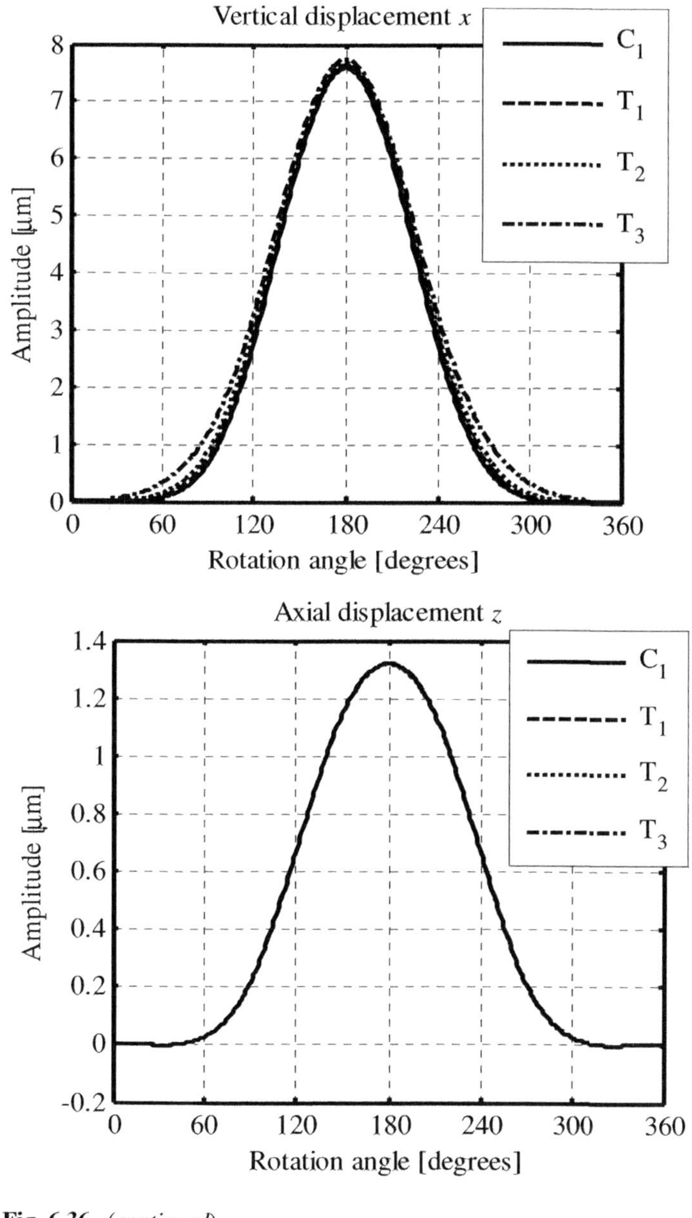

Fig. 6.36. (*continued*)

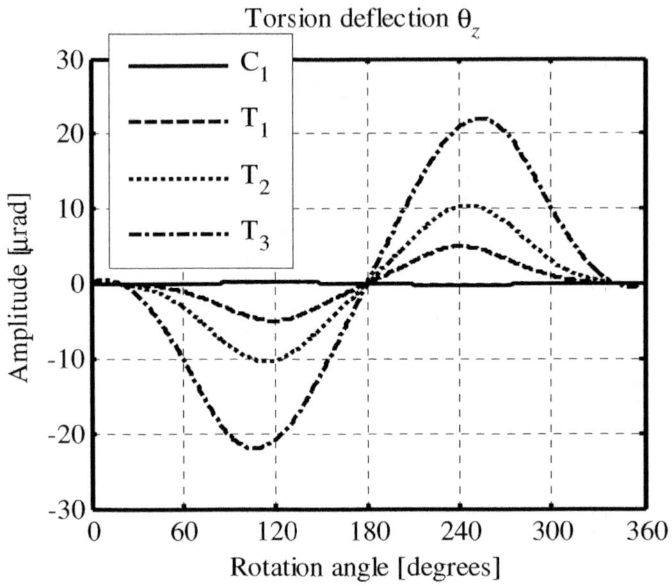

Fig. 6.36. (*continued*)

Taking instead into account friction the torsion deflection is prevented in all points of the cracked area where tangential stresses are lower than the compressive stresses multiplied by the friction coefficient. Therefore in positions between 0° and 90° (and between 270° and 360°) torsion deflections are considerably lower with respect to the sinusoidal law.

Figure 6.37 shows the results obtained with a 70% deep crack: only the vertical displacement and the torsion rotation are shown for a comparison with the 30% crack.

Displacements are magnified more than 10 times, as could be seen also from other degrees of freedom, but the behaviour is exactly the same as for the small crack. The torsion excitation due to shear forces is now rather consistent. Displacements are increasing more than proportionally with the increasing shear load. The huge excitation of torsion vibrations of a rotating shaft with such a deep crack could constitute an important symptom of the presence of a crack in positions where high shear loads combined to low bending loads are responsible for weak excitation of lateral vibrations.

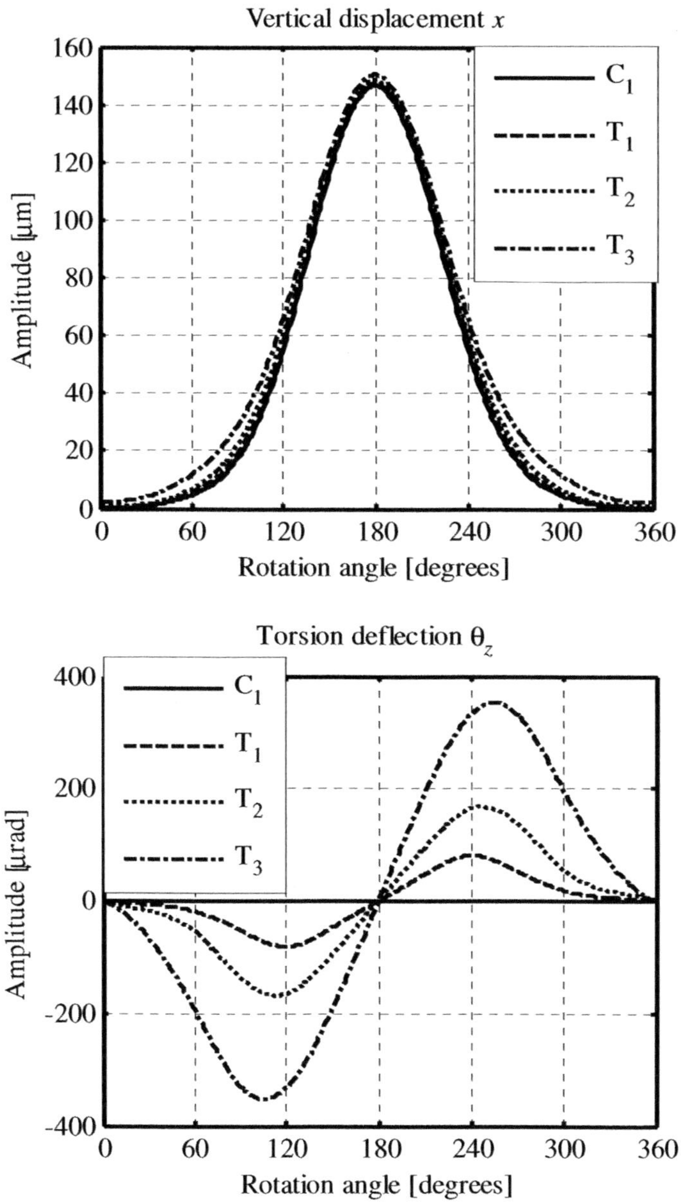

Fig. 6.37. Deflections obtained with 70% deep rotating crack and different shear loads.

From the analysis, the following results can be drawn:

1) Only small effects due to shear forces can be seen in the horizontal and vertical deflections of the rotating cracked beam. These effects are so small that they could also be neglected.
2) No effects at all are recognized in axial displacements and in angular deflections about horizontal and vertical axis.
3) Strong effects have been shown in torsion deflection, which are null when shear forces are not present.

Therefore shear flexibility can be neglected when only lateral deflections (in a static situation) or lateral vibrations (in a dynamic situation) are of interest. However shear effect is predominant and cannot be neglected when torsion deflections or vibrations due to coupling effect should be evaluated. This last consideration is important for rotordynamic analyses where vibrations are excited by bending and shear loads.

References

[6.1] Lapini GL, Zippo M, Bachschmid N, Collina A, Vallini A (1993) Experimental tests and model based calculations for the diagnosis of a crack in a 320 MW generator. Diagnostics of rotating machines in power plants CISM/IFToMM Symposium Proceedings, Oct 1993, Udine, Italy.

[6.2] Sekhar AS (1999) Vibration characteristics of a cracked rotor with two open cracks. Journal of Sound and Vibration, 223(4), 497-512.

[6.3] Henry H, Hasnaoui F, Verrier P (2006) Crack detectability using on-line vibration level shift monitoring: application to horizontal axis power generation turbines. Proc. of 7th IFToMM International Conference on Rotor Dynamics, Vienna, Sept. 2006.

Chapter 7
Some Special Effects Caused by Cracks

7.1 Effect of Transverse Cracks on Torsional and Axial Vibrations

Transverse cracks affect not only bending behaviour of rotating shafts but have effects also on torsional and axial vibrations. Torsional and axial stiffness are reduced when the crack is open. These stiffness are 1X periodical when the crack is breathing and cause parametric excitation when torsion or axial loads are applied. Moreover also coupling between shear loads and torsion and vice versa between torsion loads and bending behaviour exists. These coupling effects develop with open crack and generally disappear with closed crack. In fact when the crack is open or partially open, then the resultant of the tangential stresses due to the applied torque and/or to shear forces is eccentric with respect to the rotor axis: this causes the coupling of bending and torsional behaviour. In addition the 1X opening and closing of the crack, imposed by the bending load, generates also axial 1X displacements.

The coupling of bending and torsional vibrations in cracked rotors has been already studied by some researchers (see e.g. [7.1], [7.2] and [7.3]). All these studies are based on the strain energy release rate (SERR) approach, which allows the additional flexibility due to the presence of the crack to be evaluated. The main drawback of the SERR approach is that the true breathing behaviour of the crack, which is characteristic for heavy horizontal axis rotating shafts such as typically turbo generator shafts, cannot be accounted for. In [7.4] the improved model based on SERR approach, which has been shortly described in section 5.4.1, has been used for showing coupling effects due mainly to torsional or axial external dynamic excitation that seems difficult and unpractical to apply to industrial machinery. It should be reminded that axial excitation influences crack breathing, whilst torsion excitation does not. The authors of [7.4] claim that these additional vibrations could be useful for detecting cracks in rotating machinery. These symptoms are unfortunately so small that their use for crack detection seems still not practical. In the doctoral thesis [7.5], an extensive analysis has been made on the different available crack models and an exhaustive analysis of axial and torsional excitation generated by weight effect with the aid of a simplified breathing model has been described. Graphs show the intensity of excitations due to coupling mechanism, which allows to forecast vibration excitation if static bending moment and shaft diameter are known. Not only bending-torsion coupling effects are generated by a breathing transverse crack, but the 1X periodical torsional stiffness reduction generates also other two effects:

i) a reduction of the torsional natural frequency;
ii) static torsion load generates parametric excitation of torsional vibrations.

Furthermore, it should be noted that the reduction of torsional stiffness is generally higher than the reduction of bending stiffness, especially for smaller cracks. A small, not very deep crack affects more the second polar moment of area than its second moment of area.

In the following all effects will be shown with the aid of the FLEX model described in section 5.4.3. Some applications to test-rig models and to industrial machinery (a turbo-pump and a turbo generator unit) will show the vibration levels due to these effects which can be expected in real machinery.

7.1.1 Static Axial and Torsional Deflections due to Coupling with Bending and Torsional Loads

First static results obtained with FLEX model are compared to 3D non-linear calculation results, in order to check accuracy and validity of the proposed FLEX model. Results obtained on a beam affected by a crack with a depth of 60% of the diameter have been shown in section 5.4.5 for the 3D model and for the simplified FLEX model (see figures 5.46-5.51). The good accuracy in the comparison allowed to validate the FLEX model.

The results obtained with a elliptical crack with 24% depth are shown here compared to the FLEX model results

Figure 6.31 shows the mesh used for the 3D calculation. Only one elliptical sector is used as crack surface, although the mesh is designed to allow two or three equal elliptical cracks. The results relative to three equal elliptical cracks in the same cross section are shown in section 6.2.2.4.

The results for a load combination of shear load and torsion are shown from figure 7.1 to figure 7.3. As usual, only the effect of the crack on the deflections are shown.

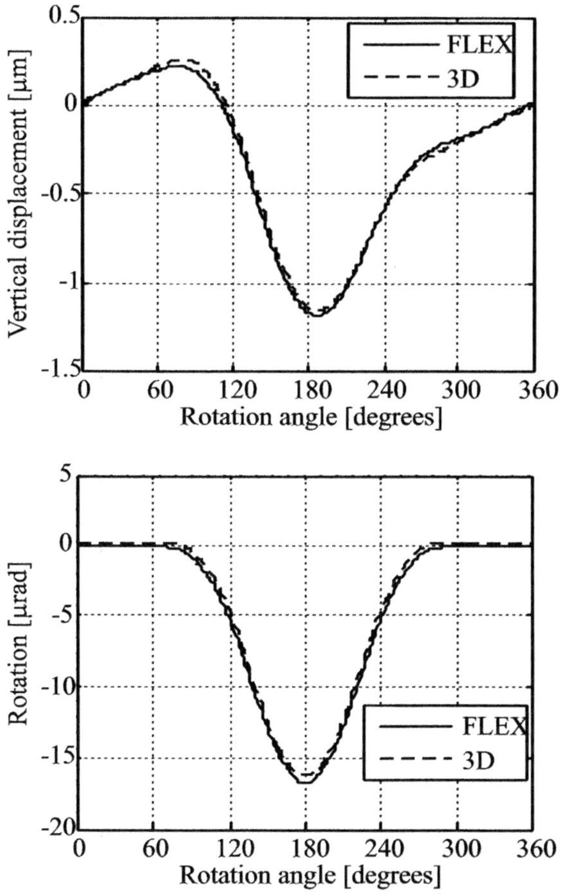

Fig. 7.1. Comparison between 3D calculation and FLEX simplified model, elliptical crack, vertical displacement (top) and angular deflection about horizontal axis (bottom).

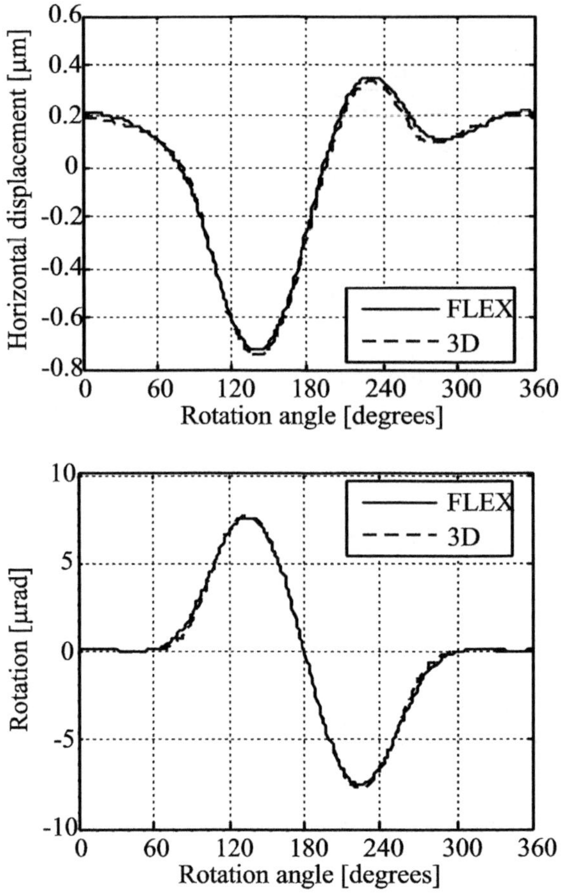

Fig. 7.2. Comparison between 3D calculation and FLEX simplified model, elliptical crack, horizontal displacement (top) and angular deflection about vertical axis (bottom).

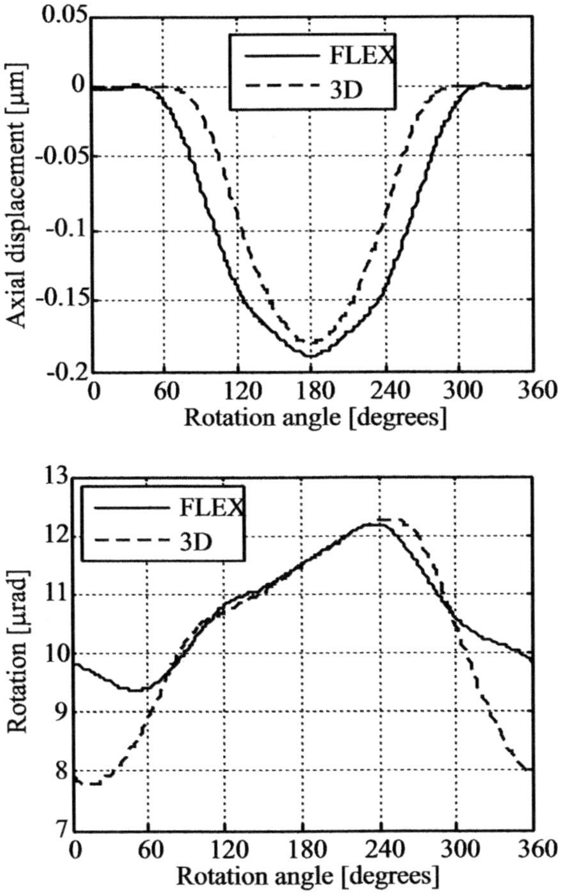

Fig. 7.3. Comparison between 3D calculation and FLEX simplified model, elliptical crack, axial displacement (top) and torsional deflection about axis of rotation (bottom).

Also in this case, the agreement of the results obtained by means of the two models seems to be very good for most of the different degrees of freedom, except for the axial displacement, which is somewhat overestimated, and for the torsional rotation, in which the effect of friction forces is probably underestimated. In fact in closed crack configuration, if the normal stresses generated by bending loads multiplied by the friction coefficient of the cracked surface are higher than the tangential stresses generated by torsional load, no micro-slipping will occur on the crack surface and the shaft will exhibit its full torsional stiffness. Instead, where micro-slipping takes place there a torsional stiffness reduction occurs. In figure 7.3-right, only in the positions close to 0° and 360°, where the crack is closed, different torsional rotations occur that are due to micro-slipping conditions in the cracked area. The differences between 3D and FLEX models could be due to the fact that the simplified model neglects completely the friction forces that are generated by the shear load, and considers only the linear distribution of friction forces due to torsional load, which is certainly unrealistic. Anyway the excitation of axial and torsional vibrations will be very weak with this small crack, since the corresponding deflections are so small.

Summarizing, the following effects of a crack on torsional vibrations can be expected:

1) the reduction of torsional stiffness will lower the torsional natural frequencies;
2) the periodical 1X variation of torsional stiffness will excite periodical torsional vibrations when constant torque is applied (parametric excitation);
3) the torsional stiffness variation will excite sideband vibration components when sinusoidal torque is applied;
4) due to bending-torsion coupling mechanism torsional vibrations are excited even if the shaft is loaded only by its weight;

Effect 1) will be shown in following section 7.1.2 where the shift of torsional natural frequency, induced by the crack, is simulated and compared to the measured value using the EURoPE test-rig.

Effect 3) will be shown in a numerical simulation of the behaviour of the same test-rig rotor in section 7.1.3.

The effects 2) and 4) will be shown in two numerical applications on industrial machinery: a vertical axis RCP pump in section 7.1.5 and a 1300 MW turbo generator in section 7.1.6.

Regarding the excitation of axial vibrations a numerical application on the same test-rig rotor will be shown in section 7.1.4.

7.1.2 Shift of Natural Torsional Frequency

Differently from the bending natural frequency shift that can be detected measuring the change of the rotating speed at which the resonance peak of the critical speed occurs, the change of the torsional natural frequency can be measured only if a torsional excitation is present or is applied to the shaft, in order to excite the torsional resonance. The advantage of exciting torsional vibrations is that torsional vibrations are in general poorly damped (because the oil film bearings do not contribute to the damping of the system), so that the resonance peaks are very sharp allowing high definition of resonance frequencies.

In EDF R&D laboratories, EURoPE test-rig was equipped by a torsional exciter (see figure 7.4) and the shift was measured. Figure 7.5 (which is reproduced from [7.6]) shows the shift of torsional natural frequency of the shaft comparing the measured peaks in the frequency response curves. The results were obtained by means of a sinusoidal torsional excitation applied to the rotating shaft equipped with a cracked specimen and with the same but un-cracked specimen. The crack had a slightly convex shape and a maximum depth of 36% of the diameter. The measured frequency shift was 0.43% of un-cracked specimen natural frequency.

Fig. 7.4. Torsional exciter installed on EURoPE test-rig.

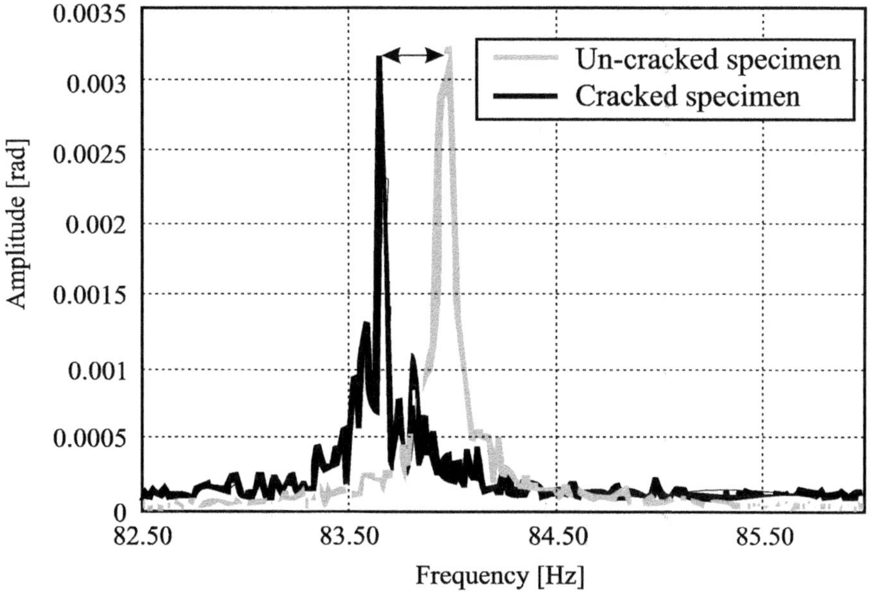

Fig. 7.5. Torsional frequency response curves obtained on cracked and un-cracked rotating test-rig shaft showing the shift of resonance peaks at its first natural torsional frequency.

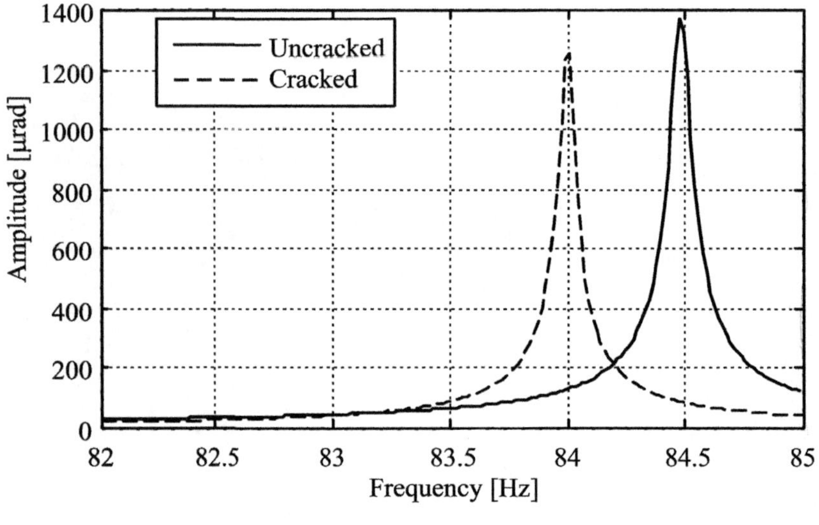

Fig. 7.6. Calculated torsional frequency response curves for cracked and un-cracked test-rig shafts.

Figure 7.6 shows the calculated results obtained with the model of the test-rig shaft and the FLEX model. The calculated frequency shift is 0.59% , somewhat higher than the measured one. Note that the torsional natural frequencies of cracked rotors must be deduced from response peaks and measured during the rotation of the shaft, when the crack is breathing and not completely open, otherwise the shift in frequency would be overestimated.

In the case considered, the crack was in a position where the mode shape of the first natural frequency shows a steep slope, which means that in this position the change in torsional stiffness will generate biggest effects, as stated in section 6.2.1.

For cracks in other positions frequency shifts of the first natural frequency could be much smaller or could even tend to vanish.

7.1.3 Excitation of Sideband Component in Torsional Vibrations of a Test-Rig Shaft

Excitation of sidebands will be described by means of a numerical simulation. Figure 7.7 shows the model of the shaft and the crack shape. The static torque M_s applied to the ends of the shaft was assumed equal to 25 Nm, to which the harmonic excitation component M equal to 12.5 Nm is superposed. Rotating speed is assumed equal to 1500 rpm (25 Hz). Crack maximum depth is 40% of shaft diameter.

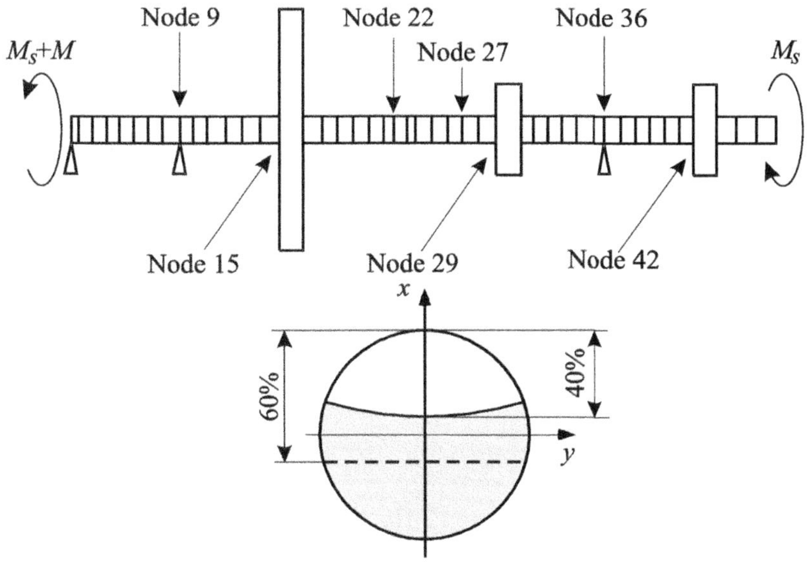

Fig. 7.7. Model of a test-rig shaft, and two types of crack. The crack is assumed to develop in node 22.

Some components at different frequencies are expected in the spectrum of torsional system response:

- at the exciting frequency f_e;
- at frequencies equal to the frequency f of rotation of the steady state system or at its integer multiples, according to the harmonic components of the stiffness matrix $[\mathbf{K}(\Omega t)]$.

Therefore, given a rotating speed equal to 1500 rpm, corresponding to the frequency $f = 25$ Hz, components at 25 Hz, 50 Hz, 75 Hz, ..., $n \times 25$ Hz, with $n \in \mathbb{N}$, will appear. Their presence is caused by the static torque applied to the system. Finally also the following components will appear:

- at the sideband frequencies, equal to the exiting frequency plus or minus the harmonics of rotation frequency f, such as $f_e \pm f$, $f_e \pm 2f$ and $f_e \pm 3f$, which can be called respectively 1X, 2X and 3X sidebands.

Figure 7.8 shows the results calculated with the harmonic excitation in resonance with the torsional first natural frequency ($f_c = 88$ Hz).

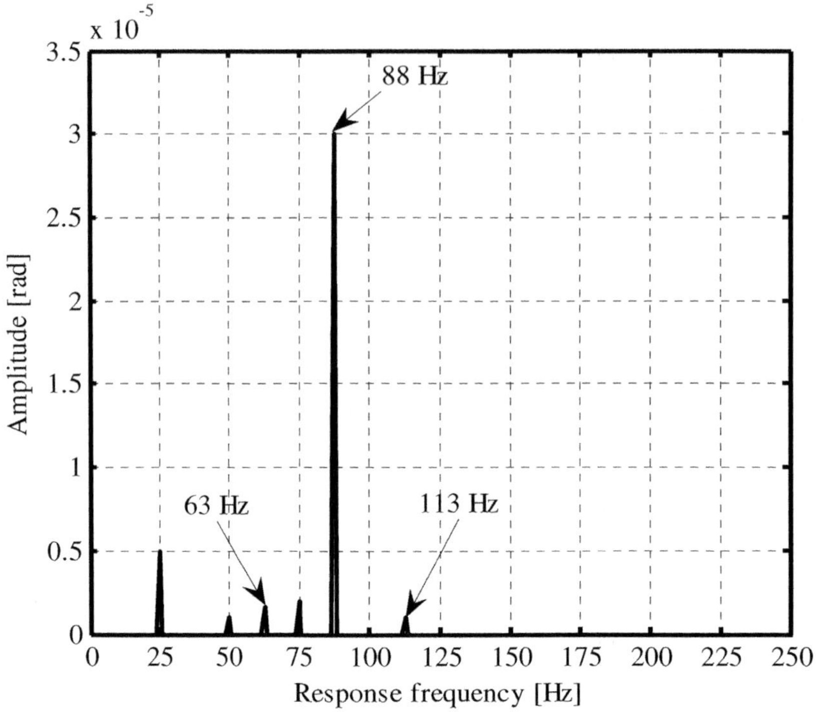

Fig. 7.8. Response calculated in node 27 with excitation in resonance, showing two sideband components (63 and 113 Hz) and parametric excitations (25, 50 and 75 Hz).

The spectrum shows:

- the amplitude of 5 μrad at 25 Hz due to the parametric excitation (1X variation of torsional stiffness) and lower amplitudes for the higher harmonics (50, 75 Hz). These excitations are directly proportional to the applied static torque, which was very small in this case compared to the shaft diameter. This shaft could afford torques up to 20-30 times the actual value: correspondingly the parametric excitation would be 20-30 times higher;
- the amplitude of 30 μrad at resonance (88 Hz), which depends on the amplitude of the harmonic excitation torque;
- the amplitude of 1.6 μrad at sideband frequency 63 Hz (88 Hz–25 Hz) and a smaller peak at 113 Hz (88 Hz+25 Hz), which also depend on the excitation torque.

If the exciting frequency is swept in the range 30-140 Hz and 1X and 2X sidebands are calculated, figure 7.9 and figure 7.10 are obtained respectively.

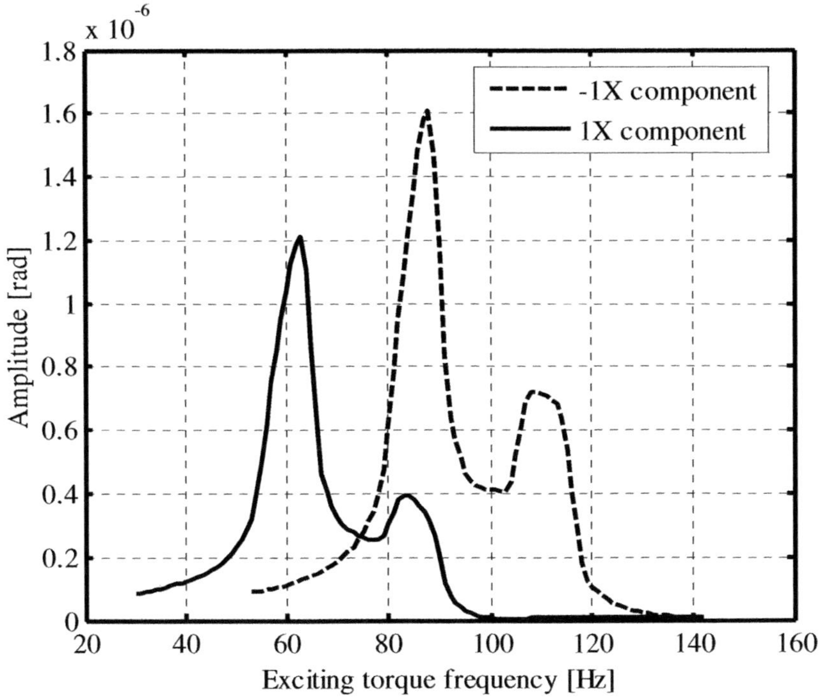

Fig. 7.9. 1X sideband components ($f_e - 25\,\text{Hz}$) and ($f_e + 25\,\text{Hz}$) when the exciting frequency sweeps from 30 to 140 Hz.

When the exciting frequency f_e is equal to $f_c \pm f$, where f_c is the natural critical frequency and f is the frequency of rotation, the ±1X sidebands are excited in resonance.

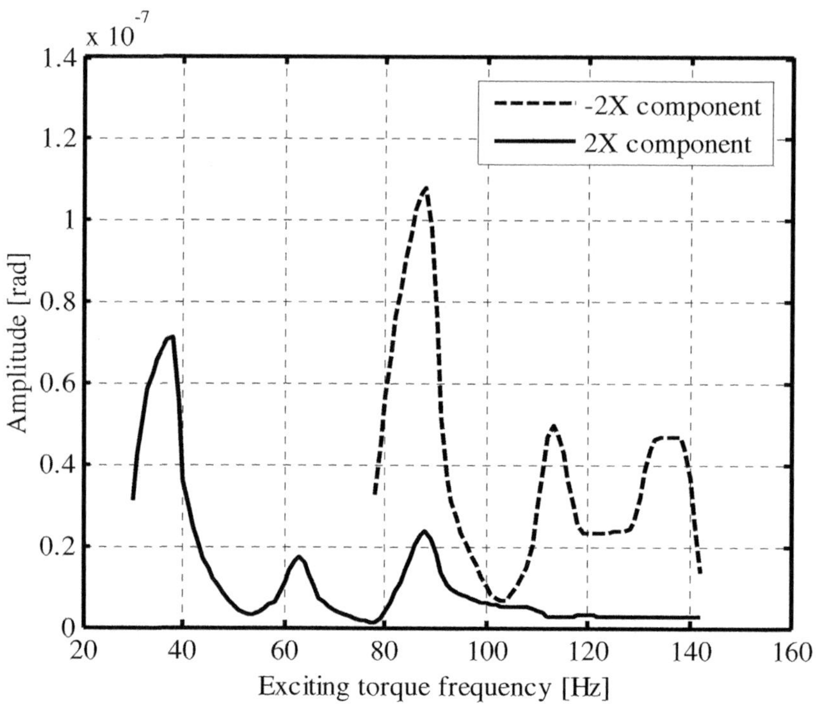

Fig. 7.10. 2X sideband components ($f_e - 50\,\text{Hz}$) and ($f_e + 50\,\text{Hz}$) when the exciting frequency sweeps from 30 to 140 Hz.

Similarly, when f_e is equal to $f_c \pm 2f$, the ±2X sidebands are excited in resonance.

The sidebands amplitudes are in general rather small, but are obviously related to the position of the crack and of the measuring station, as well as to the exciting torque amplitude. In industrial turbo-machinery the dynamic torsional excitation is generally weak, therefore sidebands will hardly be measured and only parametric excitation will effect measurable torsional vibrations, but only for breathing cracks (and not for open cracks).

7.1.4 Axial Vibration Excitation of a Test-Rig Shaft

Axial vibrations, caused by the coupling mechanism with the flexural excitation, could be measured when no other sources of axial vibration excitation is present. From the results of the static analysis (as shown in figure 5.50 and 7.3), it can be deduced that the excitation due to the coupling mechanism is rather weak. This will be shown by a numerical application on the EURoPE cracked test-rig shaft (shown in figure 7.7). The shaft is loaded by its own weight only.

Figure 7.11 shows the amplitude of the frequency response of the axial vibrations as they would be measured at the free end of the shaft. Three harmonic

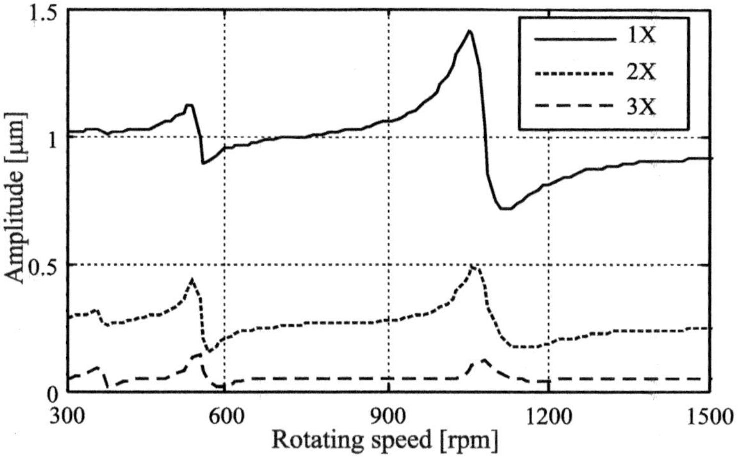

Fig. 7.11. Amplitude of the harmonic components of the axial vibration of the free end node of the shaft, excited by the coupling mechanism with bending excitation.

components are excited, but only the 1xrev component has a measurable amplitude (around 1 μm). The axial resonance is far beyond the rated rotating speed of 1500 rpm, in fact the first axial critical speed of the test-rig shaft with free ends is 28500 rpm. Some dynamic amplification is experienced in correspondence of the 1X and 2X bending critical speeds.

The axial deflection shape at rated speed has a node in correspondence of the crack, with the two parts of the shaft in opposition of phase.

If axial static load would be present, then parametric excitation due to the periodic variation of axial stiffness would be present, provided that the axial load does not prevent breathing. When the axial load holds the crack always open then the parametric excitation disappears.

If also axial dynamic excitation would be present, then sidebands in axial vibrations would be excited, in similar way as described for torsional vibrations, provided that breathing is still governed by bending.

It should be reminded that the crack induced axial excitation is proportional to the shaft diameter: industrial machines with diameters up to 10 times the diameter of the test-rig will show axial vibrations excited by a crack with the same dimensionless depth, which are 10 times the axial vibration of the test-rig. This could then be considered an additional symptom of the presence of a breathing crack.

7.1.5 Torsional Excitation in a Vertical Axis Centrifugal Pump

This analysis is aimed at evaluating the dynamic behaviour of a single stage, vertical axis centrifugal pump typically installed in the primary loops of both BWR (boiling water reactor) and PWR (pressurized water reactor) nuclear plants worldwide with the function of guaranteeing the cooling water flow from the core to the steam generator (from which the names "Reactor Coolant Pump" or "Main Coolant Pump" arise). In a number of these machines over the world the presence of transverse cracks, some of which having considerable depth - up to 70% of the diameter of the shaft -, has been detected, as reported in section 2.2.

The scope of the present analysis is to investigate the dynamical behaviour of the pump shaft, using a 6 degrees of freedom per node model (therefore considering also the torsional and axial d.o.f.s). The vibrations are excited by fluid forces on the impeller, which cause static radial forces, rotating radial forces (also called hydraulic unbalance), static torque, as well as by mechanical unbalances mainly on the motor and on the impeller. Higher harmonic components in torque and hydraulic forces as well as the random narrow band excitation from turbulent fluid flow are neglected.

In the case of a breathing crack 1X, 2X and 3X components are excited in the flexural vibrations due to the static radial force and the same components are excited in the torsional vibrations due to the static torque. In the case of an open (non-breathing) crack only 2X components are excited in the flexural vibrations due to static radial force and no component in the torsional vibrations due to static torque. Finally, due to the coupling effect between bending behaviour and torsional behaviour introduced by the crack, torsional vibrations should appear due to static bending and flexural vibrations should appear due to static torsion. Other components due to coupling effects between flexural and torsional vibrations are so small that they could hardly be detected: these components can then be neglected.

7.1.5.1 Description of the Pump

A typical RCP pump is shown in figure 2.4. Pumps of this kind are driven by induction motors and their nominal speed is 1500 rpm in Europe, or 1200 rpm in the US (sometimes even 1800 rpm). The electric motors are usually provided with a massive flywheel and two hydrodynamic lubrication radial bearings, where the upper one supports also the thrust loads in axial direction. The pump is coupled to the motor by means of a rigid coupling and supported by hydrodynamic bearings while a seal prevents the hot pressurized water from flowing along the shaft.

In correspondence of the seal a critical situation arises, because the shaft is subjected to thermal stresses in this region, deriving from the oscillation of the hot/cold interface caused by the contact of the hot pressurized water of the primary loop and the cold sealing water. This aspect is believed to be one of the causes of the initiation of micro-cracks. Once the micro-crack has appeared, it could propagate more or less rapidly by fatigue stresses generated by bending loads on the rotating shaft due to the radial hydraulic force on the impeller.

The vibrations of these machines are generally measured only by proximity probes placed in correspondence of the coupling between pump and electric engine, which happens to be rather far a position from crack location, hence the detectable vibration levels can be very insensitive to the presence of a crack.

Typical service parameters of these pumps are:

1)	pressure / water temperature	155 bar @ 290 °C
2)	sealing water temperature	35 °C
3)	motor nominal power	6500 kW
4)	nominal rotating speed	1500-1800 rpm

7.1.5.2 Description of the Model

As shown in figure 7.12, the machine has been represented by a finite beam element model consisting of a set of 80 Timoshenko beam elements. In order to evaluate also the axial and torsional vibrations, the mass, damping and stiffness matrices have been implemented in a 6 degrees of freedom per node system, instead of 4 d.o.f.s as usual in rotordynamic analysis. Moreover the non-linear simplified 1D model allows to calculate the breathing behaviour of the crack, e.g. the gradual opening and closing of its lips during one revolution of the shaft as a function of the cracked section geometry and of the applied loads.

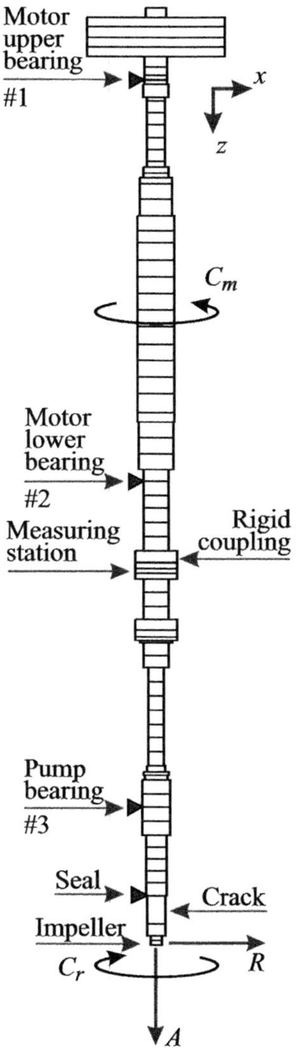

Fig. 7.12. Model of the pump shaft.

The crack, located in correspondence of the seal, is modelled as a rectilinear tip one, with depth equal to 40% of the diameter.

The assumed dynamic coefficients of the bearings at rated speed are shown in Table 7.1, while the supporting structure is considered as being rigid. The direct stiffness coefficients of bearing #3 are null, as the static loads in radial direction are rather small at this location. Also the seal provides stiffness and damping for the rotating shaft.

Table 7.1. Bearing dynamic coefficients: stiffness in N/m and damping in Ns/m.

Bearing	*Stiffness coefficients*				*Damping coefficients*			
	k_{xx}	k_{xy}	k_{yx}	k_{yy}	c_{xx}	c_{xy}	c_{yx}	c_{yy}
1	4.0e+8	1.9e+5	1.9e+5	3.8e+8	1.8e+6	4.9e+2	4.8e+2	1.7e+6
2	4.7e+8	9.4e+4	2.0e+5	1.2e+8	1.3e+6	3.0e+2	7.7e+2	4.4e+5
3	0	3.2e+8	3.1e+8	0	4.0e+6	1.1e+5	1.1e+5	4.1e+6
Seal	1.3e+7	3.8e+6	3.8e+6	1.3e+7	1.0e+5	7.9e+3	7.9e+3	1.0e+5
Thrust bearing.	k_{zz} = 4.0e+9				c_{zz} = 1.8e+6			

7.1.5.3 Definition of Loads

The loads applied to the shaft are following: on the motor a constant driving couple C_m, and on the impeller a hydraulic resisting torque $C_r = -C_m$, a stationary radial hydraulic force R, a stationary axial hydraulic force A, and finally a hydraulic U_b and a mechanical U_m unbalance rotating force, all proportional to the square of the rotating speed.

The assumed numerical values at rated speed are the following:

C_m	R	A	U_b	U_m
40.000 Nm	1.12·e+5 N	2.52·e+5 N	0.4 kgm	0.4 kgm

The opening/closing of the crack during the rotation is governed not only by the bending moment in correspondence of the crack, but also by axial loads and by the pressure of the water circulating within the pump and by the thermal effects, too. As a matter of fact, it is believed that the high pressure water could enter between the crack lips, causing their opening and preventing the breathing mechanism. Therefore two different conditions will be considered for the crack: breathing crack and always open crack.

7.1.5.4 Main Results

The frequency response of the machine is evaluated in the measuring station. The simulations have been performed by means of a frequency sweep simulating a coast-down transient after removal of the driving torque. All hydraulic forces as well as the static resisting torque are assumed to be decreasing with the square of the rotating speed. The first harmonic displacements are mainly due to the unbalances on the impeller and no visible effect is given by the presence of the crack, therefore they are not shown. Also, being the responses along the two transversal directions quite similar, only direction x and rotations along z axis will be considered: in the diagrams of figure 7.13 and figure 7.14 the vibrations of the shaft affected by a breathing crack are shown.

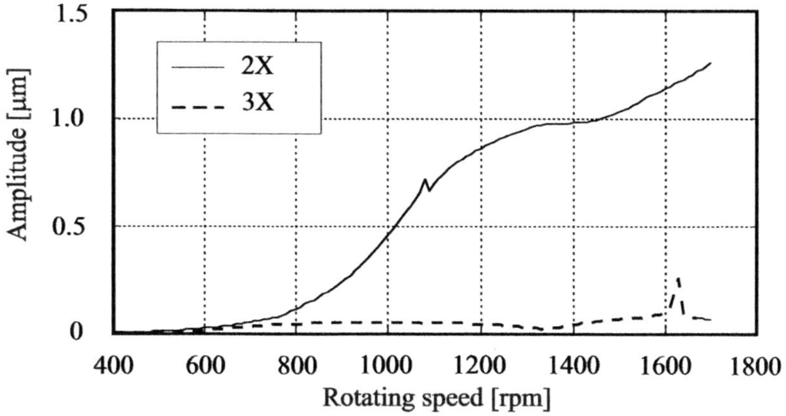

Fig. 7.13. Breathing crack – 2X and 3X harmonic lateral vibration components.

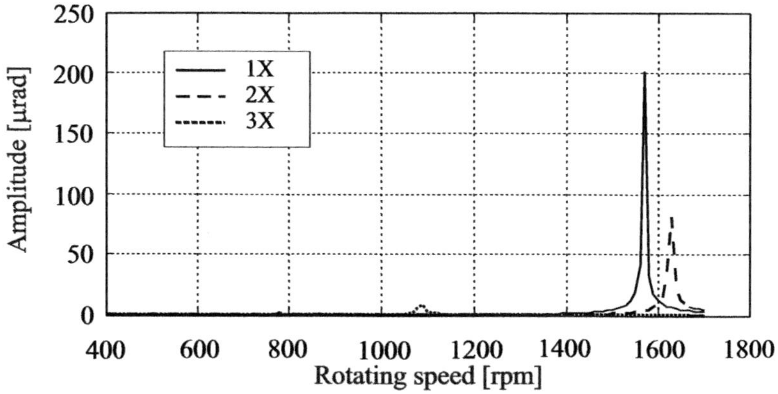

Fig. 7.14. Breathing crack –torsional vibration components.

The second and third harmonic responses of the lateral vibrations, due to the static bending moment and excited by the breathing crack, show very small amplitudes that hardly could be detected by a monitoring system. A small disturbance can be seen as the machine crosses the first torsional resonance: this is the effect of bending-torsion coupling.

The most important effect of the presence of the crack is given by the torsional response. The first harmonic component excited by the static torque and by the periodical torsional stiffness of the breathing crack becomes particularly high when the machine crosses the torsional resonance (at 1566 rpm corresponding to 26.1 Hz). Also the second harmonic component (at 1641 rpm corresponding to half of the second natural frequency of 54.7 Hz) is strongly excited by the same mechanism. Also the coupling effect, e.g. the excitation of torsional vibrations due to bending load, is present but its contribution is rather small (as will be shown by the always open crack calculation where only the coupling effect excites the torsional vibrations).

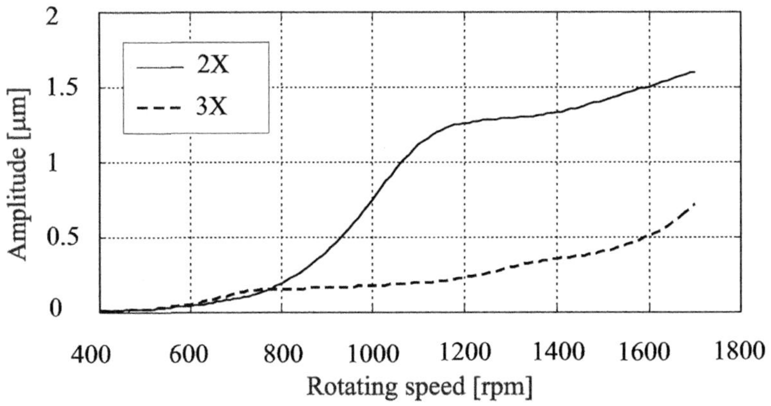

Fig. 7.15. Open crack – 2X and 3X harmonic lateral displacements.

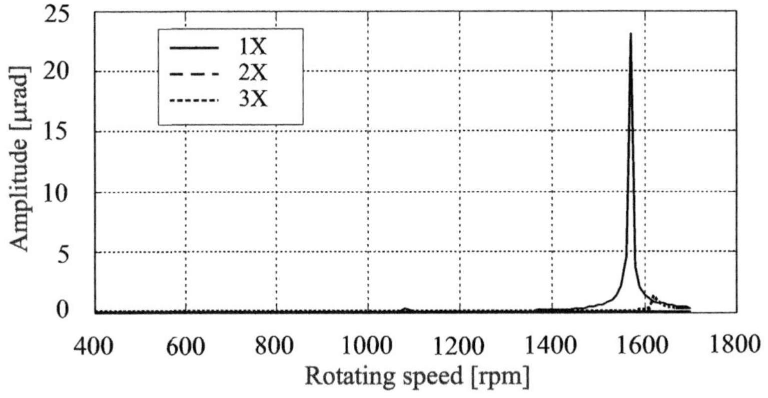

Fig. 7.16. Open crack – torsional vibration components.

The diagrams in figure 7.15 and figure 7.16 show the lateral and torsional vibrations excited in the measuring station by a non-breathing (always open) crack.

In comparison to the case of a breathing crack, the following considerations can be drawn when the crack is always open:

- the second and third harmonic components of the flexural vibration increase slightly and the small disturbance in correspondence of the lateral critical speed disappears;
- the second harmonic component is due to 2X sinusoidal bending stiffness variation and to the static bending load;
- the third harmonic component is only due to the combination of 1X unbalance load with the 2X shaft stiffness. The disturbance in correspondence of the lateral critical speed has disappeared because the lateral vibration excitation is much smaller;
- the torsional vibrations have decreased consistently (by a factor of 10) because the torsional excitation is due only to the coupling effect;
- the coupling effect is mainly a 1X effect in the case of always open crack, as can be seen from the relevant formulas defining the coupling coefficients in the flexibility matrix.

Summarizing, the main results are the following. In the measuring station the excitation of torsional vibrations due to a breathing crack could be consistent: 1X and 2X resonances are excited at rotating speeds which are close to the operating speed of the pump. Obviously they can be measured only if the operating speed is actually higher than the first torsional critical speed. The excitation of bending vibrations is instead rather small and does not constitute a certain symptom for the presence of a crack. In the case of non-breathing crack the excitation of torsional vibrations is much weaker but still recognizable in correspondence of critical speeds. The bending vibrations are a little higher but still too small to be used as a significant symptom of the presence of the crack.

These results are obviously related to the particular design of the machine, to the positions of the crack and of the measuring station, as well as to the assumed depth of the crack, to the assumed static hydraulic loads in the pump and to some extent also to the value of the assumed friction coefficient on the crack surface.

Higher loads, higher friction coefficients and deeper cracks will increase consistently the vibration levels.

7.1.6 Excitation of Torsional Vibrations in a 1300 MW Turbo Generator

The scope of this analysis is to investigate the dynamical behaviour of a 1300 MW turbo generator shaft-line with a crack in a LP turbine, using both a torsional model with only one degree of freedom per node and a model with 6 degrees of freedom per node that considers also the rotation and the displacement along rotor axis. After a short description of the machine and of the model, the model without crack is analysed for defining critical speeds (or eigenfrequencies) and mode shapes (or eigenvectors).

Then the crack is introduced and the natural frequencies and mode shapes are calculated with always open crack and are compared to the un-cracked shaft values. Also the frequency response curve due to unitary torsion applied to the generator is calculated and compared to the curve of the un-cracked shaft. Moreover a frequency response calculation is repeated introducing the breathing crack, which simulates a run-down transient in which vibrations are excited only by the weight acting on the rotating shaft and by the crack. Both lateral and torsional components are considered.

Finally the expected values of the amplitudes of vibration components, both lateral and torsional, at normal operating speed are presented, when the group is loaded by its weight and rated torque.

7.1.6.1 Description of the Unit

A sketch of the unit is shown in figure 7.17. The models of shaft, bearings and pedestals are omitted here for brevity. The crack is located in the middle of the first LP steam turbine, as shown in figure 7.17, and has a depth of only 10% of the diameter and two different angular extensions are assumed: crack A with angular extension 120° and crack B with angular extension 240° as shown in figure 7.18. A crack with similar depth had developed in this type of machine in circumferential direction reaching different angular extensions. The crack is rather small, its effects on lateral vibrations is practically un-measurable, but its effect on torsional vibrations could be a measurable symptom.

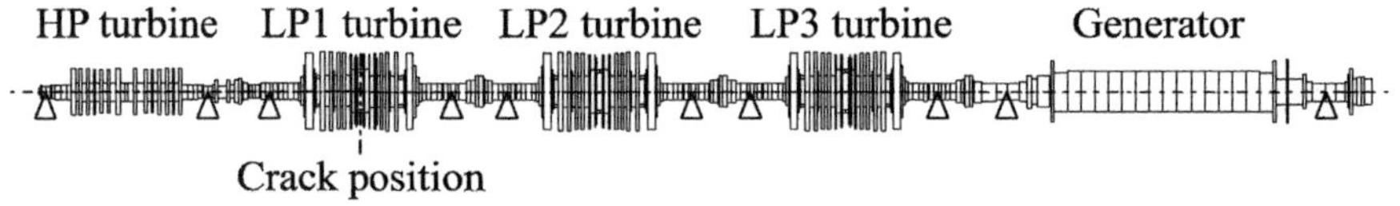

Fig. 7.17. Sketch of the turbo generator unit.

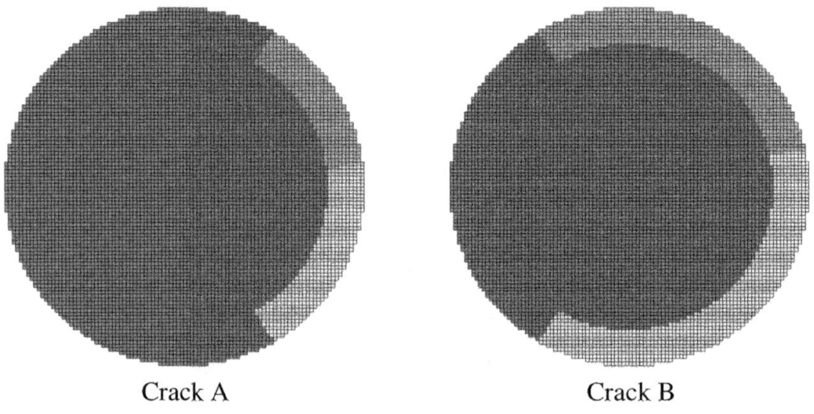

Fig. 7.18. 120° and 240° angular extension breathing cracks (in position rotated by 90° with respect to the direction of the weight).

Loads are defined as follows:

1) The bending load due to the weight is 1.47 MNm
2) The rated torque load is equal to 2.84 MNm

The bending load due to the weight corresponds to the bearing alignment condition in which the bending moments on the coupling flanges, between all shafts, is reduced to zero.

7.1.6.2 Results of the Torsional Model Only with Open Crack

First eigenfrequencies and eigenmodes of the un-cracked shaftline, using a torsional model with only one degree of freedom per node, have been calculated. Then the crack model, considered as always open crack, has been introduced and the eigenfrequencies are calculated again. Table 7.2 shows the differences introduced by the two cracks with extension 120° and 240° and depth 10% of the diameter with respect to the un-cracked shaft.

Table 7.2. Comparison between the eigenfrequencies of the un-cracked and the cracked shafts.

	Un-cracked [Hz]	Crack 120° [Hz]	Difference [Hz]	Crack 240° [Hz]	Difference [Hz]
Mode 1	5.80	5.80	0.00	5.80	0.00
Mode 2	11.05	11.04	0.01	11.04	0.01
Mode 3	15.07	15.07	0.00	15.07	0.00
Mode 4	17.49	17.49	0.00	17.49	0.00
Mode 5	42.53	42.44	0.09	42.27	0.26
Mode 6	44.04	43.95	0.09	43.82	0.22
Mode 7	45.53	45.48	0.05	45.43	0.10

The changes in eigenfrequency values of the modes 5, 6, and 7 are the most significant, but very small. The 5th mode of the cracked shaft shape is shown in figure 7.19. The differences with respect to the un-cracked mode shape are negligible. Since the 5th mode has the highest relative rotations in correspondence of the position of the crack, even if the crack is not deep, the corresponding eigenfrequency changes could be measured.

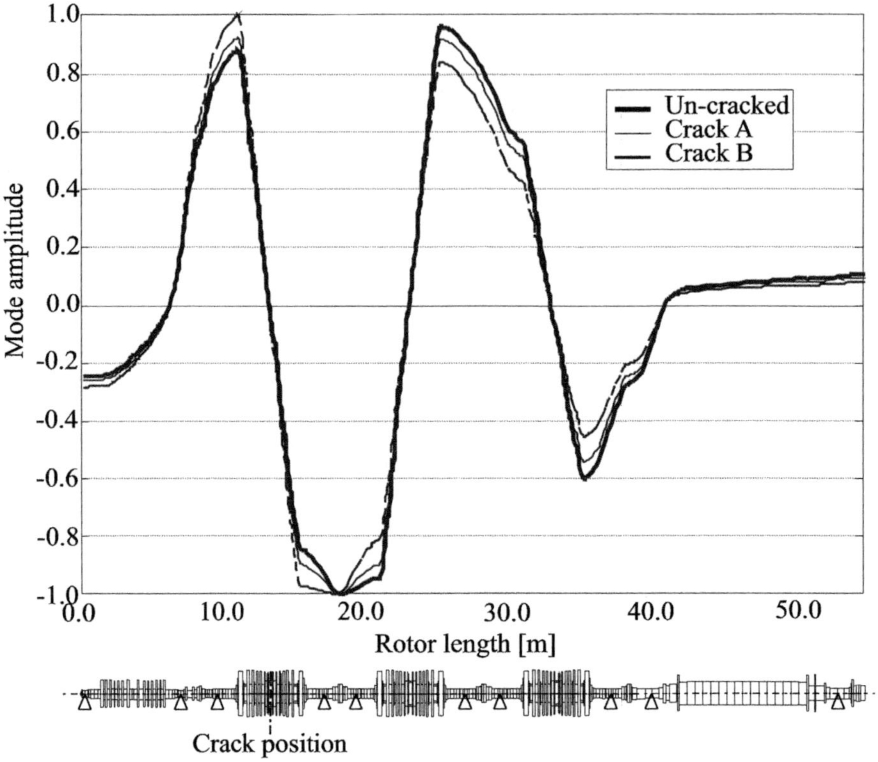

Fig. 7.19. Torsional modal deformation of the 5th mode of the cracked rotor.

7.1.6.3 Results of the complete 6 d.o.f. model with breathing crack

Now the frequency response curves of the cracked machine with breathing crack, excited by weight only, have been calculated. Both crack sizes, the shorter crack A of 120° and the larger crack B of 240° of angular extension, have been considered and the results are compared.

The calculation simulates the behaviour of the unit, during a slow run-down transient, in which the applied torque is considered zero. The speed range has been enlarged from the actual 1500 rpm rated speed up to 3000 rpm in order to check the behaviour beyond the rated speed of the machine.

Stiffness and damping coefficients of the oil-film bearings, which are function of rotating speed, are here considered constant. Flexural vibrations are shown in correspondence of the bearings and only in vertical direction, where they are actually measured in this type of machines. Torsional vibrations are shown in correspondence of the coupling flanges, where they could easily be measured. These last simulations show the torsional vibrations excited by the breathing crack, due to the coupling effect of torsional vibration with lateral loads.

The comparisons are shown for the lateral vibrations in bearing #4 of the cracked turbine in figure 7.20 for the 1X component and in figure 7.21 for the 2X component . For the torsional vibrations the comparison is shown in the coupling flange between LP1 and LP2, in the figure 7.22 for the 1X component and figure 7.23 for the 2X component.

It can be seen that the crack B with higher angular extension excites lateral vibrations with higher amplitudes for the 1X component and smaller amplitudes for the 2X component, with respect to crack A.

A shift of the lateral resonance frequency cannot be recognized because the change is very small and the resonance peak is rather wide.

The torsional vibrations, if excited by any excitation mechanism, allow to verify the shift of the torsional resonance frequency of the cracked steam turbine with crack B with respect to the smaller crack A, when the crack is truly breathing. This would allow to monitor the circumferential propagation of the crack.

A closer view of the resonance peaks of figure 7.22 is shown in figure 7.24, for highlighting the differences in resonant frequencies. The results are:

Crack 120°:	0.4 μrad	at	2543 rpm
Crack 240°:	0.6 μrad	at	2533 rpm

The resonance shift for the larger crack with respect to the smaller crack is 10 rpm equal to 0.166 Hz. A very similar frequency shift had been found in the natural frequencies of the shaft line with completely open (non-breathing) cracks.

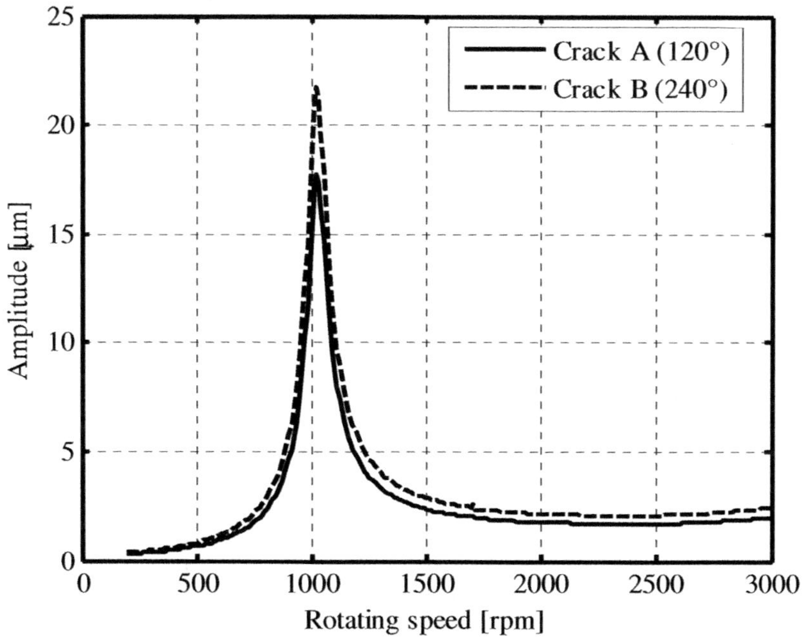

Fig. 7.20. Lateral 1X vibrations in bearing #4.

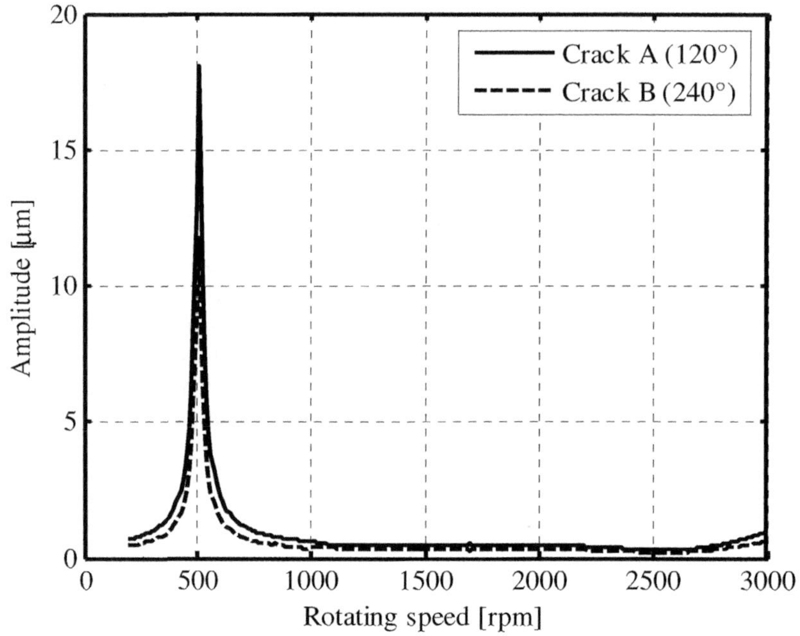

Fig. 7.21. Lateral 2X vibrations in bearing #4.

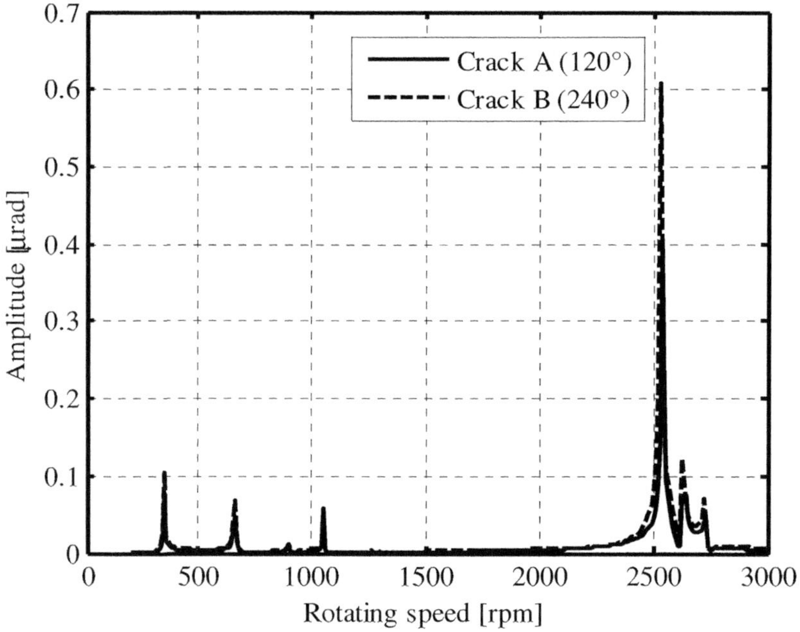

Fig. 7.22. Torsional 1X vibrations in the coupling flange between LP1 and LP2.

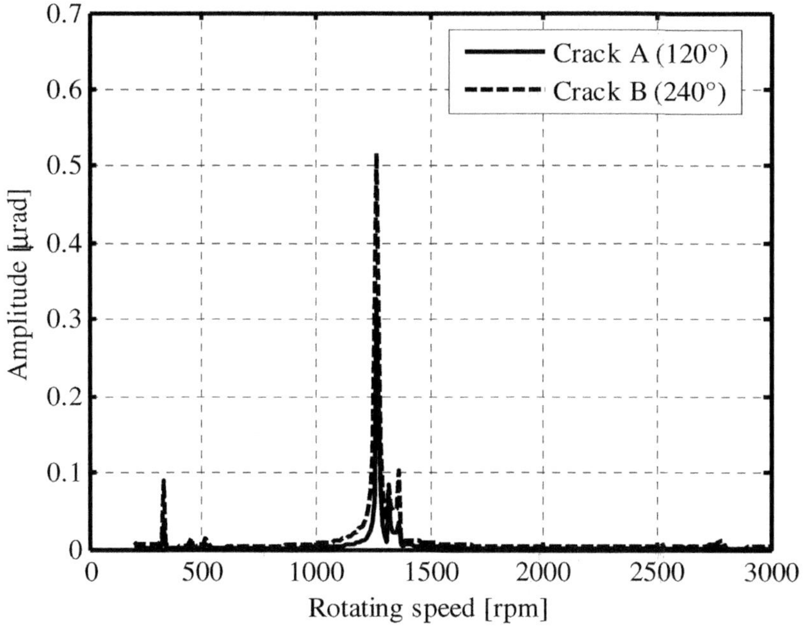

Fig. 7.23. Torsional 2X vibrations in the coupling flange between LP1 and LP2.

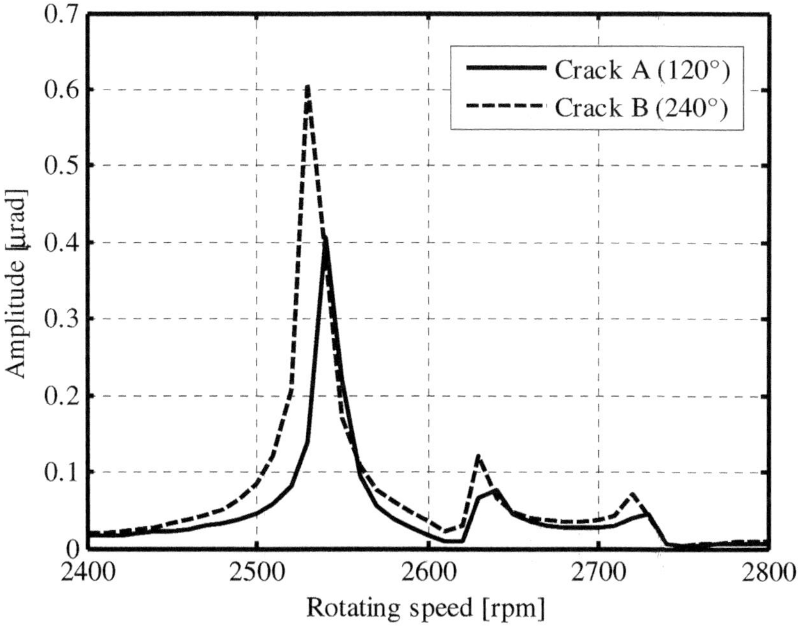

Fig. 7.24. Zoom of torsional 1X vibrations in the coupling flange between LP1 and LP2.

7.1.6.4 Results of the Complete 6 d.o.f. Model with Breathing Crack at Full Load and Nominal Speed

Finally the behaviour of the cracked shaft-line with breathing crack excited by weight and rated torque has been calculated at the operating speed of 1500 rpm. Both crack sizes, the shorter crack A (120°) and the larger crack B (240°), have been considered and the results are compared.

Vibration amplitudes at 1500 rpm have been calculated for the two cracks: these are the values that could be expected and measured in normal operating conditions, in all bearings of the turbo generator.

Table 7.3 shows the 1X components of vertical vibrations excited by the 120° crack (crack A) and by the 240° crack (crack B) at rated speed, only in the bearings where vibration amplitudes are significantly different from zero. All amplitudes are very small.

Measurable amplitudes are excited by the crack only in the bearings of the cracked turbine. As expected the 240° crack excites somewhat higher 1X vibration levels.

Table 7.3. 1X vertical vibration amplitudes in bearings #2, #3, #4 and #5 of the unit due to the crack at 1500 rpm.

	120° crack	*240° crack*
	[μm]	[μm]
Bearing #2	0.25	0.32
Bearing #3	2.37	2.92
Bearing #4	2.20	2.69
Bearing #5	0.20	0.26

Table 7.4 shows the 2X components of lateral vibrations excited by both cracks.

As already mentioned the larger crack excites in the 2X component lower amplitudes in bearing #4, but higher amplitudes in bearing #3, due to a local resonance. All amplitudes are anyway extremely low.

Table 7.4. 2X vertical vibration amplitudes in bearings #2, #3, #4 and #5 of the unit due to the crack at 1500 rpm.

	120° crack	*240° crack*
	[μm]	[μm]
Bearing #2	0.00	0.04
Bearing #3	0.35	1.35
Bearing #4	1.05	0.50
Bearing #5	0.11	0.04

Finally the torsional vibration amplitudes in all coupling flanges are shown in Table 7.5 for the 1X component and in Table 7.6 for the 2X component.

Table 7.5. 1X component of torsional vibrations in all coupling flanges due to the crack at 1500 rpm.

	120° crack [μrad]	*240° crack* [μrad]
Coupling #1	0.68	1.15
Coupling #2	0.90	1.56
Coupling #3	0.15	0.25
Coupling #4	0.01	0.02

Table 7.6. 2X component of torsional vibrations in all coupling flanges due to the crack at 1500 rpm.

	120° crack [μrad]	*240° crack* [μrad]
Coupling #1	0.20	0.09
Coupling #2	0.13	0.07
Coupling #3	0.02	0.01
Coupling #4	0.00	0.00

The torsional vibration at rated speed is rather low despite the fact that the full load torque contributes significantly to the excitation. Measurements would hardly be able to identify a crack with a small depth of 10% of the diameter, even if its angular extension reaches 240°.

In the described turbo generators affected by small cracks in the middle of the LP steam turbine, the symptoms which are useful for identifying the crack are:

1) the 2X lateral vibrations (measured in the bearings) at the 2X critical speed.
2) The shift in torsional natural frequency associated to the 5th torsional vibration mode. This requires accurate measurements and suitable torsional excitation.

7.2 Slant and Helical Cracks

Generally cracks propagate in surfaces which are roughly planar and perpendicular to the rotation axis of the shaft. Anyway, if huge torque combines with high bending loads, the crack may also propagate along an helical path, therefore these cracks are called helical or slant cracks.

In literature the dynamical behaviour of a shaft with a slant crack, a crack which has developed along a plane which forms an angle of 45° with the axis of the shaft, is analysed by Ichimonji et al. [7.7], [7.8] and Sekhar and Prasad [7.9]; probably the study was triggered by the detection of similar kind of defect in an industrial machine, but the failure has never been reported in public literature. In these studies the slant crack is assumed to open and close periodically accordingly to the direction of a sinusoidal torque that is assumed to be applied to the shaft. The bending stiffness of the shaft is changing from a maximum with closed crack to a minimum with open crack. The lateral vibrations are then modulated by the torsional frequency. This situation seems not applicable to machines of power plants during normal operating conditions when the constant driving torque is much higher than its possible small oscillations. Darpe [7.10] has studied with the SERR approach a Jeffcott rotor affected by slant cracks that have developed on planar surfaces forming different angles with the shaft axis (from 30° up to 90°: this last angle corresponds to the planar transverse crack) and different depths. He assumes that the breathing is determined by the weight and considers a gradually breathing mechanism defined by the position of the crack closure line. He compares the results with transverse cracks and finds that, assuming a reference system fixed on the shaft, the torsional d.o.f. is coupled with all other d.o.f.s, as well as with the axial degree of freedom (with the exception of one d.o.f., the displacement according to the axis of the crack, which is the axis of symmetry of the crack surface). All the flexibility coefficients (direct and cross-coupled) depend obviously on the angle of the slant crack and all increase with similar trends with crack depth.

Cracks due to torsional stresses develop most likely along helical surfaces rather than along planar inclined surfaces.

Slant cracks (with helicoid angles up to 45°) can develop only when torque is alternating, which is usually not the case in turbomachinery.

The crack studied in this section developed along a helical path with an angle of 6° only on the outer surface of the shaft, due to the combined action of bending and torsion. This kind of crack could develop at mid-span of double flow steam turbines in high power turbo generators, where the maximum bending moment due to weight combines to the transmitted torque. The aim of this analysis is to evaluate the behaviour of a shaft with this type of crack during full load operation and to check if helical cracks with so small angles might generate relevant differences with respect to the more common transverse flat crack. In order to study the behaviour of the shaft affected by a slight helical crack, a reduced scale model specimen has been designed, suitable to be used also for laboratory experimental tests. A static bending moment as well as a static torque is applied to the cracked rotating shaft. A crack depth of 42.8% of the diameter and two different angular extensions of the crack have been considered: 120° which will be called *short crack* and 240° which will be called *long crack*. This research has been performed in cooperation with R&D of EDF - Électricité de France, where also some experimental tests on a specimen affected by helical cracks have been performed. The specimen has a diameter of 70 mm, the angle of the helical curve of the crack on the cylindrical specimen surface with respect to a plane orthogonal to the cylinder axis is equal to 6°. This angle increases when moving from outside of the shaft to the inside, closer to the crack tip, where it reaches its maximum of 36°.

The helical crack has been obtained in a fatigue test machine where the rotating shaft specimen has been loaded with torsion and bending. A small helical slot had been machined in the shaft by electroerosion. The shaft specimen mounted in the fatigue testing machine, with the crack made visible by dye penetrant, is shown in figure 7.25, along with the crack depth and profile as revealed by ultrasonic tests. The crack had propagated along the helical path in both directions starting from slot, reaching a relevant depth and a wide angular extension. The obtained crack shape was rather similar to the short crack with angular extension of 120°.

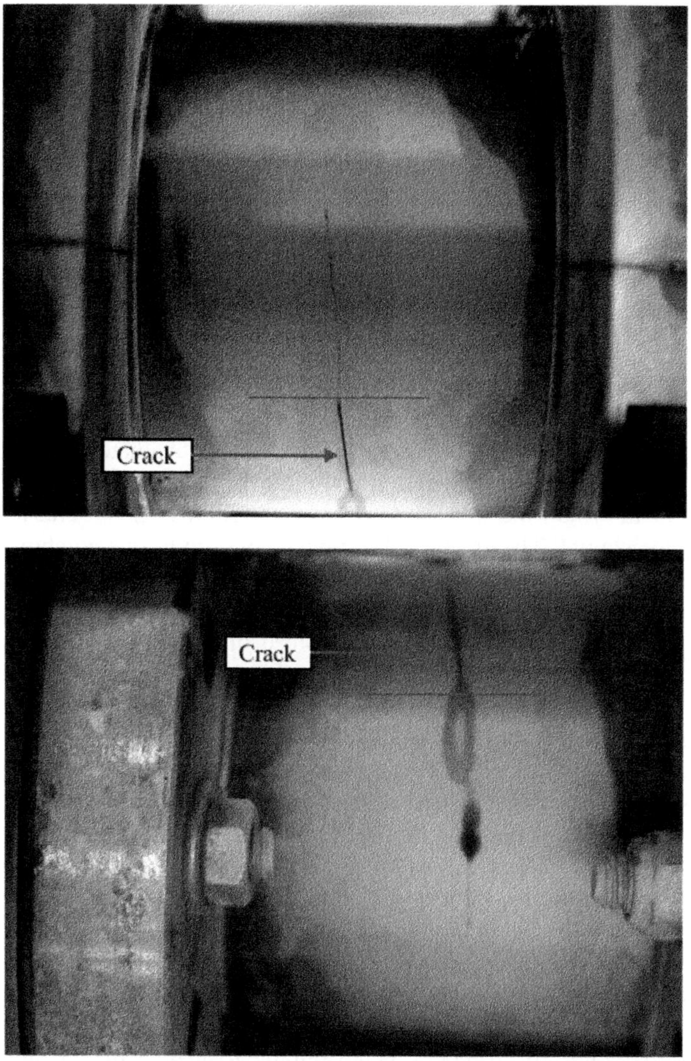

Fig. 7.25. Slightly helical crack generated on the fatigue test machine, highlighted by dye penetrant. Shape of the obtained crack as revealed by ultrasonic tests (measures in mm).

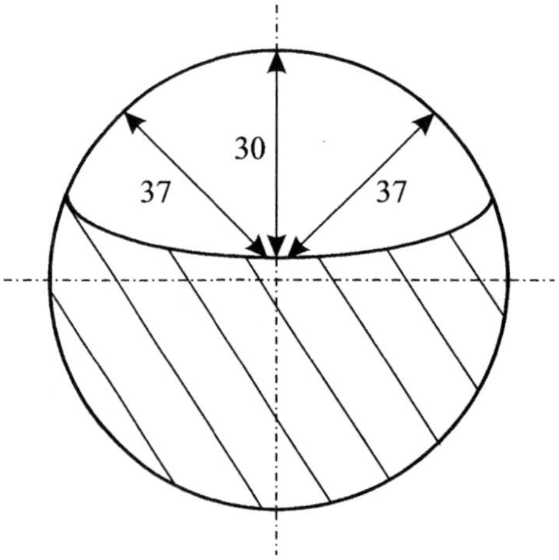

Fig. 7.25. (*continued*)

Different rotating loads have been applied to one end of the cracked specimen clamped at its other end and the breathing mechanism and the deflections according to the 6 degrees of freedom in space have been evaluated for the different angular positions of the loads with respect to the crack, by means of a 3D finite element model in which the non-linear contact conditions on the cracked surface are taken into account. The effect of the applied constant torque combined to the rotating bending load, which are responsible for generating the helical crack, is analyzed in detail. For the sake of simplicity the meshed shaft specimen has been considered fixed and the load rotating, instead of fixed load acting on rotating shaft. Dynamical loads are disregarded in this analysis. In industrial machines the static torque during normal operating conditions overcomes completely the dynamic torque components, therefore the direction of torque does not change in the full load condition. In these conditions the breathing mechanism is determined by the static bending and torsional loads only.

In order to emphasize the effect of the helical crack and to highlight the differences in its static elastic behaviour with respect to the transverse crack with the same shape and extension, the deflections of the un-cracked specimen have been subtracted from the deflections of the cracked specimens and these differences, which represent the "additional" deflections generated by the crack only, are compared in the same diagrams for the helical and for the transverse cracks.

This approach with 3D non-linear models is not suitable to analyse the dynamic behaviour of a shaft line affected by a crack and to evaluate the additional vibrations generated by the crack, due to the huge calculation effort required by the 3D non-linear analysis. Anyhow, the same differences between the statical behaviour of the shaft affected by the helical crack and that one of the transverse planar crack, will be found in its dynamical behaviour. That means that if one likes to calculate the dynamical behaviour of a shaft affected by a slightly helical crack, he can calculate the behaviour related to the transverse crack with the same shape and depth and update the results for the helical crack with some corrective factor.

7.2.1 Description of the Model

The model of the crack surface has been created by generating on the cylindrical specimen a separation surface obtained by moving a radial segment along a helical path. The obtained solid model is shown in figure 7.26 for the long crack. The two parts of the cylinder separated by the separation surface have then been meshed by an automatic procedure. The final configuration of the crack is then obtained by connecting corresponding nodes of the two parts on the separation surface where the crack has not arrived during its propagation, re-establishing the material continuity, and imposing the usual contact conditions on the remaining nodes. The obtained mesh is shown in figure 7.27 for the long crack.

The specimen has a length of twice the diameter, is clamped at one extremity, and the loads are applied at the other extremity.

This configuration has been meshed automatically. This procedure has the advantage of simplicity but has the disadvantage that the position of the nodes of the mesh on the two surfaces of the crack, which are master nodes and slave nodes, do not correspond each other. The consequence of this fact is that the interpolating surfaces which connect the nodes, are not exactly the same on both faces of the crack. Figure 7.28 shows also the superposition of master and slave nodes of the cracked surface. When contact occurs between the two surfaces, it is not continuous and some points of these surfaces are not in contact. This does not affect the overall deflections, but makes the breathing mechanism analysis less accurate. On the other hand the experimental investigation on the breathing mechanism of transverse cracks, described in section 4.2 has shown that, in closed crack configuration, only a smaller part of the cracked surface is really in contact. Therefore also in real cracks the contact does not occur in all points of a crack surface. The accuracy of the numerical results is expected to be good enough for calculating deflections due to applied loads, but will not be sufficient for checking stress intensity factors or for predicting propagation speed of the crack. The friction coefficient in the contact areas of the crack surfaces is considered equal to 0.4. The smaller crack has a similar solid model and a corresponding mesh.

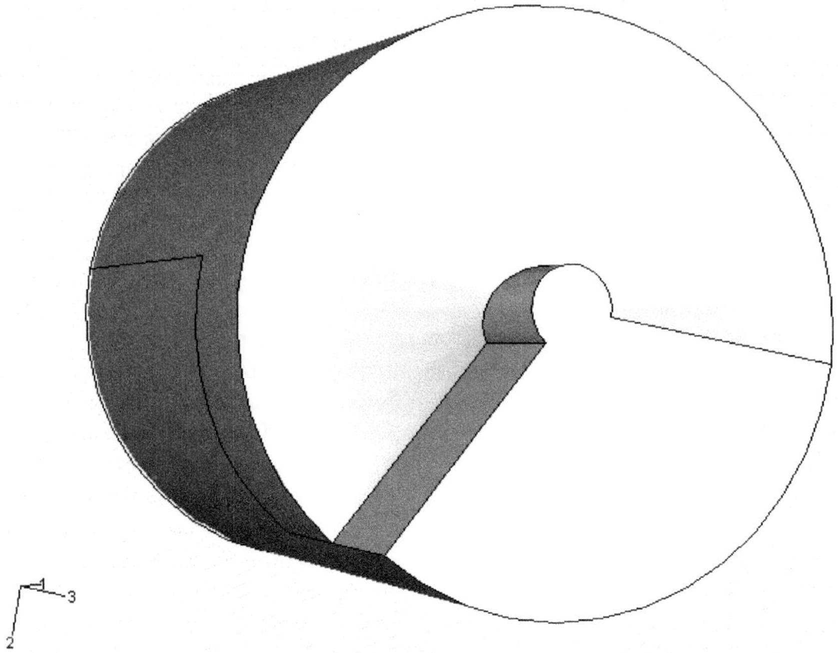

Fig. 7.26. Solid model of half specimen with a 240° long helical crack.

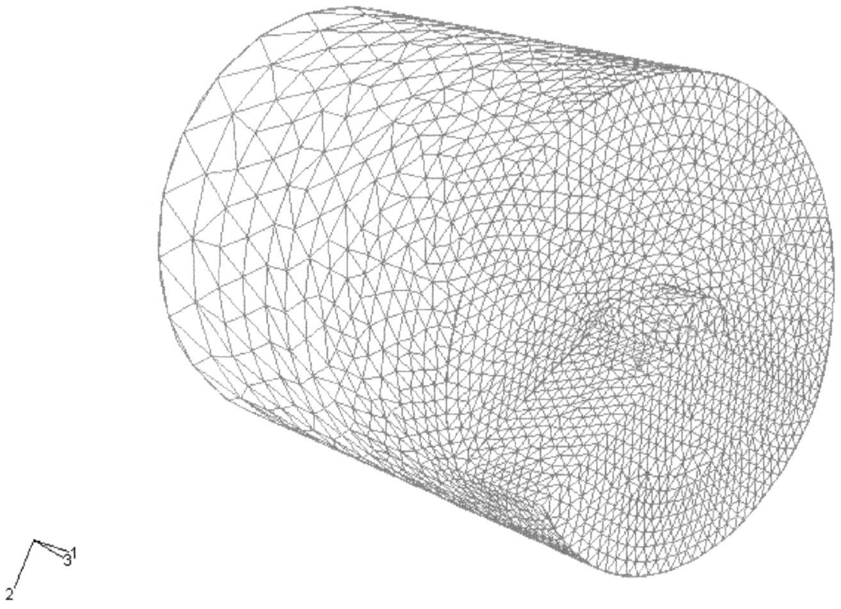

Fig. 7.27. Mesh of the half specimen with a 240° long helical crack.

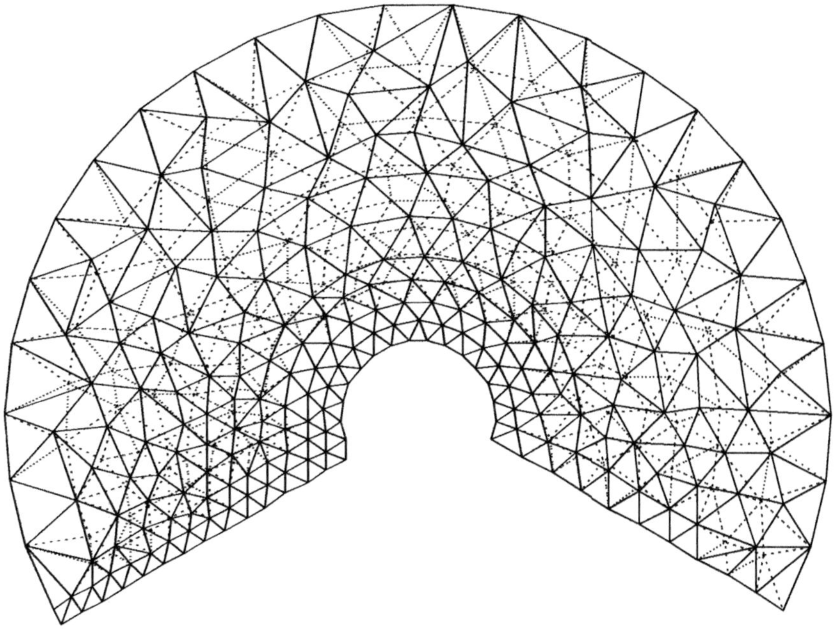

Fig. 7.28. Superposition of master and slave nodes on the cracked surface.

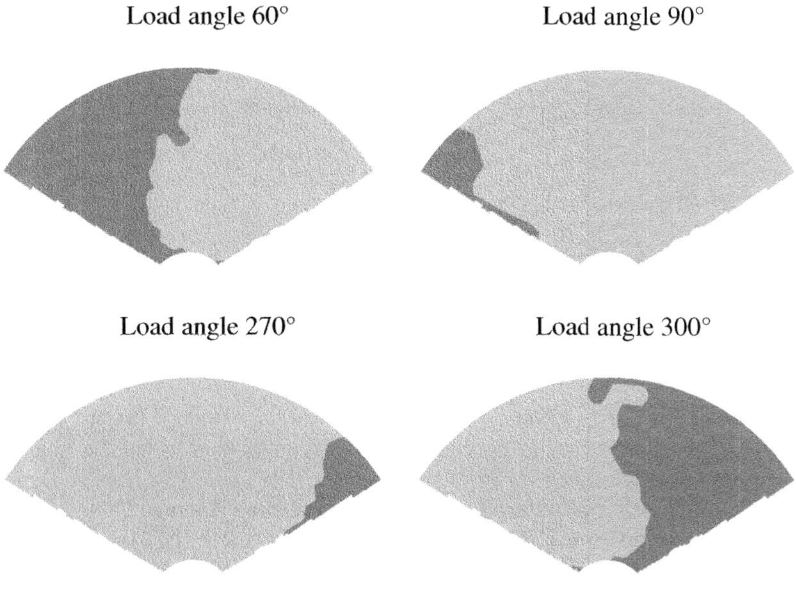

Fig. 7.29. Breathing mechanism of the short crack in some angular positions.

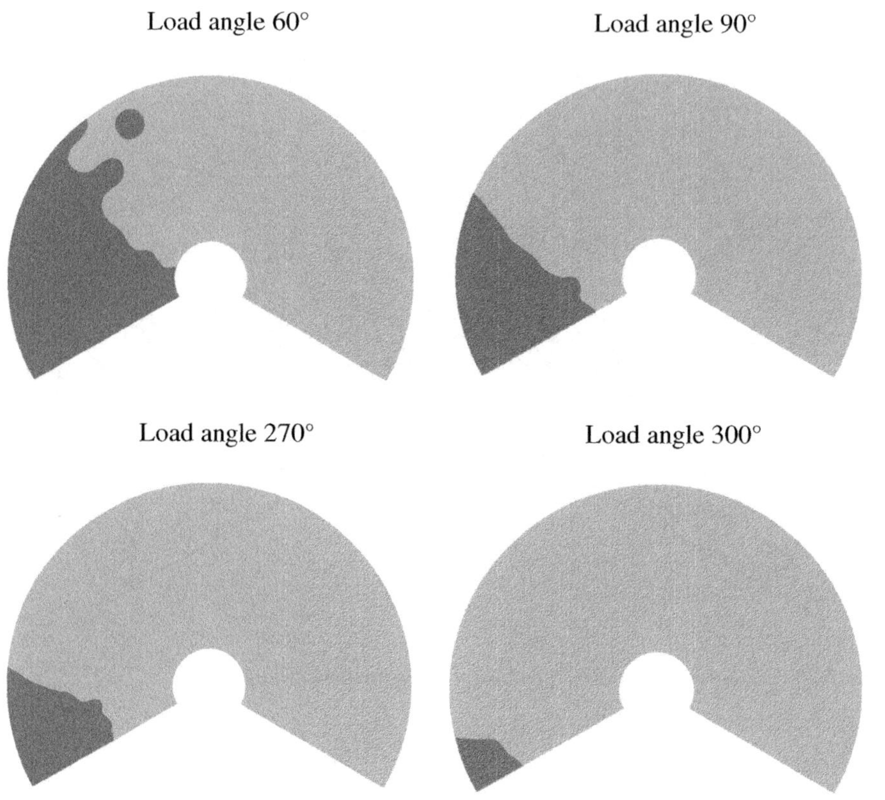

Fig. 7.30. Breathing mechanism of the long crack in some angular positions.

7.2.2 *Breathing Mechanism*

Preliminarily the contact conditions in the helical crack have been investigated determining the breathing mechanism. Taking into account the scale ratio of the specimen to a real machine and the loads on a real machine in operating conditions at full load, the scaled loads are following: bending moment $M_b = 600\,\text{Nm}$ and torsional moment $M_t = 1200\,\text{Nm}$.

Figure 7.29 shows the breathing mechanism for the short crack loaded by full bending and torsional load in some characteristic angular positions: the contact occurs between the crack lips only in the dark zones. The same is shown in figure 7.30 for the long crack. The separation lines between closed and open sectors of the crack surface are not straight lines as they should be when the crack surfaces are exactly flat, but are irregular curves as it happens when the surfaces have a random roughness.

7.2.3 Deflections

In order to provide valuable insight into the behaviour of helical cracks and to check the linearity of the overall behaviour of the cracked specimen, several different load conditions have been applied to it and the deflections have been calculated. All the different load conditions have been applied also to a planar transverse crack with the same depth and angular extensions in order to have a reference situation to which the results of the helical crack can be compared. Furthermore it should be reminded that only the effect of the crack on the deflections at the end section of the specimen is shown in all the figures. The uncracked shaft deflections are subtracted from the cracked shaft ones.

First the effect of torsion alone (no bending) has been analysed. Figure 7.31 shows the cracked specimen with its boundaries and its reference frame. Crack specimen is fixed, load is rotating and crack axis is directed according to axis y. Deflections have been calculated according to all 6 degrees of freedom. Vertical deflection, angular deflection around horizontal axis (which is shown in figure 7.32), axial deflection and torsional deflection are vanishing small for the planar transverse (flat) crack but reach consistent values for the helical crack. This means that the coupling between these degrees of freedom and the torsional degree of freedom, to which the torsional load is applied, is negligible for transverse flat cracks and relevant for helical cracks. The deflections according to the remaining degrees of freedom are smaller for the helical crack than for the flat crack.

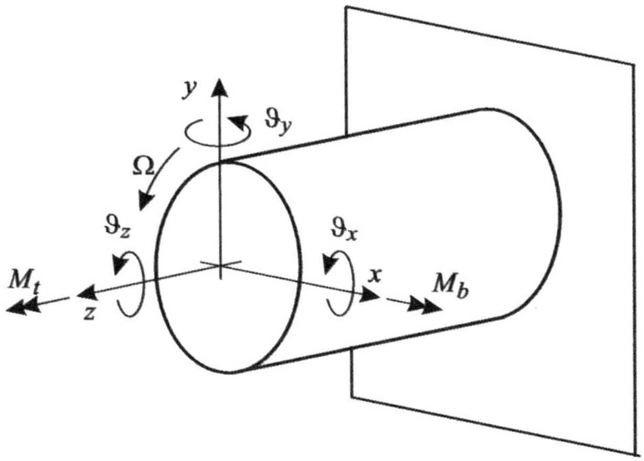

Fig. 7.31. Cracked specimen model, reference frame and loads.

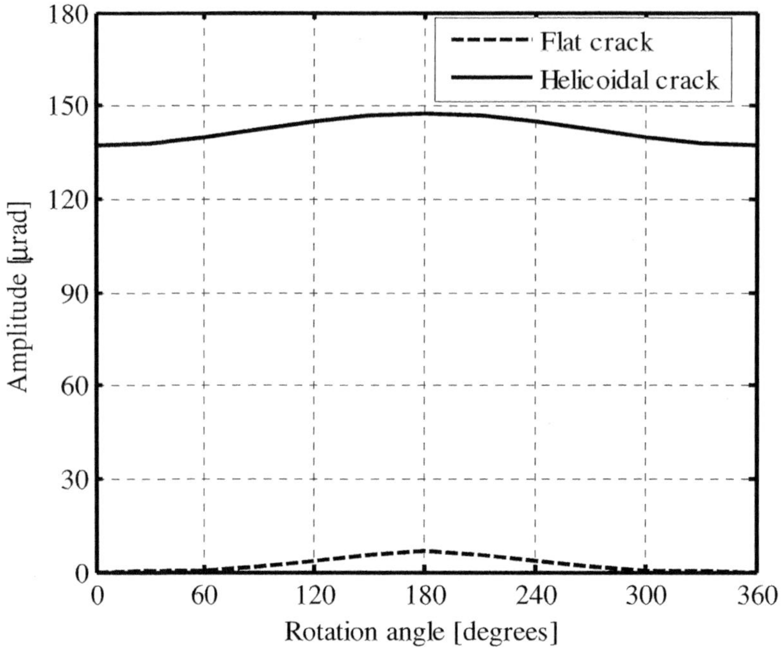

Fig. 7.32. Angular deflections ϑ_x around horizontal axis, as a function of angular position of load, due to torsional load only: comparison between helical and flat planar cracks.

A very similar behaviour has been found also for the short crack, for which roughly all displacements are reduced to ¼ with respect to those of the long crack.

At full load (bending plus torsion) the deflections according to all degrees of freedom have been calculated and compared to those of the planar crack: this allows to evaluate all coupling effects which can be expected in an industrial machine loaded by static bending and torsion and affected by a transverse crack or by a slightly helical crack.

These coupling effects generate vibrations during the rotation of the shaft that are not present in integer (non cracked) shafts and constitute symptoms of the presence of the crack. Figure 7.33 shows the diagrams for the short crack. Coupling effects are present for all degrees of freedom for both the transverse crack and the helical crack. Among them, some are relevant and some others are small.

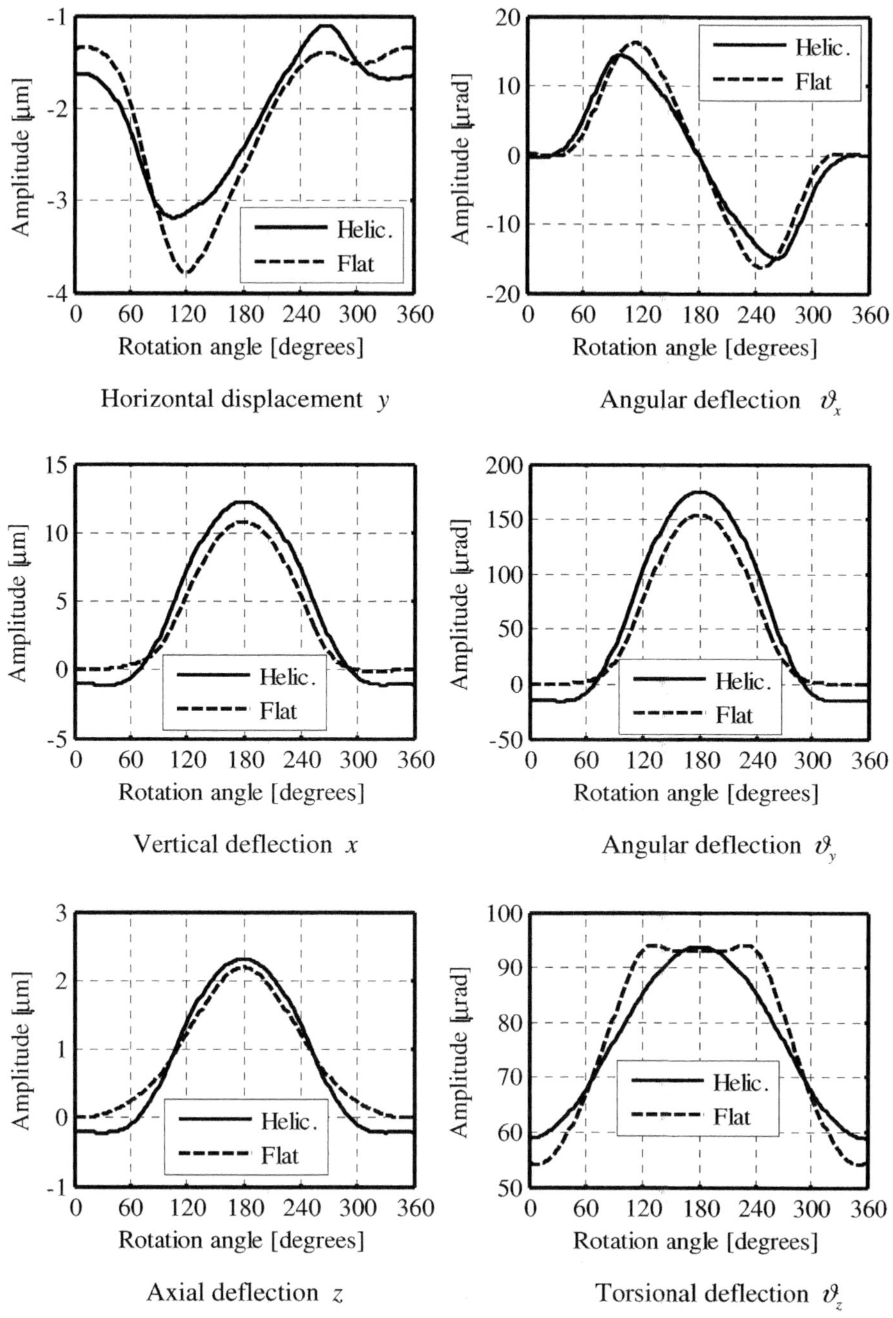

Fig. 7.33. Deflections according to all degrees of freedom for shaft specimen affected by the short crack at full load, as a function of angular position of load.

The maximum axial deflection is 2.3 μm, which is small compared to the vertical deflection of 12 μm, and the maximum torsional deflection is 90 μrad, which is comparable to the angular deflection around the horizontal axis of 175 μrad. Therefore, coupling between axial and lateral deflections exists but it is smaller than the coupling between lateral and torsion ones. This means that excitation of torsional vibrations at full load is higher than the excitation of axial vibrations.

The helical crack generates an additional flexibility somewhat higher than that of flat crack: this result could be expected because positive torsional loads tends to open crack surface and an open crack introduces a higher flexibility.

Similar results have been found for the long crack. While in case of the load condition with torsion only relevant differences have been found between the behaviour of helical and flat cracks, in full load condition the differences are so small that they can simply be neglected.

Since the behaviour of the specimen with an helical crack shows a certain sensitivity to torsion, an analysis has been made with constant bending load (as it is in real machines) and increasing torsional loads (as it occurs during the start up of the machine), also in order to check how far the overall behaviour can be considered linear or quasi-linear. Bending load M_b has been assumed constant and equal to 600 Nm and torsion M_t is respectively 0, 200, 400, 600, 1200 Nm for the loading conditions labelled respectively as LC1, LC2, LC3, LC4 and LC7. Figure 7.34 shows the angular deflection around horizontal axis and the torsional deflection.

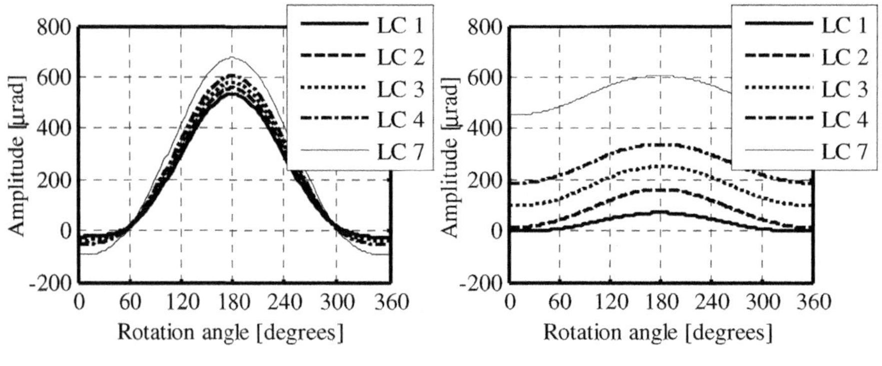

Fig. 7.34. Angular deflections ϑ_x around horizontal axis (left) and torsional deflections ϑ_z (right) due to constant bending load and increasing torsional loads, as a function of angular position of load.

Despite the fact that the stresses become highly non-linear with the deep cracks and the huge loads and reach values that overcome the elastic limit, in particular at the tip of the crack, the overall behaviour can be considered quite linear: all deflections are roughly proportional to the torsional load.

Finally, the behaviour in two different loading conditions have been compared to the behaviour of the flat crack: full load condition and no torsional load condition (that represent for instance the situation when electrical load is removed from a generator). As shown previously the differences between helical and flat crack at full load could be neglected; Figure 7.35 (left) shows these differences for the torsional degree of freedom; the differences are higher than those of other degrees of freedom for the long crack. In the same figure 7.35 (right), the differences in no torsional load conditions between the helical and the transverse crack are represented.

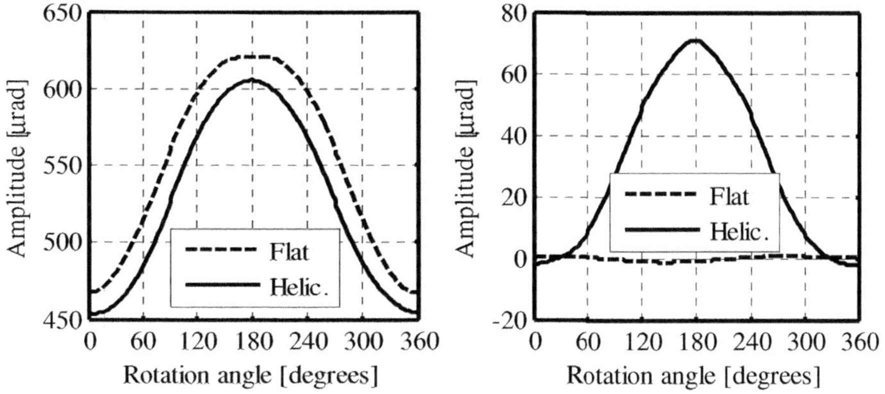

Fig. 7.35. Torsional deflection: comparison between helical and flat crack at full load (left) and at no torsional only bending load (right), as a function of angular position of load.

In the no torsional only bending load condition, some small differences are found for all degrees of freedom: the specimen with the helical crack is generally more stiff than that with the flat crack, except for the torsional degree of freedom where the bending load excites more strongly the torsional deflection, as shown in figure 7.35. Still smaller differences are found for the short crack. It seems reasonable to conclude that the only relevant difference between flat cracks and slightly helical cracks for horizontal shafts, where bending loads are always present, arise for the torsional degree of freedom in no torsional load condition. In this condition torsional deflections are generated by bending moments due to coupling effects.

7.2.4 Dynamic Tests

EDF has performed some dynamic tests on the specimen affected by the helical crack shown in figure 7.25, introducing the cracked specimen in the shaft of the

EURoPE test-rig, and measuring bending and torsional behaviour. The aim of the tests were mainly to measure the shift in bending natural frequency and in torsional natural frequency of the cracked shaft with respect to the un-cracked shaft. The results for torsional natural frequencies are shown in Figure 7.5. The excitation of axial and of torsional vibrations due to the coupling effect at the rotating speed of 1500 rpm was very weak and could not be measured. The axial and torsional natural frequencies are much higher with respect to the excitation frequency, which is mainly 1X and 2X, therefore the excitation was not amplified by dynamic effects and could not be measured.

No measurable difference with respect to planar crack behaviour could be found. This may confirm the calculated results.

7.2.5 Final Remarks about Slant and Helical Cracks

Summarizing the results it seems that at full load the differences between helical and flat crack are so small that they could be neglected.

When torsional load is removed then higher differences arise but mainly for the torsional degree of freedom, which is excited by the bending load by means of a coupling effect.

In this condition torsional deflections are generated by bending moments due to coupling effects and torsional vibrations are excited in rotating shafts. This constitutes a clear symptom of the presence of a helical crack in a shaft line: when during a run-down transient (at no torsional load) the torsional natural frequencies are strongly excited at the corresponding rotating speed, this could be due to the relevant coupling effect of the helical crack.

7.3 Non-linear Behaviours in Cracked Rotors

In order to avoid misunderstanding, the *non-linear behaviour* will be defined as the behaviour that depends not only but significantly on the response of the system. Therefore non-linear behaviour of cracked shaft occurs when the breathing of the crack is not anymore determined by the static forces, like the weight, and by the bearing reaction forces, but by the dynamic forces associated to the vibration, in other words by the response of the system.

It is well known that cracked rotors may experience different vibrational behaviours such as instabilities, sub-harmonic responses and 1X, 2X and 3X harmonic responses. The complex behaviour resulting from superposition of all these components can be found if the non-linear equation of motion are integrated in the time domain. The relative weight of the different components in the complex response depends on the values of the different parameters, like the depth of the crack, the damping, the rotating speed, the external loads and the position of the crack. To the authors' knowledge instabilities and significant sub-harmonic components have never been found in industrial rotating machinery. Cracks have been discovered before failure because the synchronous and higher harmonic vibration components had reached alarm levels, but no abnormal behaviour has been recorded. Also in case of rotor failure due to propagating crack, only increasing 1X, 2X, and 3X sometimes, vibration components have been observed before failure, no sub-harmonic components nor unstable behaviour has been measured.

The present analysis aims at evaluating all the non-linear effects due to a rather deep transverse crack that has developed in a full size shaft. Time integration has been used in [7.11] for the non-linear analysis of a heavy, horizontal axis, well damped steam turbine rotor. The results confirmed that no instabilities nor sub-harmonic components did appear and that the overall deviations from linear behaviour were rather small. In the present analysis the unbalance has been increased to overcome completely the static bending moment, as it might happen in vertical axis shafts, and also the damping has been reduced consistently, as it might occur in shafts supported by roller bearings, instead of oil film bearings.

Stable and unstable behaviours have been analysed by several different authors (see e.g. [7.12] and [7.13]). Ishida et al. [7.12] consider a Jeffcott rotor with piecewise open – closed crack model and the highlighted behaviour is the passing of the 1X resonance curve with different unbalance orientations determining stable and unstable behaviours in resonance, which are also confirmed by experimental results. Gasch et al. [7.13] consider a similar model equipped with different types of oil film bearings and investigates all different vibration components (sub- and super-harmonics) excited only by a relevant relative change of shaft stiffness (from 0 to 80%) and by no unbalance and defines their degree of stability. Non-linear effects in vibrations of stationary cracked beams have been studied extensively by Brandon [7.14], but these analysis on non rotating beams, where crack opening and closing is abrupt and governed by the vibration only, cannot catch the peculiar behaviour of rotating beams where breathing occurs that is governed by the rotation and the vibration. Simple Jeffcott rotors were used to emphasize non-linear behaviours such as apparent resonances at speeds that are twice or two-third of the critical speed [7.15] or even chaotic behaviour and bifurcation [7.16]. Wu et al. in [7.17] uses the SERR approach and calculates the condition of the crack in each time step: the breathing and the compliance is defined by the crack closing line (CCL) [7.4]. They analyse the behaviour of a Jeffcott rotor during a rotating speed transient by means of this non-linear model.

The present analysis investigates to which extent a real rotor (modelled by means of finite elements) with a deep crack (50% of the diameter) in a position not far from midspan between bearings, can experience behaviours that deviate strongly from linear steady state behaviour. The true breathing mechanism is accurately reproduced and only a small change in its overall stiffness is produced (the change in the modal stiffness of its 1st mode is only roughly 1.5%), despite the position and depth of the crack. In order to separate the crack induced non-linear effects from other non-linearities, the oil film bearing forces have been linearized. The effect of the reduction of damping on non-linear behaviour and instabilities is also investigated.

7.3.1 Cracked Rotor Linear and Non-linear Modelling

With reference to eq. (5.136) of section 5.6, the effect of the crack is to reduce the mean stiffness $[\mathbf{K}_m]$ of the rotor system and to introduce the stiffness variation $[\Delta\mathbf{K}(\Omega t, \mathbf{x})]$ in the stiffness matrix. How the variable part of the rotor stiffness $[\Delta\mathbf{K}(\Omega t, \mathbf{x})]$ varies between a minimum (for closed cracks) and a maximum (for open cracks), depends on the so-called breathing mechanism. As previously explained in section 5.3.1, the breathing mechanism is known when the open and closed parts of the cracked area are known, considering that the open and closed parts of the crack depend on static and dynamic forces and moments that act on the cracked section. Therefore $[\Delta\mathbf{K}]$ depends on the angular position Ωt of the crack with respect to the fixed frame external static loads, on the angular position of the rotating unbalance forces and on the vibration $\mathbf{x}$ itself, or more precisely on the vibration dependent inertia forces.

Periodic steady state solution dominate the vibration response of real rotors because:

1) the variation $[\Delta\mathbf{K}]$ of the stiffness matrix, to which instabilities and parametric excitation are related, is very small (generally less than 5%) with respect to the mean stiffness of the rotor system, also in case of deep cracks;
2) the damping of the oil film bearings, the internal damping of the rotor and some additional damping due to the breathing mechanism, is enough to overcome any instability and parametric excitation.

Generally the vibration response of the cracked rotor is small, so that the bending moment $\mathbf{M}_d$ due to the dynamic forces (external forces and inertia forces) is smaller than the static bending moments $\mathbf{M}_s$ due to the external, fixed in space, forces (such as the weight, the bearing reaction forces and any other stationary force), in horizontal heavy rotors of industrial plants. Therefore, the breathing behaviour is dominated by the static bending moment and the dependence of the stiffness variation $[\Delta\mathbf{K}]$ on the vibration $\mathbf{x}$ can be neglected.

On the contrary, this section aims at analysing conditions in which the influence of the vibration $\mathbf{x}$ on the breathing and on the related stiffness variation $[\Delta\mathbf{K}]$ cannot be neglected. This can occur: i) for light and lightly damped rotors when passing a critical speed (where vibrations and related dynamic bending moments overcome the static bending moment acting in correspondence of the crack) or ii) when the crack has developed in a position where the bending moment is small (e.g. close to a coupling flange) or iii) also in vertical axis machines, in which the lateral stationary forces that cause stationary bending moments are small.

In these cases [$\Delta\mathbf{K}$] depends also on $\mathbf{x}$ and eq. (5.136) of section 5.6 becomes non-linear.

7.3.2 Description of the Model and of the Method

The non-linear behaviour is calculated by integrating eq. (5.136) in the time domain. One revolution of the shaft is here divided in 12 parts, obtaining angular steps of 30°: the open and closed crack areas and the corresponding stiffness are evaluated in each angular step, taking into account the corresponding dynamic and static forces.

The cracked section is represented by a mesh of 60×60 elements for evaluating the breathing behaviour, as indicated schematically in figure 7.36.

The breathing behaviour of a shaft excited by static bending moment only is shown in figure 7.37. It will be shown that breathing can be strongly influenced by dynamic loads.

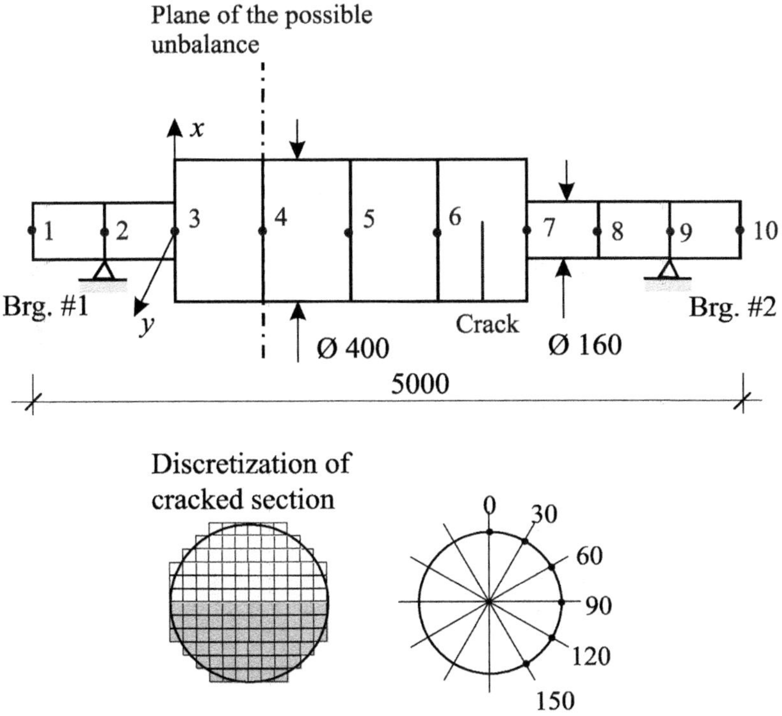

Fig. 7.36. Rotor model used for numerical simulation, discretization of cracked section and angular steps.

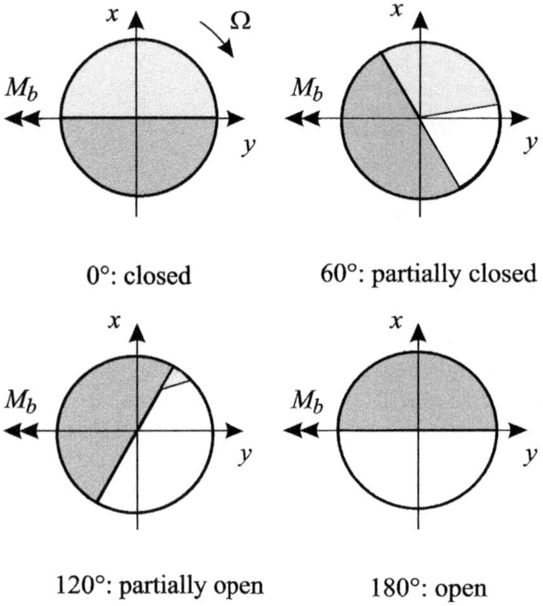

Fig. 7.37. Breathing behaviour of 50% deep rectilinear crack excited by static bending moment M_s only.

In the non-linear approach the influence of **x** on [Δ**K**] is due to the breathing mechanism: this mechanism is now governed mainly by the dynamic bending moments. These last ones are calculated from differences of rotations around the radial x and y axes, in the two nodes of the cracked element, which are obtained in each step from eq. (5.136) integrated in the time domain. Newmark β-method has been used for time integration.

Figure 7.36 shows also the FE model of the rotor used for the simulation, its overall dimensions and the position of the crack and the plane of the possible unbalance.

The overall mass of the rotor is 4000 kg. Bearings are modelled by means of their linearized stiffness and damping coefficients that are considered constant with the rotating speed and symmetrical. Their reference values are shown in table 7.7. Bearing coefficients have been considered symmetrical in order to avoid any influence on stability of unsymmetrical cross coupling coefficients. The supporting structure has been considered rigid and gyroscopic effect is taken into account. The position of the crack and of the plane of the possible unbalance are shown in figure 7.36. The 1st eigenfrequency of the un-cracked rotor is about 1700 rpm, as can be seen in table 7.8.

Table 7.7. Bearing stiffness [N/m] and damping [N s/m] coefficients.

k_{xx}	k_{xy}	k_{yx}	k_{yy}	c_{xx}	c_{xy}	c_{yx}	c_{yy}
5.0e+7	-2.0e+7	-2.0e+7	5.0e+7	3.0e+6	-1.0e+5	-1.0e+5	3.0e+6

The modal damping of the 1st mode is also shown in correspondence of the reference bearing damping coefficients. In order to let instabilities to appear, the damping coefficients have been changed in order to reduce modal dampings that are listed in table 7.8 along with the eigenfrequencies of open and closed crack. Four different conditions are shown: the reference one with more realistic stiffness and damping coefficients and three others with decreasing modal damping (LD1, LD2 and LD3).

Table 7.8. Eigenfrequencies and modal damping in open and closed crack configuration as a function of bearing damping.

Damping configuration		Crack configuration			
		Closed		Open	
		1st eigen-frequency [rpm]	Modal damping [%]	1st eigen-frequency [rpm]	Modal damping [%]
Normal	BW	1692	8.66	1616	8.13
	FW	1697	9.36	1681	8.89
LD1	BW	1700	4.44	1621	4.09
	FW	1701	4.60	1687	4.60
LD2	BW	1701	3.81	1622	3.51
	FW	1702	3.93	1688	3.81
LD3	BW	1701	3.34	1622	3.07
	FW	1702	3.43	1688	3.33

LD1: modal damping reduced of 50%;
LD2: modal damping reduced of 65%;
LD3: modal damping reduced of 78%;
BW: backward whirl, FW: forward whirl.

As will be shown in reference conditions non-linear effects are not recognizable. By reducing damping of about 50% (LD1), non-linear effects arise but rotor is stable for all rotating speeds. The further reduction of about 15% (LD2) brings the system very close to instability and the final reduction of about 13% (LD3) makes the system definitely unstable. The presence of unbalance can change the stability threshold.

7.3.3 Results Obtained with Shaft Loaded by Weight Only

In order to emphasize the non-linear behaviour, the damping has to be reduced: this way the vibration amplitude increases whilst the static deflection remains constant and the breathing mechanism is governed by the vibration rather than by the weight. The following figures show the vibration amplitude in the speed range close to resonance.

As much the damping is reduced, as much the non-linear approach deviates from linear, in which breathing is governed by weight only. The linear behaviour used for the comparison is evaluated as usual in the frequency domain, using eq. (5.144). Figure 7.38 shows the results for 2 different values of damping. Figure 7.39(left) shows that the further reduction of damping allows the instability to arise. The speed range in which instability occurs is in between 1640 and 1670 rpm, which is inside the spread of the two natural frequencies corresponding to the open crack 1622 and 1688 rpm.

7.3.4 Results Obtained with Shaft Loaded by Weight and Unbalance

The presence of the unbalance may restore the stability of the system. Figure 7.39 (right) shows that a suitable unbalance (0.15 kgm at 180° phase shift with respect to the crack) is able to stabilize the shaft again. Similar results have been found by Ishida et al. [7.8] theoretically and experimentally for a Jeffcott rotor. In order to check if instability is related to the breathing or non-breathing mechanism, the breathing behaviour in both cases of figure 7.39 is reproduced in figure 7.40.

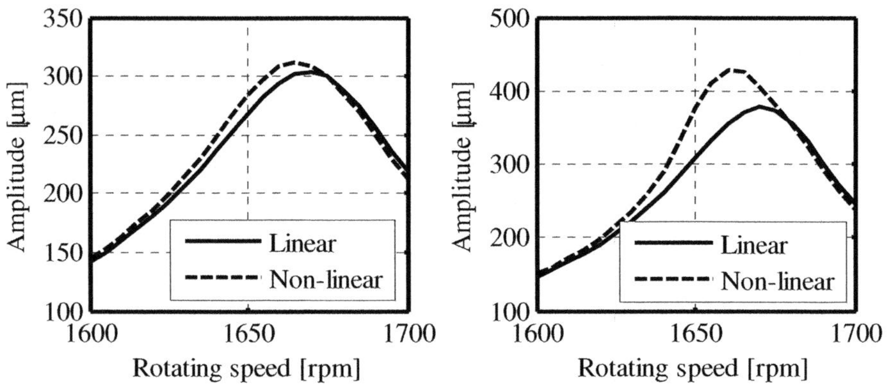

Fig. 7.38. Resonance curves with decreasing damping: LD1-left, LD2-right.

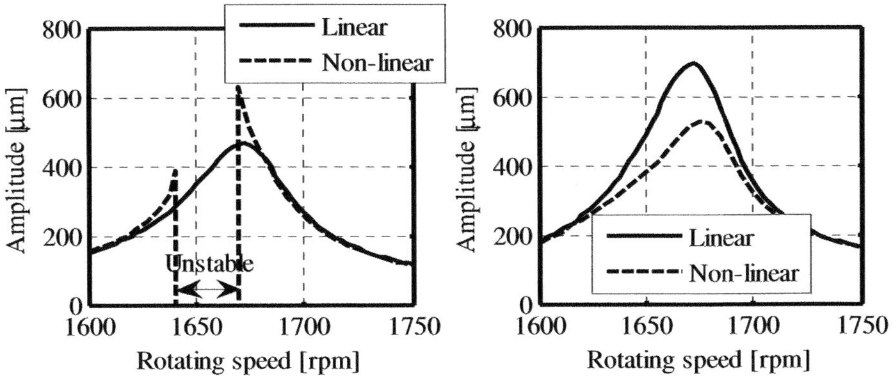

Fig. 7.39. Damping reduction to LD3 (on the left) allows instability excitation. With the same damping, the unbalance stabilizes the vibrations of the shaft (on the right).

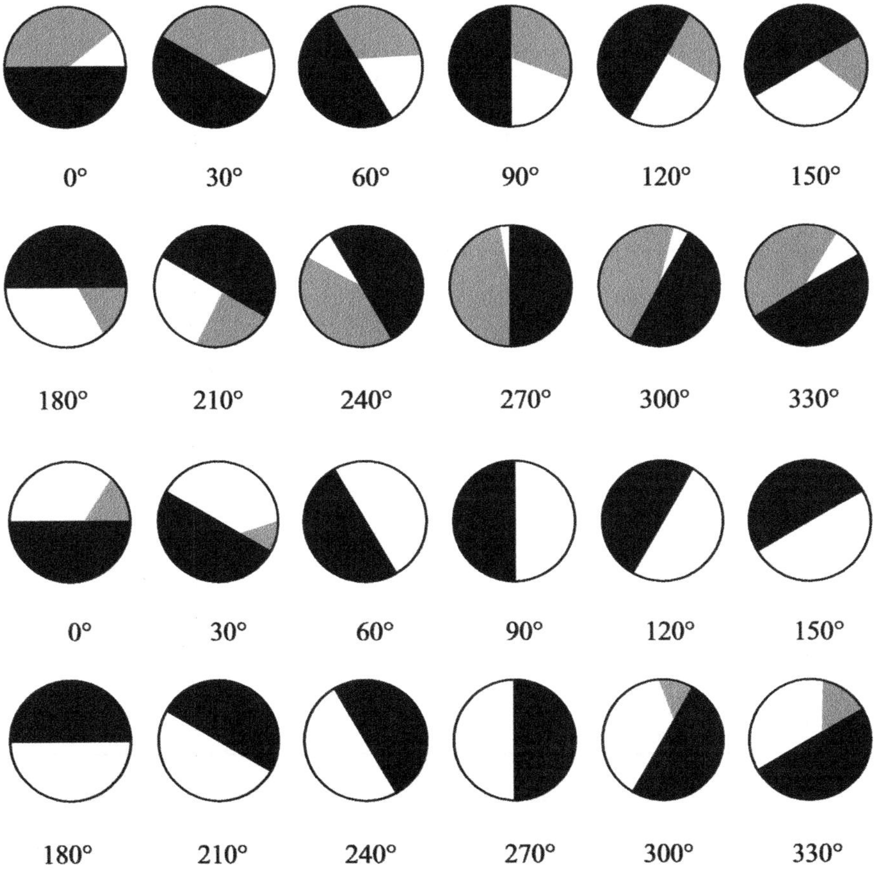

Fig. 7.40. Breathing behaviour during stable vibration (top) and during unstable vibration (bottom).

As can be seen from figure 7.40, by comparing the breathing during unstable vibration to the breathing of stable vibration, the crack in unstable condition is almost always open. The effect of unbalance is to hold the crack more closed: this is sufficient to avoid instability. If the degree of instability is directly related to the stiffness asymmetry, as it is generally believed, then the breathing crack, for which the stiffness changes are higher than that of the open crack, should have a higher degree of instability. The main difference with respect to the open crack is that the periodical stiffness variation in the breathing crack is mainly 1X with also 2X components, whilst an open crack has only 2X stiffness periodicity. Therefore it seems that the instability is related to the amount of 2X asymmetry rather than to the overall amount of stiffness asymmetry.

Now the effects of the unbalance are analyzed more in detail, considering the simulated vibrations in node 5 of the model (see figure 7.36): non-linear results are compared with the approximated linear approach results. The unbalance value has been chosen in order that the excited vibrations (also with reference damping) overcome the static deflections and govern the breathing mechanism, at least in resonance. Figure 7.41 shows the values of the bending moment due to the weight (which is obviously constant) compared to that one due to dynamic loads (unbalance plus equivalent crack forces): the simulation with reference damping is on the top, the simulation with reduced damping LD1 is on the bottom. Both simulations are with the same unbalance directed according to crack axis.

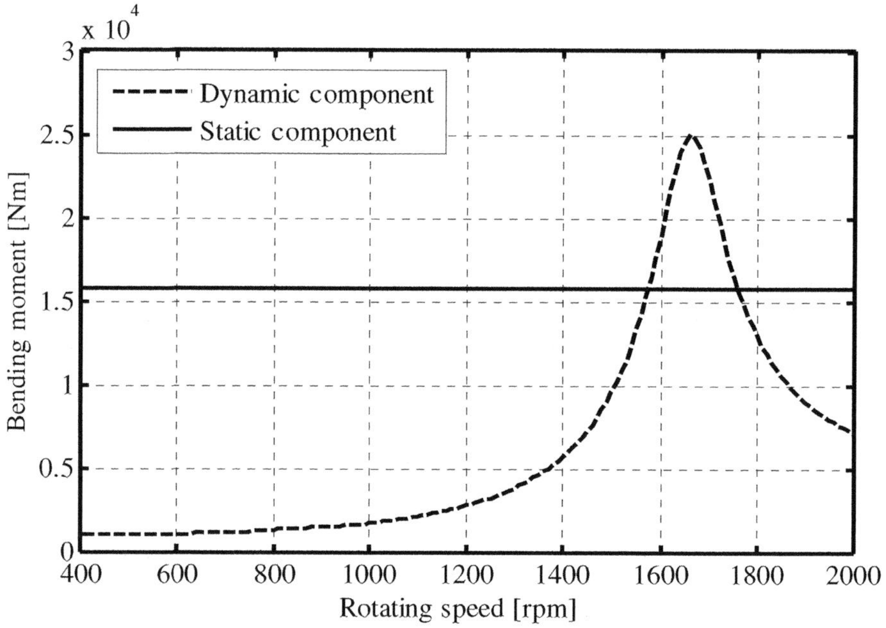

Fig. 7.41. Static and dynamical bending moments in correspondence of the crack, governing the crack behaviour, with reference damping on the top, with reduced damping LD1 on the bottom.

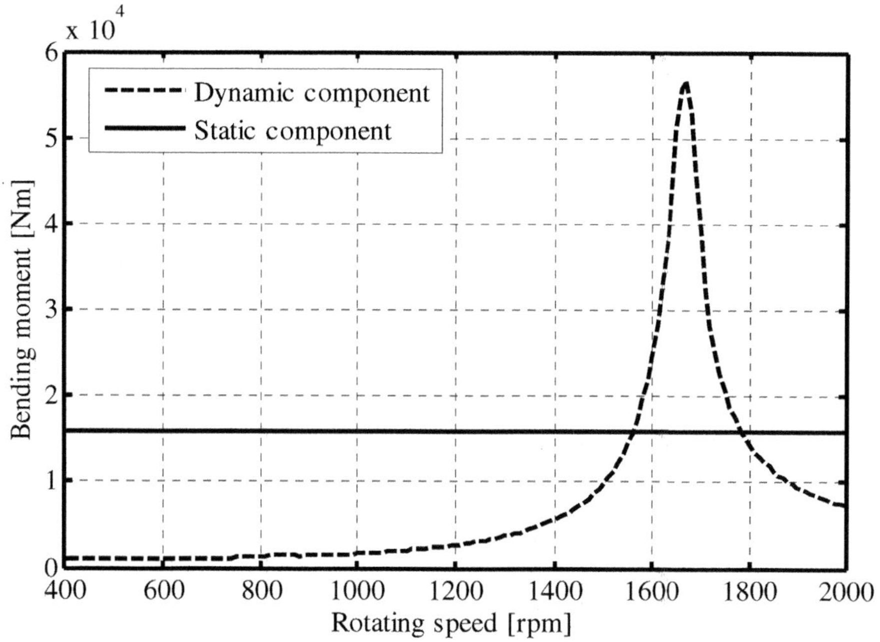

Fig. 7.41. (*continued*)

The vibration dominates and governs the breathing mechanism close to the resonance; lower damping increases the speed range where dynamic behaviour dominates. In the speed range below resonance the vibration is low and breathing is governed by weight.

Figure 7.42 shows the results obtained when the unbalance is directed according to the crack axis, so that its effect tends to hold the crack open and crack induced vibration sums up to unbalance induced vibration.

Left curve with reference damping shows that the linear approach is a very good approximation of the non-linear behaviour. Right curve shows that the linear amplitude increase from 450 μm up to 1000 μm due to lower modal damping (condition LD1), but the non-linear amplitude is much higher. An explanation of this behaviour could be that the non-linear amplitude is composed by the steady state solution plus a free motion component, which is highly excited because close to the instability threshold.

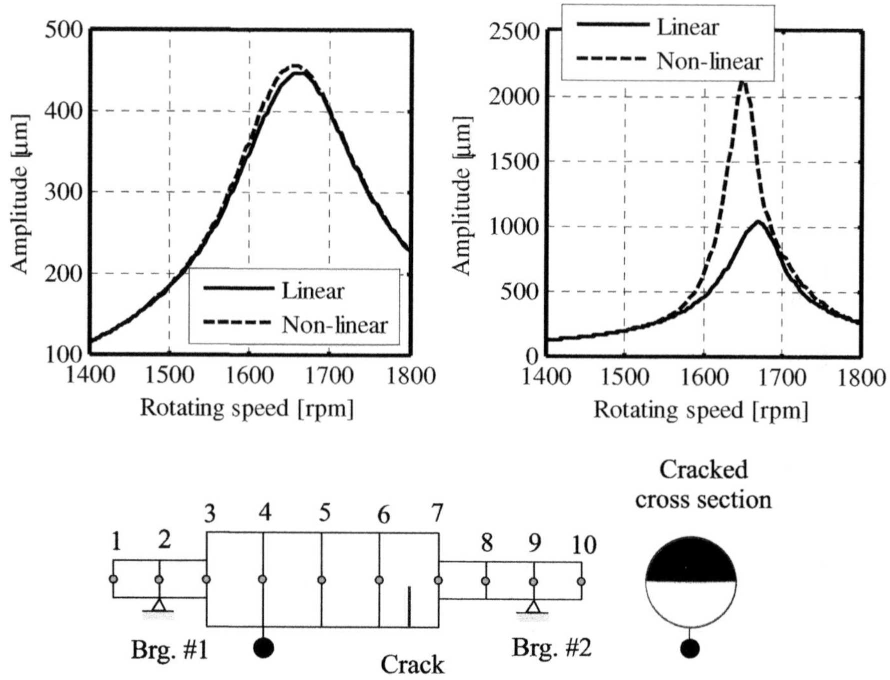

Fig. 7.42. Response curve with unbalance directed according crack axis: reference damping on the left, reduced damping LD1 on the right.

Figure 7.43(top) shows that the breathing behaviour is strongly influenced by the vibration: the crack does not close completely in any angular position. Nevertheless the linear approach, in which the crack closes completely once per revolution, as shown in Figure 7.37, gives rather accurate results. Figure 7.43(bottom) shows that the always open crack brings the system closer to instability.

Summarizing the results about breathing behaviour it seems that breathing prevents instability and non breathing open cracks are most likely subject to instability if damping is suitably small.

The effect of the position of the unbalance on the behaviour has been also investigated with reference damping: Figure 7.44 shows the results as 1X and 2X components when the unbalance is rotated by +90°, and figure 7.45 when unbalance is rotated by +180°. In this last situation vibrations due to unbalance are mainly in opposition with respect to crack induced vibrations, so that the dynamic loads are smaller than the static loads (due to weight): the breathing is governed again by weight.

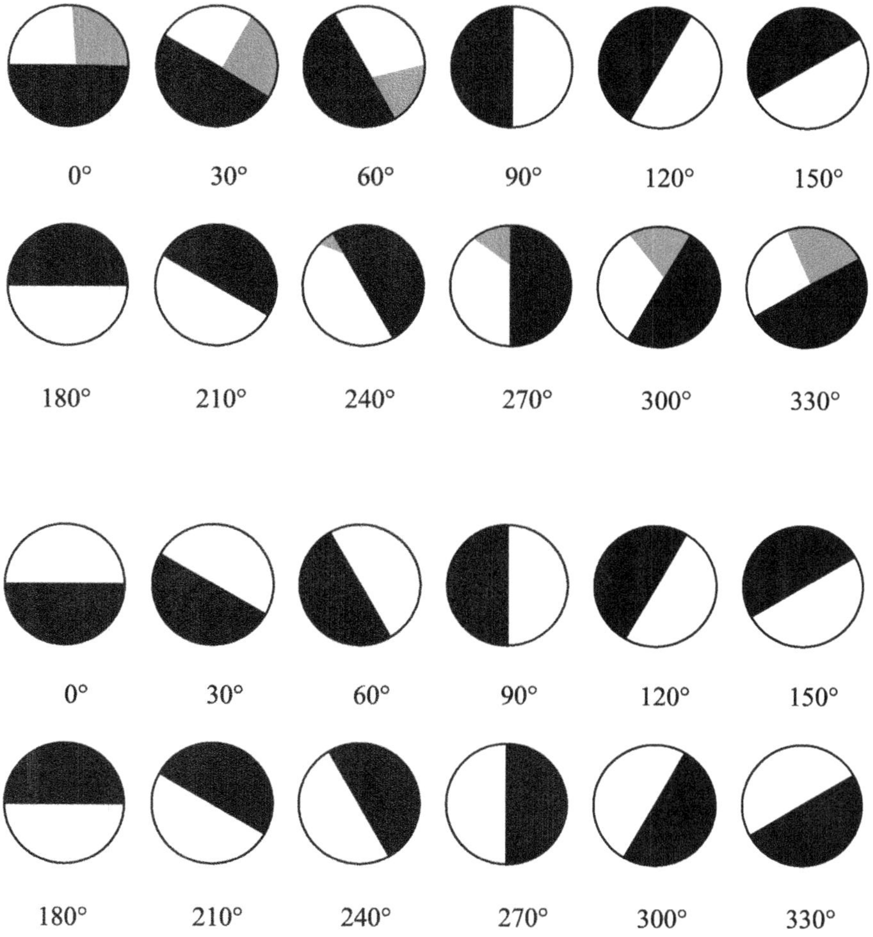

Fig. 7.43. Breathing behaviour in resonance at 1650 rpm: reference damping on the top, reduced damping LD1 on the bottom.

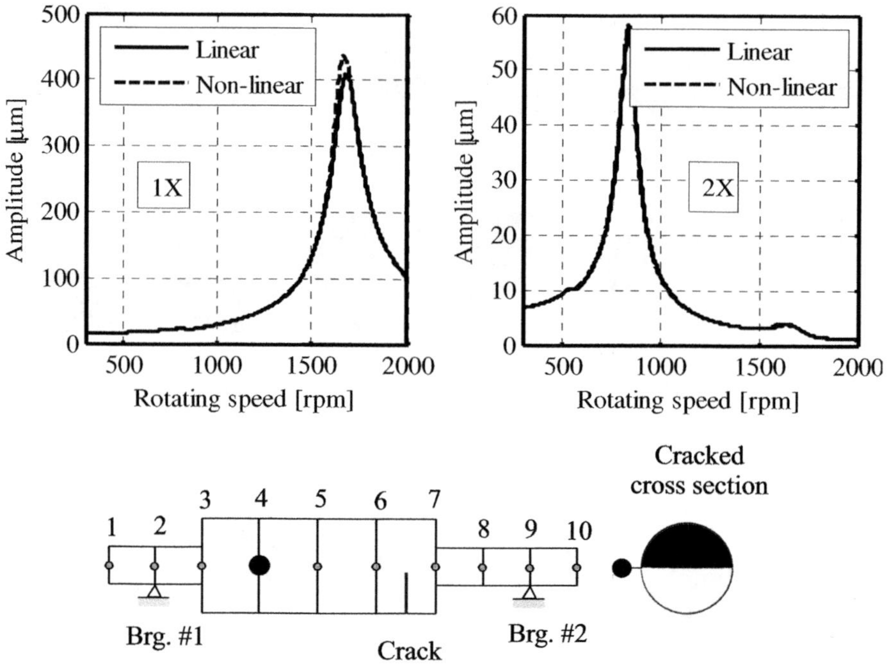

Fig. 7.44. 1X and 2X response with unbalance rotated by +90° with respect crack axis.

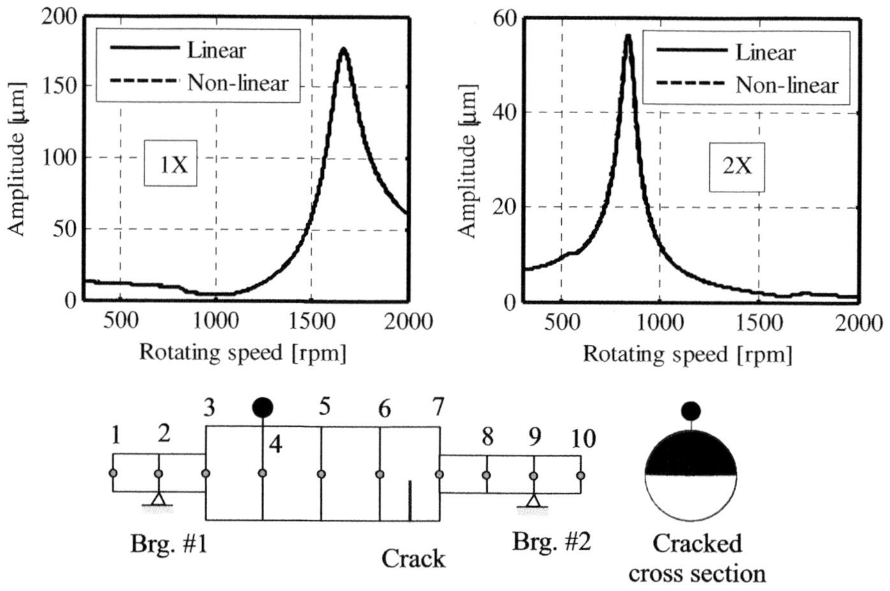

Fig. 7.45. 1X and 2X response with unbalance rotated by +180° with respect crack.

The linear approach is again very accurate, despite the fact that the breathing is governed by the vibration and not by the static deflection (except situation shown in figure 7.45). The 2X component seems completely unaffected by the non-linearity.

These results confirm the validity of the linear approach in the analysis of cracked shafts of real machines (such as turbines and generators of power plants).

7.3.5 Sub-harmonics

In order to check if also sub-harmonic components, which have not been recognized in all time histories from which 1X amplitudes as a function of rotating speed of figure 7.44 and figure 7.45 have been calculated, can be excited by the crack in this type of machine, damping has been reduced further on and speed range of investigation has been enlarged. Figure 7.46 shows the results in terms of FFT of time histories for LD3 damping at a rotating speed of 3300 rpm (twice 1st critical speed). This is a rotating speed where an instability region could appear, as shown in figure 5.64 for the simple Jeffcott / de Laval rotor.

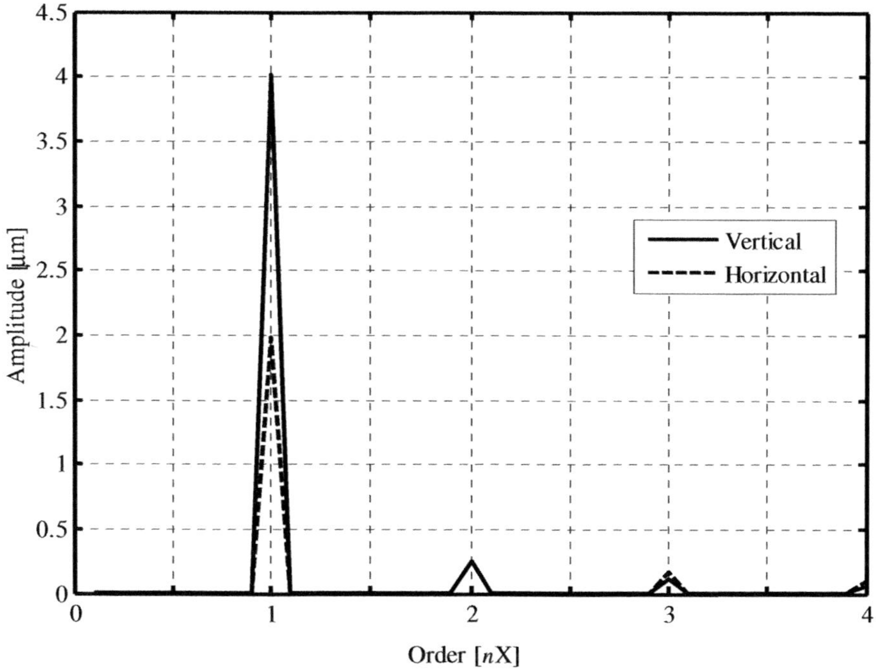

Fig. 7.46. FFT spectra of vertical and horizontal vibrations at 3300 rpm.

The damping has also been reduced further on and also the weight has been reduced to 1/100 of its value in order to avoid masking of possible sub-harmonic components by the forced motion components, but again similar results have been found, with all amplitudes scaled 1/100. All peaks in the spectra are due to the forced vibration components at 1X, 2X and 3X due to crack and weight.

The main results of this numerical investigation are the following:

a) with realistic damping values the linear approach is very accurate: instabilities, sub-harmonics do not appear and effect of unbalance is correctly predicted, even in speed ranges where the breathing behaviour is governed by the vibration;
b) a consistent reduction of damping allows the instability to be excited;
c) unstable vibrations can be stabilized applying suitable unbalances that influence the breathing behaviour;
d) unstable vibrations are most likely to appear when breathing is prevented and crack is always open;
e) sub-harmonic components do not appear even if damping is reduced further on.

Above results might explain why instabilities and sub-harmonics have never been found in industrial machines with cracked rotors to authors' knowledge. Instability and sub-harmonic excitation increase with stiffness asymmetry, but in industrial machines the modal stiffness variation is very small, the breathing mechanism and the damping prevents unstable behaviours, and sub-harmonics do not appear even if damping is consistently reduced.

All the listed results are related to the speed range of interest of the machine (0-3000 rpm), in which only the 1st critical speed is included.

Behaviours close to higher critical speeds have still to be investigated.

References

[7.1] Papadopoulos CA, Dimarogonas AD (1987) Coupled bending and torsional vibration of a cracked timoshenko shaft. Ing. Arch, 57, 257-266.
[7.2] Papadopoulos CA, Dimarogonas AD (1987) Coupled longitudinal and bending vibrations of a rotating shaft with an open crack. Journal of Sound and Vibration, 117, 81-93.
[7.3] Ostachowiz WM, Krawczuk M (1992) Coupled torsional and bending vibrations of a rotor with an open crack. Archive of Applied Mechanics, 62, 191-201.
[7.4] Darpe AK, Gupta K, Chawla (2004) A coupled bending longitudinal and torsional vibrations of a cracked rotor. Journal of Sound and Vibration, 269, 33-66.
[7.5] Theis W (1990) Laengs- und Torsions-schwingungen bei quer angerissenen Rotoren. VDI Fortschrittberichte Reihe 11 , Nr. 131.
[7.6] Stoisser CM, Hasnaoui F, Henry H, Bachschmid N, Tanzi E (2006) The resonance frequency shift in cracked rotors: numerical and experimental results. Paper ID 266, Proceedings of 7th IFToMM Conference on Rotor Dynamics, Vienna Austria, September 25-28.
[7.7] Ichimonji M, Watanabe S (1988) The dynamics of a rotor system with a shaft having a slant crack. JSME International Journal, Series 3: Vibration, Control Engineering, Engineering for Industry, 31(4), 712-718.

[7.8] Ichimonji M, Kazao Y, Watanabe S, Nonaka S (1994) The dynamics of a rotor system with a slant crack under torsional vibrations. Proc. of the 1994 International Mechanical Engineering Congress and Exposition, Chicago (IL), Nov 6-11, Vol. 192, 81-90.

[7.9] Sekhar AS, Prasad PB (1997) Dynamic analysis of a rotor system considering a slant crack in the shaft. Journal of Sound and Vibration, 208(3), 457-474.

[7.10] Darpe AK (2007) Dynamics of a Jeffcott rotor with slant crack. Journal of Sound and Vibration, 303(1-2), 1-28.

[7.11] Bachschmid N, Diana G, Pizzigoni B (1984) The influence of unbalance on cracked rotors. Vibrations in Rotating Machinery Conference, IMechE paper C304/84, 1984, 193-198.

[7.12] Ishida Y, Yamamoto T, Hitokawa K (1994) Vibrations of a rotating shaft containing a transverse crack (major critical speed of a horizontal shaft). IV Int. IFToMM Conference on Rotor Dynamics, Chicago, Sept. 7-9, 47-52.

[7.13] Gasch R, Meng G. (2000) Stability and stability degree of a cracked flexible rotor supported on journal bearings. ASME Journal of Vibration and Acoustics, 122, 116-125.

[7.14] Brandon JA (2000) Nonlinear vibrations of cracked structures: perspectives and horizons. Shock and Vibration Digest, 32(4), 273-280.

[7.15] Leng X, Meng G, Zhang T (2007) Bifurcation and chaos response of a cracked rotor with random disturbance. Journal of Sound and Vibration, 299(3), 621-632.

[7.16] Gao J-M, Zhu X-M (1992) The resonance of a vertical rotor containing a transverse crack. Acta Mecanica Solida Sinica, 5(2), 167-174.

[7.17] Wu X, Friswell MI, Sawicki JT, Baaklini JY (2005) Finite element analysis of coupled lateral and torsional vibrations of a rotor with multiple cracks. Proceedings of the ASME Turbo Expo 4, vol. 4, 841-850.

Chapter 8
Crack Diagnosis in Rotating Shafts

8.1 Diagnosis

Diagnosis is a procedure that allows identifying the impending fault or defect from symptom analysis. In the rotordynamics field, symptoms are generally vibrations, which are measured in correspondence of the bearings in industrial machines.

A diagnosis can be performed by skilled human operators, as it was common in the past, or automatically by means of suitable software. Different approaches can be used for automatic diagnostic procedures. In the following, two approaches are presented, along with some applications: i) a qualitative one, which does not allow defining severity and position of the identified fault; ii) a model based approach, which on the contrary identifies also severity and position of the fault.

8.2 Qualitative Approach

This section presents one of the possible qualitative approaches to the crack diagnose in a rotating shaft. Since the method is aimed at the identification of the most likely fault among others, which typically affects rotating machinery, it is a nonsense to explain it limiting only to cracks.

The approach consists in a *knowledge based* system, based on a *fault matrix*, in the form defined in [8.1]. This fault matrix is an underlying rule based system that includes also the conversion of measured data to *values of certainty*, which indicate whether the physical values of the symptoms are high or low. Certainties are then combined by using one of several algorithms available for handling *fuzzy logic*. The cumulative certainty, combined with a user selected value of prior probability to indicate the likelihood of a particular fault occurring on a given machine, is used to make a ranked list of fault probabilities. The scheme of the approach is shown in figure 8.1.

The idea of a fault matrix is practically the ancestor of every diagnostic method in rotating machinery starting from the classification given by Sohre [8.2] about vibration characteristics and the related faults. Some other fault matrices have been developed in literature (see for instance [8.3][8.4]) and can be used as a starting point for a knowledge base. The knowledge base cannot be developed generally for all types of rotating machinery, some specific details of each type of machine must be taken into consideration, for example the type and size of bearings. The development of a knowledge based system for diagnosis using vibration measurements, depends strongly on the cooperation of an experienced diagnostician and has to be tailored to some particular machines in order to be successful. Also in the case presented here, even if it was tried to keep it more general as possible, some characteristics are necessarily machine specific.

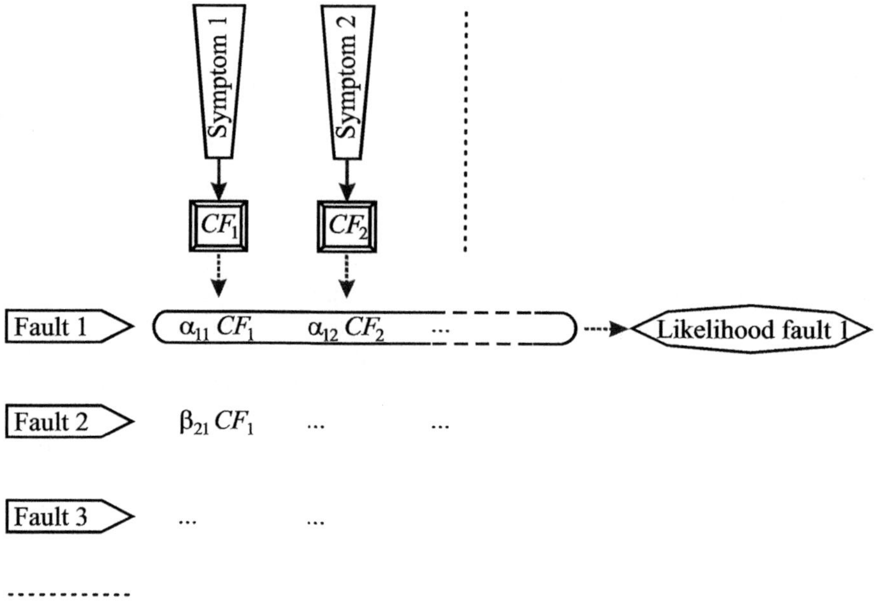

Fig. 8.1. Fault matrix approach.

8.2.1 Definition of the Fault Matrix

A fault matrix is a very compressed form of knowledge representation. It consists of a 2D array of *symptoms* (columns) and *faults* (rows). In diagnostics based on vibrations, inference chains are usually kept very short, connecting a symptom directly to a fault. A very simple example is: "if 1X vibration is very high, then the machine is probably unbalanced". In case of several symptoms, a short inference chain is a disadvantage for rule based systems, whilst the use of a matrix representation is advantageous. High level of 1X vibrations may be caused by other faults as shown hereafter in section 8.1.5.

In its simplest form, the fault matrix forms a conjunction of all symptoms to a fault and concludes with either *true* or *false*. All characteristic symptoms of a fault must be presented in order to be able to detect a fault. In case of diagnostics of machinery this is a too restrictive precondition, since often some of the symptoms could not be evident or measurable at all.

In particular, it is desirable that the fault matrix has the following properties:

- Symptoms may take any truth value, ranging from *false* over *unknown* to *true*.
- The presence of a symptom may strengthen the belief in a hypothesis.
- The absence of a symptom may strengthen the belief in a hypothesis.

- The presence of a symptom may weaken the belief in a hypothesis.
- The absence of a symptom may weaken the belief in a hypothesis.
- The use of absolute values should be avoided. The relative change of a value should be used instead, in order to facilitate tailoring of the system to a particular machine.
- It should be possible to assign prior beliefs to each possible fault, reflecting the experienced distribution of faults for a particular machine.
- Under development of a new fault matrix, it should be possible to apply different treatments of uncertainty, using the same knowledge representation in order to find the one, which is best suited for the problem.

To accomplish the above constraints, the notion of fault matrix has to be extended to be machine specific and handle uncertain symptoms and inexact reasoning.

In order to express belief in the absence or presence of symptoms and hypotheses on a scale ranging from *false* over *unknown* to *true* it is necessary to introduce the certainty factors CF. The certainty factors may take values in the range $-1.00 \leq CF \leq +1.00$. Their use is to map the qualitative natural language expression or attribute into a quantitative value as shown in table 8.1.

Table 8.1. Correspondence between certainty factors and symptom attributes.

Natural language expression (attribute)	*Certainty factor* (CF)
Definitely false	$-1.00 \leq CF \leq -0.95$
Almost false	$-0.95 \leq CF \leq -0.50$
Slightly false	$-0.50 \leq CF \leq -0.10$
Unknown	$-0.10 \leq CF \leq +0.10$
Slightly true	$+0.10 \leq CF \leq +0.50$
Almost true	$+0.50 \leq CF \leq +0.95$
Definitely true	$+0.95 \leq CF \leq +1.00$

Each symptom in a fault matrix is related to one or more faults. A symptom may have both a strengthening and a weakening effect on the diagnosis, independently of being false or true. This is taken into account by introducing two weights α_{ij} and β_{ij} that are assigned to each fault/symptom pair, where i is the fault index, j is the symptom index, independently from the CF_j value. Weight α_{ij} is used when symptom j-th has some degree of truth ($0.00 \leq CF_j \leq +1.00$), whilst β_{ij} is used when symptom j-th has some degree of falsity ($-1.00 \leq CF_j \leq 0.00$).

Table 8.2. Example of a fault matrix.

	Symptom No1	*Symptom No2*	*Symptom No3*	...
Fault no 1	20 / -15	0 / 0	-45 / 30	...
Fault no 2	35 / -30	0 / 0	20 / -15	...
Fault no 3	0 / 0	80 / -45	0 / 0	...
Fault no 4	30 / 0	0 / 0	30 / -30	...
...	...	...	...	α_{ij} / β_{ij}

Table 8.2 shows an example of building up of a fault matrix. The weights can take values in the range $[-100,+100]$: a positive weight strengthens the belief in a fault and negative weight weakens it. A zero weight denotes that the absence and/or the presence of the symptom in question is irrelevant to the fault. The weights are always defined in pairs. Two simple examples can explain how the weight pair is attributed to a fault.

Example 1. In order to express that the presence of high value of the 1X vibration amplitude is an indication for unbalance and its absence has on the contrary a weakening effect on the diagnose, the weights introduced into the fault matrix can be:

...	High value 1X vibration amplitude	...
Unbalance	$80/-35$	⋮
...	...	...

(8.1)

This means that in case of increase of the 1X vibration amplitude, a strengthening weight of 80 is given, otherwise in case of its absence a weakening weight of -35 is assigned.

Example 2. Contrarily to the previous example, high value of the 0.5X vibration amplitude should weaken the likelihood of unbalance whilst its absence should not have any influence on it. The corresponding weight pair can be:

...	High value 0.5X vibration amplitude	...
Unbalance	$-30/0$	⋮
...	...	...

(8.2)

However, the assignment of negative values to the first weight α_{ij} must be given carefully when the symptom has some degree of truth and negative values in the range -5 to -30 are recommended in general. The use of large negative values may produce misleading diagnosis in case multiple faults are present. Similarly, in general, negative value of the second weight β_{ij} should be avoided.

For each machine, the prior likelihood of the known faults may be defined. The likelihood is expressed using certainty factors in the range $[-1,+1]$.

In order to make the fault matrix applicable to a wide family of machines it is possible first to relate measured values to a *baseline value* and then to apply a membership function to the relative change of the variable. This is done in order to be able to use the fault matrix for different but related machine. The algorithms used are discussed in the following sections.

8.2.2 Inference

Two points have to be addressed. First, symptoms are evaluated often on the basis of numerical data coming from condition monitoring systems, but they must be translated into statements related to certainty factors. Second, the weighted combination of the certainty factors and the weights give the partial probability of a fault, given a symptom. When the fault condition influences more than one symptom, the partial probabilities for each individual symptom have to be combined to give the cumulative probability that a particular fault is present. Several methods can be used, the Bayesian method is used in this example.

In the qualitative approach, measured values cannot be used directly for diagnostic purposes. Indeed, rather than the measured value itself, the focus is on the relation between the measured value and other values or former values. There are generally four conclusions, which can be drawn from a measurement:

1) the value is high;
2) the value is low;
3) the value is normal;
4) the value is changed.

On the basis of these considerations, several methods can be used to define fault attributes.

The first is a Boolean answer, i.e. a conclusion *Yes / No* (true/false) is drawn depending on the measured value. This corresponds for instance to the following example:

B Is the 3X phase locked with the 1X? *"Ph_1_3x"* (8.3)

in which the field B indicates the type of the attribute (Boolean in this case), then the symptom is described, while the statement between the quotation marks is the symbolic label of the symptom.

The second way is to define directly the certainty factor using the correspondence with the natural language attributes as done in table 8.1.

The last way is to obtain certainty factors directly from measured values, considering them as fuzzy sets and defining the membership functions to which they belong. Actually, in this section the membership functions differs from those that are normally used in fuzzy logic, because they vary between $[-1,+1]$ instead of $[0,+1]$. Moreover, due to the arbitrariness in the selection of the membership function in fuzzy logic, in this application piecewise functions are used and they are characterised by three parameters i.e. *low* l, *turnover* c and *high* h. The certainty factor CF is defined for each curve as a function of the position x in relation of these parameters. The corresponding equation to each membership function assigns a certainty factor in interval $[-1,+1]$ to the value x if $l<x<h$. Certainty factor of -1 or 1 are assigned outside that range, depending on the membership function type. For the purposes of the application presented, four types of membership functions (*Up slope, Down slope, Peak* and *Sink.*) are introduced and they are defined to suit from the point of view of simplicity, convenience, speed and efficiency the use of the three parameters l, c and h. Their expressions are:

$$CF_U = \begin{cases} -1 & \text{if } x \le l \\ -\sqrt{1-\left(\dfrac{x-l}{c-l}\right)^2} & \text{if } l < x \le c \\ \sqrt{1-\left(\dfrac{x-h}{h-c}\right)^2} & \text{if } c < x \le h \\ +1 & \text{if } x > h \end{cases} \tag{8.4}$$

$$CF_D = \begin{cases} 1 & \text{if } x \le l \\ \sqrt{1-\left(\dfrac{x-l}{c-l}\right)^2} & \text{if } l < x \le c \\ -\sqrt{1-\left(\dfrac{x-h}{h-c}\right)^2} & \text{if } c < x \le h \\ -1 & \text{if } x > h \end{cases} \tag{8.5}$$

$$CF_P = \begin{cases} -1 & \text{if } x \le l \\ -\sqrt{1-\left(2\dfrac{x-l}{c-l}\right)^2} & \text{if } l < x \le \dfrac{l+c}{2} \\ \sqrt{1-\left(2\dfrac{x-c}{c-l}\right)^2} & \text{if } \dfrac{l+c}{2} < x \le c \\ \sqrt{1-\left(2\dfrac{x-c}{h-c}\right)^2} & \text{if } c < x \le \dfrac{c+h}{2} \\ -\sqrt{1-\left(2\dfrac{x-h}{h-c}\right)^2} & \text{if } \dfrac{c+h}{2} < x \le h \\ -1 & \text{if } x > h \end{cases} \tag{8.6}$$

$$CF_S = \begin{cases} 1 & \text{if } x \le l \\ \sqrt{1-\left(2\dfrac{x-l}{c-l}\right)^2} & \text{if } l < x \le \dfrac{l+c}{2} \\ -\sqrt{1-\left(2\dfrac{x-c}{c-l}\right)^2} & \text{if } \dfrac{l+c}{2} < x \le c \\ -\sqrt{1-\left(2\dfrac{x-c}{h-c}\right)^2} & \text{if } c < x \le \dfrac{c+h}{2} \\ \sqrt{1-\left(2\dfrac{x-h}{h-c}\right)^2} & \text{if } \dfrac{c+h}{2} < x \le h \\ 1 & \text{if } x > h \end{cases} \tag{8.7}$$

Figure 8.2, figure 8.3, figure 8.4 and figure 8.5 give a graphical representation of the membership functions.

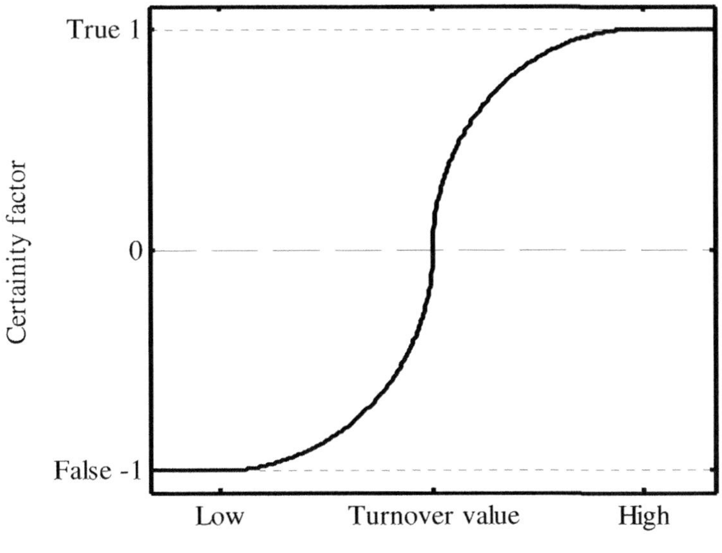

Fig. 8.2. Up slope curve (membership function).

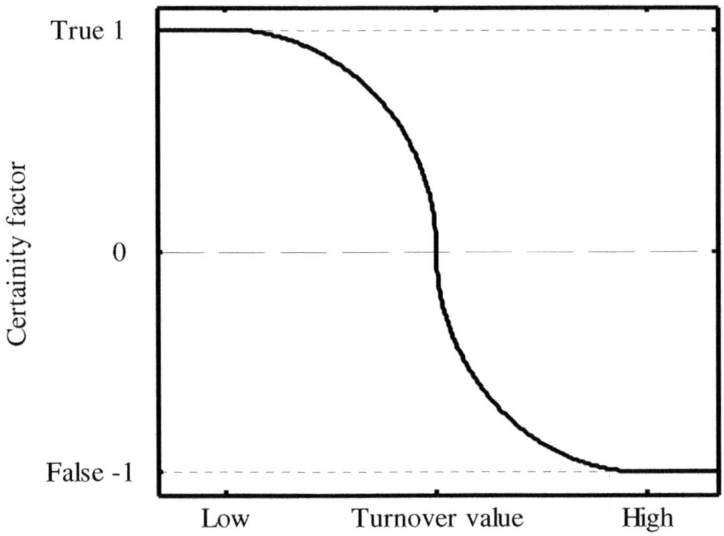

Fig. 8.3. Down slope curve (membership function).

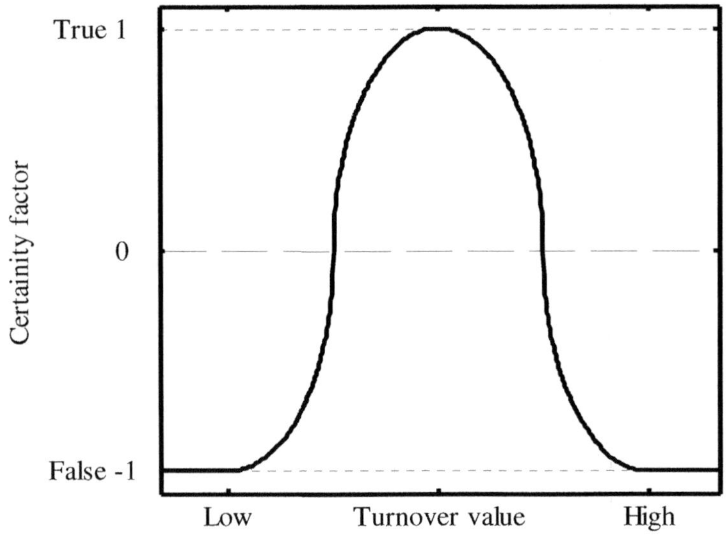

Fig. 8.4. Peak curve (membership function).

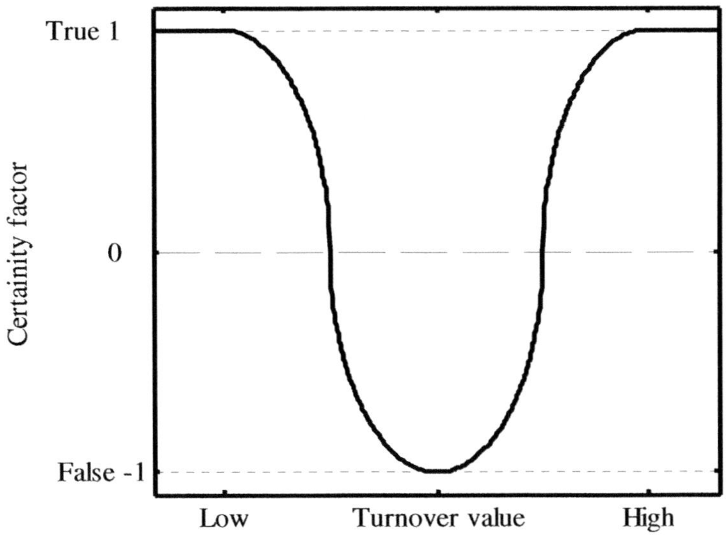

Fig. 8.5. Sink curve (membership function).

8.2.3 Knowledge Representation

Once the basics of the fault matrix is drawn, it is necessary to consider its implementation. This is more simple to be done by means of an example, rather than in the abstract. Obviously the complex relationships, which are behind the scenes of fault diagnosis using the fault matrix, exploit "naturally" computers, so the example presentation is strictly connected to computer based implementation. A possible implementation is by means of a tabular form, that can be directly transferred in a ASCII file for the input in a software, similarly to that is reported in table 8.3 (lines starting with a # are only comments and are ignored).

8.2.4 Symptom Generation

The definitions of the symptoms are the first statements of table 8.3, each one on a separated line, with the syntax explained in this section. The symptom generation is a very important task in building expert systems used for diagnostics. In the case of vibration analysis the amount of raw data is so overwhelming that data reduction is required. Note that this does not occur only in qualitative methods: data analysis in the frequency domain instead of in the time domain is a way to reduce the amount of information to be analyzed. The main aim of symptom generation is to find a transformation, which maps all observations and measurements into a smaller set of variables. This set will, hopefully, contain all the relevant and discriminatory information needed to solve the overall diagnostic problem. Since these methods are used also in other fields rather than rotor dynamics, the solution is also called *solution to the pattern recognition problem.*

Table 8.3. Example of a tabular form for a fault matrix.

```
# Symptoms
#
D0.50 1.00 1.50  Major amplitude of radial vibration A_1_1 high "A_1_1"
D0.50 1.00 1.50  Major amplitude of axial vibration A_1_3 high "A_1_3"
#
U0.10 0.40 1.40  Frequency Components  at (0.10X - 0.45X]   "A_1_sd"
U0.10 0.40 1.40  Frequency Components  at (0.45X - 0.55X]   "A_1_sc"
U0.10 0.40 1.40  Frequency Components  at (0.55X - 0.98X)   "A_1_su"
B Is subsynchron component unaffected by load drop   "SuC_1_ld"
#
U0.30 1.00 1.60  Frequency Component at (1X) high  "A_1_1x"
U0.30 1.00 1.60  Frequency Component at (2X) high  "A_1_2x"
U0.30 1.00 1.60  Frequency Component at (3X) high  "A_1_3x"
U0.30 1.00 1.60  Frequency Component at (4X) high  "A_1_4x"
U0.30 1.00 1.60  Frequency Component at (5X) high  "A_1_5x"
U0.30 1.00 1.60  Frequency Component at (6X) high  "A_1_6x"
U0.30 1.00 1.60  Frequency Component at (7X) high  "A_1_7x"
U0.30 1.00 1.60  Frequency Component at (8X) high  "A_1_8x"
#
. . .

# Faults
#
Normal operating condition
Impending fault
Axial rotor asymmetry
Rigid coupling misalignment
Flexible coupling misalignment
. . .

#
# Fault Matrix
#
100/0   0/0    0/0    0/0    0/0    0/0    0/0    ...
0/100   0/0    0/0    0/0    0/0    0/0    0/0    ...
0/0     0/0    0/0    0/0    0/0    0/0    0/0    ...
0/0     0/0    0/0    0/0    0/0    0/0    0/0    ...
0/0     20/0   0/0    0/0    0/0    0/0    0/0    ...
...     ...    ...    ...    ...    ...    ...    ...
```

The point is to find as many symptoms as necessary to distinguish between faults and preferably no more than these. However, the more faults one wants to diagnose the more symptoms one needs, while several symptoms are common to several faults. If the distinguishing symptoms are overestimated in their importance, then a fault will possibly be diagnosed even when only these few symptoms are known. If a relatively low weight is assigned to them, it will be impossible to distinguish between similar patterns.

For instance, the situation in which three symptoms are used to distinguish between three faults is considered. Each of the faults is characterised by the presence of two symptoms (see figure 8.6). The presence of the third should have no influence on the fault.

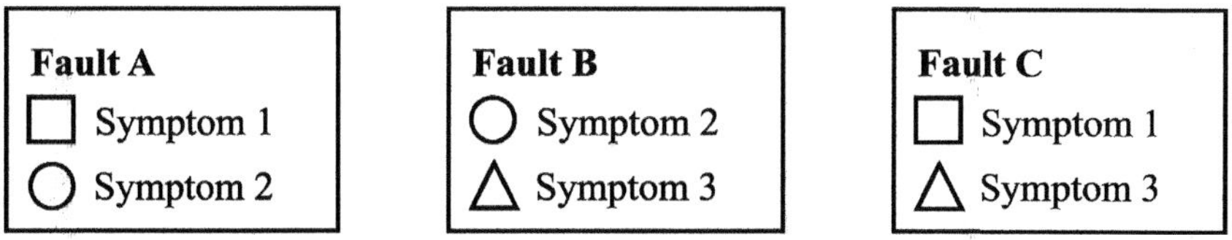

Fig. 8.6. Three faults, each characterized by two symptoms.

If two symptoms are true and one false, the diagnosis is clear and simple. If all the three symptoms are true, then there would be either a fault, which manifests itself with all the three symptoms, or two faults or even three of them. The addition of more unique symptoms – if there are any more – will help to some degree.

Claiming that the empty places in the fault matrix mean negative evidence against a fault, the presence of one of the additional symptoms would strengthen the hypothesis that none of the faults is present. This simple example suggests that:

- the number of symptoms must be sufficient to characterise each fault;
- the number of symptoms common to several faults should be minimal;
- there must be a sufficient number of symptoms, which will give negative evidence when present.

The term *sufficient* is rather difficult to be quantified in the rotating machinery field and depends on experience to a great degree. The decision about which symptoms to use is made by applying the "method" of trial and error considering the restrictions imposed by the specific data acquisition system, which could not allow some symptoms to be evaluated. Most of the symptoms can be obtained automatically, some additional information is subjective or could be difficult to acquire, such as orbit shape, etc. The selected symptoms may be logically grouped in several blocks as shown in table 8.4.

Table 8.4 refers to general condition monitoring systems of rotating machinery in which the raw data are collected by 90° apart proximity probes that are situated in each measuring plane. In addition, there is usually a proximity probe for the axial displacement and a key-phasor to obtain phase related information and to synchronous average of data.

In each line of table 8.3 where a symptom is defined, the first character determines the type of the attribute, accordingly to table 8.5. Either Boolean attributes (true, false), likelihood (see table 8.1) or numerical value may be used. If the numerical value has to be input, then the corresponding certainty factor is calculated using a membership function. The first letter determines the membership function type, while the subsequent numerical values are the *low*, *turnover* and *high* parameters. Then the symptom is described along with its symbolic name. The complete list of the 64 symptoms used in this example is reported hereafter. It is not superfluous to remark that other machines may have different number and types of faults and that not all the listed symptoms are strictly related to crack diagnosis.

Table 8.4. Grouping of the symptoms.

Group	*Symptom*	*Symbolic label*
Frequency components	Frequency components at (0.45X-0.55X)	*"A_1_sc"*
	Frequency component at (1X) high	*"A_1_1x"*
	Frequency component at (2X) high	*"A_1_2x"*
	Frequency component at (3X) high	*"A_1_3x"*
	Frequency component at (4X) high	*"A_1_4x"*
	...	...
Phase	Is the 2X phase locked with the 1X?	*"Ph_1_2x"*
	Is the 3X phase locked with the 1X?	*"Ph_1_3x"*
	...	...
	The 1X vector changed in amplitude/phase	*"RAF_1_1x"*
	...	...
Orbit	Orbit is elliptical	*"Or_1_el"*
	Orbit is flat	*"Or_1_ft"*
	Orbit is 8-shaped or banana shaped	*"Or_1_bf"*
	...	...
Speed related	Vibration increases with increasing speed	*"V_1_rtos"*
	...	...
Load related	Is vibration related to load?	*"V_1_rtol"*
	...	...
General information	Major amplitude of radial vibration A_1_1 high	*"A_1_1"*
	Major amplitude of axial vibration A_1_3 high	*"A_1_3"*
	...	...
	First critical speed	*"CS_1_1"*
	Second critical speed	*"CS_1_2"*
	...	...
	Vibration pattern changes after start/stop	*"ChV_1_ss"*
	Vibration change with time	*"ChV_1_t"*
	...	...

Table 8.5. Symptom types identifiers.

Identifier	*Selected method*
B	Boolean
L	Likelihood
U	Up slope
D	Down slope
P	Peak
S	Sink

8.2.4.1 Description of Symptoms Used in Knowledge Base

1) *D* 0.25 1.00 1.50 Major amplitude of radial vibration high *"A_1_r"*

For relative vibration of shaft with respect to bearings this symptom is defined as the higher value of the peak displacement measured in two selected perpendicular measurement directions.

For the absolute vibration of bearing pedestal or casing, it is defined as the maximum RMS vibration velocity.

2) *B* Change of major amplitude of radial vibration *"A_1_rch"*

This symptom contains information about increase or decrease of vibration level over a period of time, defined by expression (8.8). The change can be instantaneous or progressive with time, and the symptom will be triggered if the change is greater than 25% of baseline (see section 8.1.7 for the baseline definition).

$$A\_1\_rch = 1 \text{ if} \left|\frac{(A\_1\_r)_n - (A\_1\_r)_{n-1}}{(A\_1\_r)_{n-1}}\right| \cdot 100[\%] > 0.25 \cdot \mathit{Baseline} \tag{8.8}$$

where $n-1$ is the time when the reference measurement was made and n is some time later.

Note that these first two symptoms refer to broad-band vibrations and are only used for determining the general state of the machine.

3) *U* 0.25 1.00 1.75 Major amplitude of axial vibration high *"A_1_a"*

Major amplitude of axial vibration is the amplitude of total vibration level (0-pk) measured in a direction parallel to the shaft axis.

4) *U* 0.25 1.00 4.50 Frequency Components at (0.10X, 0.45X] *"A_1_sd"*
5) *U* 0.25 1.00 4.50 Frequency Components at (0.45X, 0.55X] *"A_1_sc"*
6) *U* 0.25 1.00 4.50 Frequency Components at (0.55X, 0.95X) *"A_1_su"*

These symptoms define whether sub-synchronous components in a frequency spectrum are high compared with a specified baseline level (square brackets indicate the closed interval).

7) *B* Is sub-synchronous component affected by load change *"A_1_lch"*

This symptom contains information about the relation between sub-synchronous harmonics and load drop or load change.

8) *U* 0.25 1.00 3.00 Frequency Component at (1X) high *"A_1_1x"*

9) *U* 0.25 1.00 3.00 Frequency Component at (2X) high *"A_1_2x"*
10) *U* 0.25 1.00 3.00 Frequency Component at (3X) high *"A_1_3x"*
11) *U* 0.25 1.00 3.00 Frequency Component at (4X) high *"A_1_4x"*

These symptoms define whether amplitudes for the four first harmonics in the frequency spectrum are high compared with specified baseline levels.

12) *B* The 1X is dominant *"A_1_1xd"*
13) *B* The 2X is dominant *"A_1_2xd"*
14) *B* Dominant frequency 0.5X not locked *"A_1_05nld"*
15) *B* Dominant frequency 0.5X locked on *"A_1_05lod"*
16) *B* Dominant frequency is system critical speed *"A_1_scsd"*

These symptoms define the largest peak for some vibration components. "Locked" means that the phase is locked to the shaft speed. True values are defined as follows:

$$\begin{aligned}&A\_1\_xd = 1\\&\text{if } \mathit{amplitude\ of}\ 1\text{X} > 1.41 \cdot \mathit{amplitude\ of\ all\ other\ frequency\ componer}\end{aligned} \tag{8.9}$$

$$\begin{aligned}&A\_2\_xd = 1\\&\text{if } \mathit{amplitude\ of}\ 2\text{X} > \mathit{amplitude\ of\ all\ other\ frequency\ components}\end{aligned} \tag{8.10}$$

$$\begin{aligned}&A\_1\_05nld = 1\\&\text{if } (\mathit{phase\ is\ not\ locked\ with}\ 1\text{X}) \text{ and}\\&(\mathit{amplitude\ of\ sub\text{-}harmonic}\ 0.5\text{X} > 0.5 \cdot \mathit{amplitude\ of\ all\ other}\\&\mathit{frequency\ components})\end{aligned} \tag{8.11}$$

$$\begin{aligned}&A\_1\_05lod = 1\\&\text{if } (\mathit{sub\text{-}harmonic\ is\ at\ exactly}\ 0.5\text{X}) \text{ and } (\mathit{phase\ is\ locked\ with}\ 1\text{X})\ \text{a}\\&(\mathit{amplitude\ of\ sub\text{-}harmonic}\ 0.5\text{X} > \mathit{amplitude\ of\ all\ other}\\&\mathit{frequency\ components})\end{aligned} \tag{8.12}$$

$$\begin{aligned}&A\_1\_scsd = 1\\&\text{if } (\mathit{amplitude\ at\ frequency\ of\ system\ critical\ speed} >\\&\mathit{amplitude\ of\ all\ other\ frequency\ components})\end{aligned} \tag{8.13}$$

17) *B* Did 1X change over a short period of time *"A_1_1xtcs"*
18) *B* Did 2X change over a short period of time *"A_1_2xtcs"*
19) *B* Did 1X change over a medium period of time *"A_1_1xtcm"*
20) *B* Did 2X change over a medium period of time *"A_1_2xtcm"*
21) *B* Did 1X change over a long period of time *"A_1_1xtcl"*
22) *B* Did 2X change over a long period of time *"A_1_2xtcl"*

These symptoms contain information about increase in level or a change in the vector of the two first harmonics over a period of time. The two measurements should be made with similar operating conditions. The time period is defined as:

- a short period of time is less than 1 minute or a stepwise change,
- a medium period of time is between a few minutes and a few hours,
- a long period of time is more than 12 hours.

The change of vibration amplitude should be at least 25% (25% is given as a guideline, but other values may be used based on experience with a specific machine).

23) *B* Did vibration level change in one bearing only *"A_1_bVc"*
The information about the fact that the vibration level change is predominantly in one bearing only or not is helpful information for diagnosis of *loosening bearing bushes.*
24) *B* Trend in 1X is accelerating *"A_1_1xT"*
25) *B* Trend in 2X is accelerating *"A_1_2xT"*
26) *B* Trend in 3X is increasing or accelerating *"A_1_3xT"*
27) *B* Trend in gear passing Frequency (1tX) *"A_1_txT"*
If the level of the three first harmonics, or their vectors, show an increasing rate of change during several measurements over a period of time, these symptoms contain information about kind of change for these harmonics. The total change should be at least 25%.
28) *B* Run-down test. Change of 1X on rundown over time *"A_1_1xRdc"*
29) *B* Run-down test. Change of 2X on rundown over time *"A_1_2xRdc"*
30) *B* Run-down test. Change of 3X on rundown over time *"A_1_3xRdc"*
These symptoms contain information about level change of the first three harmonics during run-down from hot conditions with respect to a reference or to the last comparable run-down.
31) *U* 0.30 1.00 3.00 Half Harmonic Frequencies (1.5, 2.5, 3.5X) *"A_1_n05x"*
32) *U* 0.30 1.00 3.00 Blade passing Frequency (1bX or 2bX) *"A_1_bx"*
33) *U* 0.30 1.00 3.00 Gear tooth passing Frequency (1tX or 2tX) *"A_1_tx"*
These symptoms contain information about other harmonics in the frequency spectrum, which are related to number of blades 'b', or number of gear teeth 't', etc.
34) *U* 0.30 1.00 3.00 Background level increases (high frequencies) *"A_1_b"*
This symptom contains information about frequency ranges which are not specified above, for example due to changes in non-synchronous resonant response.
35) *U* 0.25 1.00 3.50 Vibration level at line frequency (50 Hz) *"A_1_ef"*
36) *U* 0.25 1.00 3.50 Vibration level at slip frequency *"A_1_sf"*
37) *U* 0.25 1.00 3.50 Vibration level at harmonic of line frequency ($n \times 50$ Hz) *"A_1_nef"*
38) *B* Is line frequency equal rotating speed? *"A_1_LfRs"*

The information about vibration components at line frequency and slip frequency are indispensable for diagnosis of electrical faults (line frequency is 50 Hz or 60 Hz). To distinguish the electrical fault and other faults, which contain first harmonic one needs to have information about relation between rotating speed and line frequency.

39) *U* 0.40 1.00 4.00 Relation (2X / 1X) [%] *"A_1_2xR1x"*

Information about the ratio between second order components and first order components, for evaluating severity of misalignment.

40) *B* Is the 2X phase locked with the 1X? *"A_1_2xL1x"*
41) *B* Is the 3X phase locked with the 1X? *"A_1_3xL1x"*
42) *B* Vibration at both ends of shaft in phase *"A_1_endsf"*

These symptoms contain values of phase angle of first harmonic and information about phase angle of second and third harmonics against phase angle of first harmonic. In addition there is information about ratio between phase angle at both ends of shaft.

43) *B* The 1X vector changed in amplitude/phase *"A_1_1xCaf"*
44) *B* The 2X vector changed in amplitude/phase *"A_1_2xCaf"*
45) *B* The 3X vector changed in amplitude/phase *"A_1_3xCaf"*

These symptoms consider if amplitudes and phase angles of the first three harmonics are steady or have changed. These symptoms look for a difference in the magnitude of the complex vector (amplitude and phase) between present and previous measurements. Such changes could be detected automatically by the condition monitoring system.

46) *U* 0.30 1.00 2.60 Component *A_1_1x* at slow roll *"A_1_1xsr"*
47) *U* 0.30 1.00 2.60 Component *A_1_1x* at slow roll after cool down *"A_1_1xsrc"*

These are levels of first harmonics at slow roll (approximately at 100 – 200 rpm) for two different conditions, at operating temperature and at low temperature after cool down.

48) *B* Relation (Axial / Radial) > 0.5 *"A_1_05R"*
49) *B* Relation (Axial / Radial) > 1 *"A_1_1R"*

These symptoms define the ratio between maximum shaft displacement in axial and radial directions.

50) *D* 0.875 1.00 1.125 Reduction of first critical speed *"A_1_CS1"*
51) *D* 0.875 1.00 1.125 Reduction of second critical speed *"A_1_CS2"*
52) *D* 0.875 1.00 1.125 Reduction of third critical speed *"A_1_CS3"*

These symptoms contain information about possible reduction of the three first critical speeds.

53) *U* 0.25 1.00 3.00 Frequency Components at first critical speed *"A_1_1csx"*
54) *U* 0.25 1.00 3.00 Frequency Components at second critical speed *"A_1_2csx"*
55) *U* 0.25 1.00 3.00 Frequency Components at third critical speed *"A_1_3csx"*
56) *B* Change in vibration components at first critical speed *"A_1_scC"*

These symptoms contain values of vibration level at critical speeds for the first three critical speeds. Symptom *"A_1_scC"* is true if the vibration level at first critical speed has increased at least of 10%.

57) *U* 0.75 1.00 1.25 Bearing temperature high *"A_1_Tb"*
This symptom defines operating temperature at the bearing, which will be compared with a reference temperature for normal operating condition.
58) *B* Vibration increases with increasing speed *"A_1_Vrts"*
This symptom is analyzed for variable speed machines, like pumps.
59) *B* Vibration related to load *"A_1_Vrtl"*
60) *B* Vibration pattern changes after start/stop *"A_1_CVss"*
These symptoms contain information about changes in total vibration level versus speed, or load, and after stop and re-start.
61) *B* Vibration steady after change *"A_1_VsaC"*
This symptom contains the information about form and trend of vibration after changing of total vibration level, is it same or not.
62) *B* Similar fractional change x and y *"A_1_SCxy"*
This symptom contains information about similarity of changing vibration level in *X* and *Y* directions (e.g. 50% increase in both directions).
63) *B* Disks shrunk on the shaft *"A_1_Dsh"*
This symptom is decided once and always for a given machine and increases the likelihood of detecting shaft bow.
64) *B* H_2 seal oil temperature changes *"A_1_oTC"*
For H_2 cooled generators, the relation with the temperature of the oil in the H_2 seals can be a symptom for rub on these seals.

8.2.5 Fault Description

The next block in table 8.3 consists of the list of possible faults, one per line. In addition to the machine state, the characteristics of eleven different faults are defined in the knowledge base presented in the table. The state of machine is defined as:

- Normal operating condition
- Impending fault

The list of faults defined in the knowledge base is:

- Axial rotor asymmetry
- Flexible coupling misalignment
- Rigid coupling misalignment
- Bearing misalignment
- Transverse crack
- Rotor to stator rub
- Loosening bearing bushes

- Bow and set
- Unbalance
- Electrical faults
- Instability

In the following, the state of the machine and the different faults are described briefly, along with some of the primary characteristics of each fault that are commonly used for diagnosis. For a more detailed description, refer to [8.5][8.6][8.7][8.8].

1) **Normal operating condition**
Of course, conditions in which a machine has very low level of vibration, indicate that no significant mechanical or operational problems are present, but sometimes considerable faults in the machine do not cause high vibration levels. For this reason it is recommended to make additional measurements, for example bearing oil temperature or noise control. However, if the level of vibration exceeds defined limits this is a reliable symptom that significant mechanical or operational problems are present in the machine.

2) **Impending fault**
When the level of vibration exceeds defined limits for *Normal operating condition* an impending fault is probable. Anyway if one is not only interested in the likelihood of a certain faults but also in its severity, qualitative methods are not obviously able to evaluate quantitatively the severity: for instance they are not able to detect the depth of a crack, while they are able to diagnose the presence of a crack.

3) **Axial rotor asymmetry**
Some rotating machinery, like two-poles generators are intrinsically axially asymmetric due to the geometry of the slots for the windings, see figure 3.6. These machines present 2X native vibrations. In other machines, a form of axial asymmetry can be the effect of a fault.

4) **Flexible coupling misalignment**
It is very difficult to align two shafts so that no forces exist that will cause vibration. For this reason vibration due to misalignment is almost as common as vibration due to unbalance. The flexible coupling misalignment results in two forces, axial and radial, which cause axial and radial vibration. These forces and therefore the level of generated vibration will increase with increased misalignment.

Misalignment occurs when two shafts are not aligned properly and the three basic types of misalignment are: *angular*, *radial* and a *combination of these two* [8.9].

Misalignment will also cause a predominant vibration at a frequency of 1X in many cases, but unlike unbalance, misalignment will often be accompanied by harmonically related frequencies, including 2X and 3X and occasionally higher orders of rotating speed. In severe cases, the second-order component can exceed the first-order component.

As a result of a significant misalignment, the orbit of the shaft centre may not be circular or slightly elliptical. The orbit may have a "banana form". As the higher order frequencies take on greater significance, the shaft orbit may become a figure of eight. Flexible coupling misalignment is described in detail in [8.10].

5) **Rigid coupling misalignment**
The rigid coupling misalignment can be due to manufacturing errors of the two surfaces of the flanges, assembling errors (different tightening forces in the connecting bolts) or corrosion. An asymmetrical strain distribution is caused by a radial or angular misalignment of a rigid coupling between two rotors of a shaft line. The strain distribution does not influence the system parameters and may be considered as created by an external system of forces, provided that the associated vibrations in the bearings are small enough to consider the system as linear. In this case, the predominant vibration is 1X.

6) **Bearing misalignment**
In addition to coupling misalignment, it is possible to have misalignment between one of the bearings and the shaft-line [8.11][8.12]. In this case of a misalignment, no vibration will result unless there is also unbalance. Misalignment will change load on bearing and accordingly also its oil-film parameters (stiffness and damping). Also alternating stresses on the shaft are imposed by bearing misalignment. With unbalance, a radial vibration will be present as well as an axial vibration. The vibration may occur at a frequency of 1X, 2X or perhaps 3X. The real cause of this vibration is unbalance and both the axial and radial vibration level will be reduced when the rotor is balanced.

7) **Transverse crack**
The dynamic phenomena associated with the appearance of shaft crack are related to a decrease of the shaft stiffness in correspondence of the crack and are described in detail in previous chapters.

The dynamic response of the shaft depends significantly on the stiffness characteristics of the shaft and supporting pedestal, as well as on the damping. The most evident response is for 1X and 2X vibrations.

8) **Rotor to stator rub**
The most common types of rubs in rotating machinery are blade tip and seals rubs where clearances are small [8.13][8.14][8.15]. Both of these can be caused by thermal expansion, rotor unbalance, fluid dynamic forces, misalignment, etc.

Depending on the nature of the contact between a rotor and its bearing or casing the rub generates very complex rotor vibration, even the so-called spiral vibrations [8.16][8.17]. Depending on the conditions of full annular or of partial arc rub, it produces unsteady sub-harmonics or orders of operating speed of the rotor.

Depending on its severity, rub can be intermittent and so is classified as one of the parametric excitations. The vibration characteristic of an intermittent rub is sub-critical.

9) Loosening bearing bushes

The mechanical looseness typically results in a vibration at a frequencies of 2X rotating speed, but may also result in higher order frequencies such as 3X, 4X, 5X or perhaps 6X. The second harmonic component is not in phase with the first harmonic and phase of second harmonic is variable. The amplitude of vibration resulting from mechanical looseness will often be unsteady. Comparing amplitude and phase of vibration on various points of the machine or structure, any significant difference in amplitude or phase clearly indicates relative motion and a source of mechanical looseness.

Of course, the vibration characteristic of mechanical looseness will not occur unless there is some other exciting force such as unbalance, misalignment, etc. [8.17] However, when there is excessive looseness, even relatively small levels of unbalance or misalignment can result in large amplitude of looseness vibration.

10) Bow and set

A bow in the rotor can occur during operation as a result of unequal thermal conditions [8.18] or rotor rubbing [8.14]. A rotor bow, or a bent shaft, usually causes a preload on the bearings. The mass centre of a bent shaft can be sufficiently removed from the geometric centre to cause mass unbalance, 1X frequency component and balancing can reduce the vibration level [8.19]. However, balancing cannot straighten the shaft. The rotor bow prevents adequate alignment in some cases and causes clearance problems in others, depending on where the bend occurs. That means that characteristic spectrum of a bow shaft contain the expected 1X frequency component, but sometimes 2X frequency component too. The second harmonic is in phase with the first harmonic.

At low speed, for instance below 200 rpm for a turbo generator unit, where the dynamic effects are negligible, the measured absolute or relative shaft vibrations are given by the contribution of the slow roll eccentricity, that is the contribution of the deformation shape. A distinct value of this slow roll eccentricity can be used as a symptom of a bent shaft. Phase delay close to 90° in the readings of two transducers installed along two orthogonal directions on the same shaft section is another check of the reliability of the diagnosis of a shaft bow.

11) Unbalance

The unbalance is the most common cause of vibration at rotating machine. According to the International Standards Organisation (ISO), unbalance can be defined as: "... that condition which exists in a rotor when vibratory force or motion is imparted to its bearings as a result of centrifugal forces".

A more intuitive definition of unbalance is an unequal distribution of weight of a part about its rotating centreline. Another common and useful definition for unbalance is: "... a condition which exists whenever the rotating centreline (shaft axis) and central principal axis of a rotor do not coincide".

Types of unbalance can be classified according to the geometric relationship between the shaft axis and principal axis of a rotor. Essentially, there are four types of rotor unbalance: *static, couple, quasi static* and *dynamic*.

If a rotor is out of balance, it should be out of balance by the same quantity, after 360° of rotation. For this reason unbalance is characterised by a frequency component of 1X [8.9]. However, some other faults have dominant first order components too, such as shaft bow, bearing misalignment, etc. For this reason it is difficult to distinguish unbalance from these faults.

The following condition can be helpful to diagnose unbalance in a system:

- frequency component at half shaft or fractional shaft speed cannot be present;
- high frequency components (2X, 3X, ...) are very small;
- the unbalance has a fixed phase angle with respect to a shaft phase reference;
- vibration is related to speed but is not related to load.

Under normal conditions, 1X vibrations are obtained [8.14][8.20]. When excessive mass unbalance is present, bearings can behave non-linearly and induce higher-order vibrations (2X, 3X, ...) with amplitudes lower than the 1X vibration [8.5][8.6].

12) Electrical faults

Vibration of electrical machinery such as motors, generators etc. can be due to either mechanical or electrical origin. The vibrations caused by electrical problems are normally the results of unequal magnetic forces acting on the rotor or stator. These unequal magnetic forces may be due to open or shorted winding, a broken rotor bar [8.21], unbalanced phases, unequal air gap [8.22], etc. Generally, the frequency of vibration resulting from the electrical problem will be equal line frequency (50/60 Hz) or line frequency multiplied by a whole number [8.23].

13) Instability

Oil whirl occurs primarily on machines equipped with pressure-lubricated sleeve bearings and operating at a relatively high speed, normally above the critical speed of the rotor [8.6][8.12][8.24]. The frequency of oil whirl vibration is slightly less than one half the rotation speed of the rotor (0.42-0.48X).

Hysteresis or internal friction induced whirl is a sub-synchronous vibration. The forces originate from friction either between rotor components or in the material of the rotor. Hysteretic whirl can be observed generally in a rotating machine that operates above the first critical speed. The vibration excited by hysteresis whirl is similar in many respects to oil whirl described earlier. However, hysteresis whirl will always occur at a frequency equal to the first critical speed of system rotor-bearing [8.25].

Instabilities in turbomachinery can also be caused by fluid mechanical interactions in seals and aerodynamic cross-coupling in areas with long narrow clearances between rotating and stationary components, like balance pistons and impellers in pumps, and bladed disks in turbomachines [8.26]. Sub-synchronous vibrations may occur in machines operating above the first critical speed. Such phenomena are described further in [8.25].

8.2.6 Fault Matrix

Finally, the fault matrix follows the list of faults in table 8.3. As described in section 8.1.1, each pair fault/symptom must have two weights associated with it (α_{ij} / β_{ij}) that are separated in the ASCII format used by a slash "/". Each row contains the weights associated with a particular fault, and each column relates to a particular symptom. The complete knowledge base is presented in table 8.6.

Table 8.6. Knowledge base.

Faults \\ Symptoms	Major amplitude of radial vibration high "A_1_r"	Change of amplitude of radial vibration [%] "A_1_rch"	Major amplitude of axial vibration high "A_1_a"	Frequency components at (0.10X - 0.45X] "A_1_sd"	Frequency components at (0.45X - 0.55X] "A_1_sc"	Frequency components at (0.55X - 0.95X) "A_1_su"
Normal operating condition	100/0	0/20	0/0	0/0	0/0	0/0
Impending fault	0/100	20/0	0/0	0/0	0/0	0/0
Axial rotor asymmetry	0/0	0/0	0/0	0/0	0/0	0/0
Rigid coupling misalignment	0/0	0/0	0/0	-10/0	-10/0	-10/0
Flexible coupling misalign.	0/0	0/0	30/0	0/0	0/0	0/0
Bearing misalignment	0/0	0/0	0/0	10/0	0/0	0/0
Transverse crack	0/0	0/0	0/0	0/0	0/0	0/0
Rotor to stator rub	0/0	0/0	0/0	20/0	20/0	10/0
Loosening bearing bushes	0/0	0/0	0/0	10/0	10/0	5/0
Bow and set	0/0	0/0	0/0	-10/0	-10/0	-10/0
Unbalance	0/0	0/0	0/0	-10/0	-10/0	-10/0
Electrical faults	0/0	0/0	0/0	0/0	0/0	0/0
Instability	0/0	0/0	0/0	60/-5	60/-5	60/-5

Table 8.6. (*continued*)

Faults \\ Symptoms	Is sub-synchron component unaffected by load change *"A_1_lch"*	Frequency component at (1X) high *"A_1_1x"*	Frequency component at (2X) high *"A_1_2x"*	Frequency component at (3X) high *"A_1_3x"*	Frequency component at (4X) high *"A_1_4x"*	The 1X is dominant *"A_1_1xd"*
Normal operating condition	0/0	0/0	0/0	0/0	0/0	0/0
Impending fault	0/0	0/0	0/0	0/0	0/0	0/0
Axial rotor asymmetry	0/0	10/0	30/-20	10/-5	0/0	20/0
Rigid coupling misalignment	0/0	70/-50	20/0	10/0	0/0	30/-30
Flexible coupling misalign.	0/0	30/-10	40/-20	20/-10	10/-5	0/0
Bearing misalignment	0/0	20/-10	30/-10	10/-10	0/0	0/0
Transverse crack	0/0	40/-10	20/0	20/0	10/0	20/-10
Rotor to stator rub	0/0	20/-10	10/0	10/0	10/0	25/0
Loosening bearing bushes	0/0	25/-10	20/0	15/0	10/0	20/0
Bow and set	0/0	70/-50	30/-10	0/0	0/0	30/-30
Unbalance	0/0	80/-50	20/-10	0/0	0/0	35/-30
Electrical faults	0/0	20/-10	20/0	0/0	0/0	0/0
Instability	-20/0	-10/0	-10/0	-10/0	0/0	-30/0

Table 8.6. (*continued*)

Faults \\ Symptoms	The 2X is dominant "*A_1_2xd*"	Dominant frequency 0.5X not locked "*A_1_05nld*"	Dominant frequency 0.5X locked on "*A_1_05lod*"	Dominant frequency is system critical speed "*A_1_scsd*"	Did 1x change over a short period of time "*A_1_1xtcs*"	Did 2x change over a short period of time "*A_1_2xtcs*"
Normal operating condition	0/0	0/0	0/0	0/0	0/0	0/0
Impending fault	0/0	0/0	0/0	0/0	0/0	0/0
Axial rotor asymmetry	20/0	-30/0	-25/0	-30/0	0/0	20/0
Rigid coupling misalignment	-30/0	-30/0	-25/0	-30/0	20/0	0/0
Flexible coupling misalign.	20/0	-20/0	-20/0	-20/0	0/0	0/0
Bearing misalignment	10/0	15/0	-20/0	-20/0	0/0	0/0
Transverse crack	20/0	-20/0	-20/0	-20/0	0/0	0/0
Rotor to stator rub	15/0	-20/0	20/0	-20/0	0/0	0/0
Loosening bearing bushes	10/0	10/0	10/0	0/0	0/0	0/0
Bow and set	-30/0	-20/0	-20/0	-30/0	0/0	0/0
Unbalance	-30/0	-25/0	-25/0	-30/0	0/0	0/0
Electrical faults	0/0	0/0	0/0	0/0	0/0	0/0
Instability	-10/0	30/-20	15/-10	20/-20	0/0	0/0

Table 8.6. (*continued*)

	Did 1X change over a medium period of time *"A_1_1xtcm"*	Did 2X change over a medium period of time *"A_1_2xtcm"*	Did 1X change over a long period of time *"A_1_1xtcl"*	Did 2X change over a long period of time *"A_1_2xtcl"*	Did vibration level change in one bearing only *"A_1_bVc"*	Trend in 1X is accelerating *"A_1_1xT"*
Normal operating condition	0/0	0/0	0/0	0/0	0/0	0/0
Impending fault	0/0	0/0	0/0	0/0	0/0	0/0
Axial rotor asymmetry	0/0	0/0	0/0	0/0	-20/0	-10/0
Rigid coupling misalignment	0/0	0/0	0/0	0/0	-30/0	-20/0
Flexible coupling misalign.	10/0	10/0	10/0	10/0	-20/0	5/0
Bearing misalignment	10/0	10/0	10/0	10/0	-10/0	10/0
Transverse crack	0/0	0/0	10/-10	20/-10	-20/0	40/-20
Rotor to stator rub	10/0	0/0	0/0	0/0	-20/0	20/0
Loosening bearing bushes	0/0	0/0	0/0	0/0	20/0	5/0
Bow and set	10/0	0/0	0/0	0/0	-30/0	10/-5
Unbalance	0/0	0/0	0/0	0/0	-30/0	-10/0
Electrical faults	0/0	0/0	0/0	0/0	0/0	0/0
Instability	0/0	0/0	0/0	0/0	0/0	-10/0

Table 8.6. (*continued*)

Faults \\ Symptoms	Trend in 2X is accelerating *"A_1_2xT"*	Trend in 3X is increasing or accelerating *"A_1_3xT"*	Trend in gear passing frequency (1tX) *"A_1_txT"*	Change 1X on rundown over time *"A_1_1x RdC"*	Change 2X on rundown over time *"A _1_2x RdC"*	Change 3X on rundown over time *"A_1_3x RdC"*
Normal operating condition	0/0	0/0	0/0	0/0	0/0	0/0
Impending fault	0/0	0/0	10/0	0/0	0/0	0/0
Axial rotor asymmetry	10/0	10/0	0/0	0/0	20/0	10/0
Rigid coupling misalignment	-10/0	-10/0	0/0	0/0	0/0	0/0
Flexible coupling misalign.	10/0	5/0	0/0	0/0	0/0	0/0
Bearing misalignment	10/0	10/0	0/0	0/0	0/0	0/0
Transverse crack	30/-10	20/-10	0/0	20/-20	30/-30	50/-30
Rotor to stator rub	10/0	10/0	0/0	25/0	25/0	25/0
Loosening bearing bushes	10/0	10/0	0/0	0/0	0/0	0/0
Bow and set	10/-5	10/-5	0/0	0/0	0/0	0/0
Unbalance	-5/0	-5/0	0/0	0/0	0/0	0/0
Electrical faults	0/0	0/0	0/0	0/0	0/0	0/0
Instability	-10/0	-10/0	0/0	0/0	0/0	0/0

Table 8.6. (*continued*)

Faults \\ Symptoms	Half harmonic frequencies (1.5, 2.5, 3.5X) *"A_1_n05x"*	Blade passing frequency (1bX or 2bX) *"A_1_bx"*	Gear passing frequency (1tX or 2tX) *"A_1_tx"*	Background level increases (high freq.) *"A_1_b"*	Vibration level at line frequency (50 Hz) *"A_1_ef"*	Vibration level at slip frequency *"A_1_sf"*
Normal operating condition	0/0	0/0	0/0	0/0	0/0	0/0
Impending fault	10/0	10/0	10/0	0/0	0/0	0/0
Axial rotor asymmetry	0/0	0/0	0/0	0/0	0/0	0/0
Rigid coupling misalignment	0/0	0/0	0/0	-10/0	0/0	0/0
Flexible coupling misalign.	0/0	0/0	0/0	0/0	0/0	0/0
Bearing misalignment	0/0	0/0	0/0	0/0	0/0	0/0
Transverse crack	0/0	0/0	0/0	-10/10	0/0	0/0
Rotor to stator rub	0/0	0/0	0/0	-35/35	0/0	0/0
Loosening bearing bushes	0/0	0/0	0/0	-20/0	0/0	0/0
Bow and set	0/0	0/0	0/0	0/0	0/0	0/0
Unbalance	0/0	0/0	0/0	-40/0	0/0	0/0
Electrical faults	0/0	0/0	0/0	0/0	60/0	40/0
Instability	0/0	0/0	0/0	10/0	0/0	0/0

Table 8.6. (*continued*)

Faults \\ Symptoms	Vibration level at harmonic of line frequency (n×50 Hz) "*A_1_nef*"	Is line frequency equal rotating speed "*A_1_LfRs*"	Relation (2X/1X)[%] "*A_1_2xR1x*"	The 1X vector changed in amplitude/phase "*A_1_1xCaf*"	Is the 2X phase locked with the 1X "*A_1_2xL1x*"	The 2X vector changed in amplitude/phase "*A_1_2xCaf*"
Normal operating condition	0/0	0/0	0/0	0/0	0/0	0/0
Impending fault	0/0	0/0	0/0	0/0	0/0	0/0
Axial rotor asymmetry	0/0	0/0	30/0	5/0	30/-20	20/-10
Rigid coupling misalignment	0/0	0/0	30/0	30/-30	0/0	10/0
Flexible coupling misalign.	0/0	0/0	60/0	10/0	0/0	30/-10
Bearing misalignment	0/0	0/0	0/0	25/-10	0/0	10/-10
Transverse crack	0/0	0/0	0/0	15/-20	0/0	20/-10
Rotor to stator rub	0/0	0/0	0/0	25/-20	0/0	20/-10
Loosening bearing bushes	0/0	0/0	0/0	10/0	20/0	20/0
Bow and set	0/0	0/0	0/0	30/-40	20/0	10/0
Unbalance	0/0	0/0	0/0	30/-40	0/0	10/0
Electrical faults	30/0	-30/10	0/0	0/0	0/0	0/0
Instability	0/0	0/0	0/0	10/0	0/0	10/0

Table 8.6. (*continued*)

Faults \\ Symptoms	Is the 3X phase locked with the 1X *"A_1_3xL1x"*	The 3X vector changed in amplitude/phase *"A_1_3xCaf"*	Vibration at both ends of shaft in phase *"A_1_endsf"*	Component A_1_1x at slow roll *"A_1_1xsr"*	Component A_1_1x at slow roll after cool down *"A_1_1xsrc"*	Relation (Axial / Radial) > 0.5 *"A_1_05R"*
Normal operating condition	0/0	0/0	0/0	0/0	0/0	0/0
Impending fault	0/0	0/0	0/0	0/0	0/0	0/0
Axial rotor asymmetry	0/0	10/0	0/0	0/0	0/0	0/0
Rigid coupling misalignment	0/0	10/0	0/0	0/0	0/0	0/0
Flexible coupling misalign.	10/-10	10/-5	0/0	0/0	0/0	10/0
Bearing misalignment	0/0	10/-10	0/0	0/0	0/0	0/0
Transverse crack	0/0	0/0	0/0	10/0	0/0	0/0
Rotor to stator rub	0/0	10/0	0/0	0/0	0/0	0/0
Loosening bearing bushes	0/0	10/0	0/0	0/0	0/0	0/0
Bow and set	0/0	0/0	10/0	20/-10	20/-10	0/0
Unbalance	0/0	0/0	10/0	0/0	0/0	0/0
Electrical faults	0/0	0/0	0/0	0/0	0/0	-10/0
Instability	0/0	10/0	10/0	0/0	0/0	10/0

Table 8.6. (*continued*)

Faults \\ Symptoms	Relation (Axial / Radial) > 1 "*A_1_1R*"	Reduction of first critical speed "*A_1_CS1*"	Frequency components at first critical speed "*A_1_1csx*"	Reduction of second critical speed "*A_1_CS2*"	Frequency components at second critical speed "*A_1_2csx*"	Reduction of third critical speed "*A_1_CS3*"
Normal operating condition	0/0	0/0	0/0	0/0	0/0	0/0
Impending fault	0/0	0/0	0/0	0/0	0/0	0/0
Axial rotor asymmetry	0/0	0/0	0/0	0/0	0/0	0/0
Rigid coupling misalignment	0/0	0/0	0/0	0/0	0/0	0/0
Flexible coupling misalign.	5/0	0/0	0/0	0/0	0/0	0/0
Bearing misalignment	0/0	0/0	0/0	0/0	0/0	0/0
Transverse crack	0/0	30/-10	0/0	20/-10	0/0	10/-10
Rotor to stator rub	0/0	0/0	0/0	0/0	0/0	0/0
Loosening bearing bushes	0/0	0/0	0/0	0/0	0/0	0/0
Bow and set	0/0	0/0	20/-5	0/0	20/-5	0/0
Unbalance	0/0	0/0	0/0	0/0	0/0	0/0
Electrical faults	-5/0	0/0	0/0	0/0	0/0	0/0
Instability	5/0	0/0	35/-5	0/0	30/-5	0/0

Table 8.6. (*continued*)

Faults \\ Symptoms	Frequency components at third critical speed "*A_1_3csx*"	Change in vibration components at first critical speed "*A_1_scC*"	Bearing temperature high "*A_1_Tb*"	Vibration increases with increasing speed "*A_1_Vrts*"	Vibration related to load "*A_1_Vrtl*"	Vibration pattern changes after start/stop "*A_1_CVss*"
Normal operating condition	0/0	0/0	0/0	0/0	0/0	0/0
Impending fault	0/0	0/0	0/0	0/0	0/0	0/0
Axial rotor asymmetry	0/0	-40/0	0/0	0/0	-25/0	-10/0
Rigid coupling misalignment	0/0	-30/0	0/0	15/-10	-10/0	-10/0
Flexible coupling misalign.	0/0	-30/0	0/0	0/0	10/-5	10/0
Bearing misalignment	0/0	20/0	30/0	0/0	0/0	0/0
Transverse crack	0/0	10/0	0/0	0/0	-20/0	0/0
Rotor to stator rub	0/0	10/0	0/0	0/0	20/0	0/0
Loosening bearing bushes	0/0	10/0	0/0	0/0	-20/0	20/0
Bow and set	20/-5	-30/0	0/0	0/0	-10/0	0/0
Unbalance	0/0	-30/0	0/0	20/-20	-40/0	0/0
Electrical faults	0/0	0/0	0/0	0/0	0/0	0/0
Instability	30/0	10/0	0/0	0/0	10/0	15/0

Table 8.6. (*continued*)

Faults \\ Symptoms	Vibration steady after change "*A_1_VsaC*"	Similar fractional change *X* and *Y* "*A_1_SCxy*"	Disks shrunk on the shaft "*A_1_Dsh*"	H_2 seal oil temperature changes "*A_1_oTC*"
Normal operating condition	0/0	0/0	0/0	0/0
Impending fault	0/0	0/0	0/0	0/0
Axial rotor asymmetry	0/0	0/0	0/0	0/0
Rigid coupling misalignment	20/-20	10/-20	0/0	0/0
Flexible coupling misalign.	0/0	0/0	0/0	0/0
Bearing misalignment	0/0	0/0	0/0	0/0
Transverse crack	10/-10	15/-15	0/0	0/0
Rotor to stator rub	-35/35	25/-20	10/0	15/0
Loosening bearing bushes	0/0	0/0	0/0	0/0
Bow and set	0/0	10/-20	15/0	0/0
Unbalance	20/-20	10/-20	0/0	0/0
Electrical faults	0/0	0/0	0/0	0/0
Instability	0/0	0/0	0/0	0/0

8.2.7 Baseline

If numerical values are used, together with the membership functions to obtain certainty factors, then the measured value must be related to a baseline value that has to be defined for each symptom. One way to define the baseline value is on the basis of the data that were measured when the machine was new and in good condition.

If a reference condition is not available, the other way is to define baseline values as a function of the limit in μm for the good state of the machine as defined by international standards like ISO 7919 or 10816 or national ones like VDI 2059. Considering for instance the last one, the baseline value for shaft relative vibration, as function of rotating speed in rpm is:

$$G_{limit} = \frac{2400}{\sqrt{Speed}} \tag{8.14}$$

The baseline for each symptom of section 8.1.4.1 with a numerical value is defined in table 8.7 as a function of the limit for total vibration G_{limit} (note that the symbolic name in the table is the same of that in section 8.1.4.1 with the addition of suffix 'g'). If bearing pedestal vibration is measured then the baseline values can be determined e.g. from VDI 2056 standard for casing vibration. The G_{limit} would then be expressed in mm/s, according to this standard.

Table 8.7. Baseline values.

Value	***Baseline***
$A\_1\_1g = 1.00 \cdot G_{limit}$	Baseline for total radial vibration
$A\_1\_3g = 1.00 \cdot G_{limit}$	Baseline for total axial vibration
$A\_1\_1xg = 0.50 \cdot G_{limit}$	Baseline for 1X component
$A\_1\_2xg = 0.35 \cdot G_{limit}$	Baseline for 2X component
$A\_1\_3xg = 0.15 \cdot G_{limit}$	Baseline for 3X component
$A\_1\_4xg = 0.15 \cdot G_{limit}$	Baseline for 4X component
$A\_1\_5xg = 0.075 \cdot G_{limit}$	Baseline for 5X component
$A\_1\_6xg = 0.075 \cdot G_{limit}$	Baseline for 6X component
$A\_1\_7xg = 0.075 \cdot G_{limit}$	Baseline for 7X component
$A\_1\_8xg = 0.075 \cdot G_{limit}$	Baseline for 8X component
$A\_1\_1csg = 0.25 \cdot G_{limit}$	Baseline for component at first critical speed
$A\_1\_2csg = 0.25 \cdot G_{limit}$	Baseline for component at second critical speed
$A\_1\_3csg = 0.25 \cdot G_{limit}$	Baseline for component at third critical speed
$A\_1\_1xsrg = 0.05 \cdot G_{limit}$	Baseline for component at slow roll
$A\_1\_1xsrcg = 0.05 \cdot G_{limit}$	Baseline for component at slow roll after cool down
$A\_1\_xbg = 0.10 \cdot G_{limit}$	Baseline for component at blade passing frequency
$A\_1\_xtg = 0.10 \cdot G_{limit}$	Baseline for component at gear passing frequency
$A\_1\_n05xg = 0.10 \cdot G_{limit}$	Baseline for component at half harmonic frequency
$A\_1\_sdg = 0.15 \cdot G_{limit}$	Baseline for subharmonic (0.10-0.45) X
$A\_1\_scg = 0.15 \cdot G_{limit}$	Baseline for subharmonic (0.45-0.55)X
$A\_1\_sug = 0.15 \cdot G_{limit}$	Baseline for subharmonic (0.55-0.95)X
$A\_1\_bg = 0.10 \cdot G_{limit}$	Baseline for background component
$A\_1\_efg = 0.50 \cdot G_{limit}$	Baseline for component at line frequency
$A\_1\_sfg = 0.35 \cdot G_{limit}$	Baseline for component at slip frequency
$A\_1\_nefg = 0.35 \cdot G_{limit}$	Baseline for ($n \times 50$)Hz component

Example. For a rotor with operating speed of 1500 rpm, G_{limit} given by eq. (8.14) is:

$$G_{limit} = \frac{2400}{\sqrt{Speed}} = \frac{2400}{\sqrt{1500}} = 61.96\mu\text{m} \tag{8.15}$$

Considering the symptom "Frequency component at (1X) high", its baseline value is (see table 8.7):

$$A\_1\_1xg = 0.5 \cdot G_{limit} = 30.98\mu\text{m} \tag{8.16}$$

it has *Up slope* as membership function (see section 8.1.4.1, symptom 8) and the parameters are:

$$\begin{aligned} l &= 0.30 \cdot A\_1\_1xg = 9.29\mu\text{m} \\ c &= 1.00 \cdot A\_1\_1xg = 30.98\mu\text{m} \\ h &= 1.60 \cdot A\_1\_1xg = 49.75\mu\text{m} \end{aligned} \tag{8.17}$$

while the weights associated to this symptom and the fault "Unbalance" are $\alpha_{ij} = 80$ and $\beta_{ij} = -50$, as shown in the knowledge basis of table 8.6.

Considering figure 8.7, if the symptom "Frequency component at (1X) high":

- is less than 9.29 μm, then the likelihood of the fault "Unbalance" is 50% false;
- is more than 49.57 μm, then the likelihood of the fault "Unbalance" is 80% true;
- is between 9.29 μm and 49.57 μm, the CF is given by figure 8.7 and the likelihood is obtained multiplying the CF times the weights α_{ij} or β_{ij}.

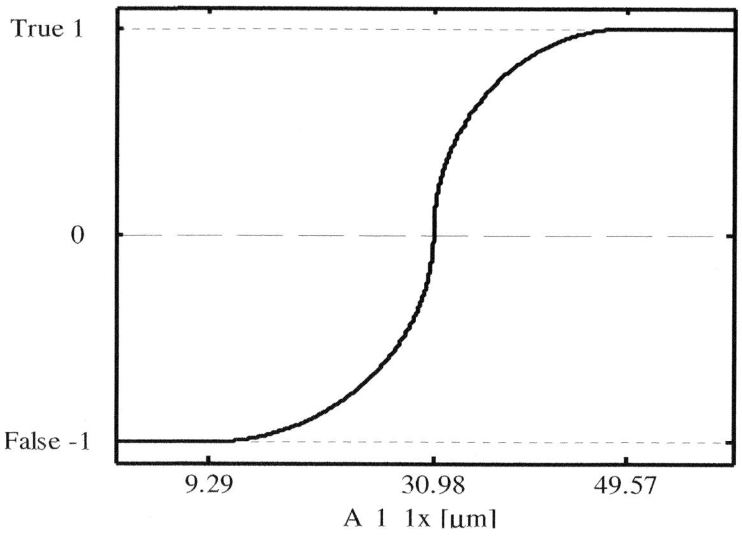

Fig. 8.7. Up slope curve for symptom "Frequency component at (1X) high" for a rotor operating at 1500 rpm.

8.3 Results for the Qualitative Approach

The first experimental case presented here describes a case in which a cracked shaft with a crack depth of about 50% of the diameter is installed in the EURoPE test-rig. The description of the test-rig is presented in detail in section 8.4.2. The raw measurement data are obtained when the rotor is operating at 1500 rpm and the relative vibrations are measured not in bearing but in a section close to the shaft midspan. Table 8.8 shows the data and it is possible to observe that some of the symptoms were not measured nor evaluated. Moreover, the rotor vibrates severely: the higher value of the major amplitude of the radial relative vibration is 262 μm that should be compared with the G_{limit} value, which is equal to that of eq. (8.15).

Using the knowledge base of table 8.6, the data of table 8.8 and the symptoms description of section 8.1.4.1, the likelihood of the faults are combined using a Bayesian method and the results are shown in figure 8.8.

Table 8.8. Raw measurement data for the symptoms.

Symptom	*Symbolic name*	*Measured*	*Value*
Major amplitude of radial vibration high	*"A_1_r"*	1	262.0
Change of major amplitude of radial vibration [%]	*"A_1_rch"*	0	0
Amplitude of axial vibration high	*"A_1_a"*	0	0
Frequency Components at 45X]	*"A_1_sd"*	0	0
Frequency Components at 55X]	*"A_1_sc"*	0	0
Frequency Components at 98X)	*"A_1_su"*	0	0
Is subsynchron component unaffected by load change	*"A_1_lch"*	0	0
Frequency Component at (1X) high	*"A_1_1x"*	1	262.0
Frequency Component at (2X) high	*"A_1_2x"*	1	13.0
Frequency Component at (3X) high	*"A_1_3x"*	1	3.8
Frequency Component at (4X) high	*"A_1_4x"*	1	3.7
The 1X is dominant	*"A_1_1xd"*	1	1
The 2X is dominant	*"A_1_2xd"*	1	-1
Dominant frequency 05X not locked	*"A_1_05nld"*	0	0
Dominant frequency 05X locked on	*"A_1_05lod"*	0	0
Dominant frequency is system critical speed	*"A_1_scsd"*	1	-1
Did 1X change over a short period of time	*"A_1_1xtcs"*	0	0
Did 2X change over a short period of time	*"A_1_2xtcs"*	0	0
Did 1X change over a medium period of time	*"A_1_1xtcm"*	0	0
Did 2X change over a medium period of time	*"A_1_2xtcm"*	0	0
Did 1X change over a long period of time	*"A_1_1xtcl"*	0	0
Did 2X change over a long period of time	*"A_1_2xtcl"*	0	0

Table 8.8. *(Continued)*

Symptom	*Symbolic name*	*Measured*	*Value*
Did vibration level change in one bearing only	*"A_1_bVc"*	1	-1
Trending in 1X is increasing or accelerating	*"A_1_1xT"*	1	1
Trending in 2X is increasing or accelerating	*"A_1_2xT"*	1	1
Trending in 3X is increasing or accelerating	*"A_1_3xT"*	1	1
Trending in gear passing Frequency (1tX)	*"A_1_txT"*	0	0
Rundown test Change 1X on rundown over time	*"A_1_1xRdc"*	0	0
Rundown test Change 2X on rundown over time	*"A_1_2xRdc"*	0	0
Rundown test Change 3X on rundown over time	*"A_1_3xRdc"*	0	0
Half Harmonic Frequencies n05X	*"A_1_n05x"*	0	0

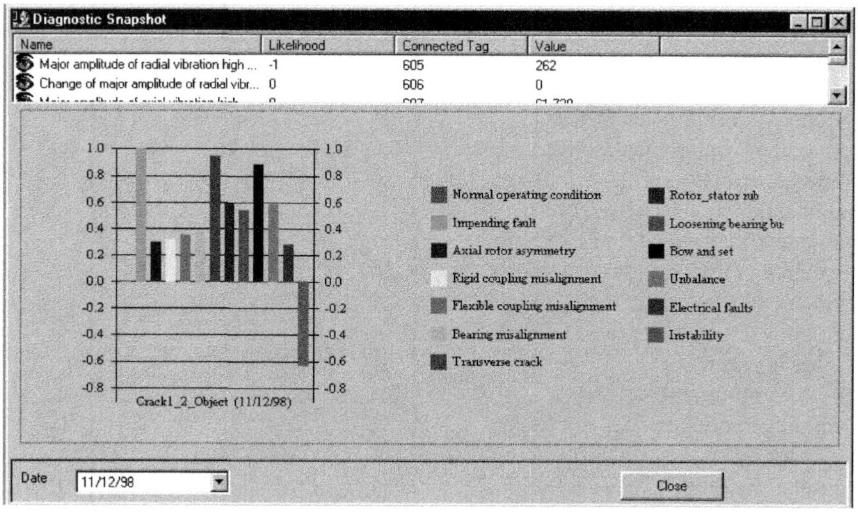

Fig. 8.8. Results for the likelihood of the faults in case of 50% crack depth at 1500 rpm.

The total level of 262 μm is much greater than the vibration limit for "Good state", for that reason the "Impending faults" has very large likelihood and "Transverse crack" has the largest likelihood, i.e. 0.94, among all the modelled faults. All faults with dominant 1X component have large value of the likelihood, in particular "Bow and set" has the likelihood of 0.88 and this could be plausible, since the cracked shaft may have a residual bow due to the process used to make the crack (alternate fatigue).

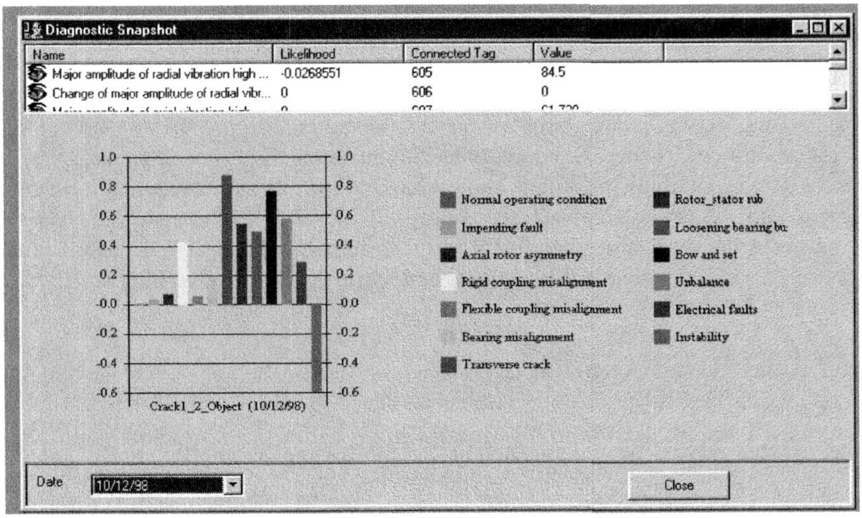

Fig. 8.9. Results for the likelihood of the faults in case of 34% crack depth at 850 rpm.

In the second example, for the same test-rig, the shaft has a crack of 34% depth and the rated speed is 850 rpm. As explained in section 8.4.2, this is not actually a breathing crack, but a slot. The measured vibration has the total level of 84.5 μm . Using again VDI 2059 standard, the G_{limit} value results 82.31 μm . The total vibration level is close to vibration limit for "Good state". Considering the results of figure 8.9, the likelihood of "Impending faults" is much less than the previous example and "Transverse crack" has again the largest likelihood, equal to 0.87.

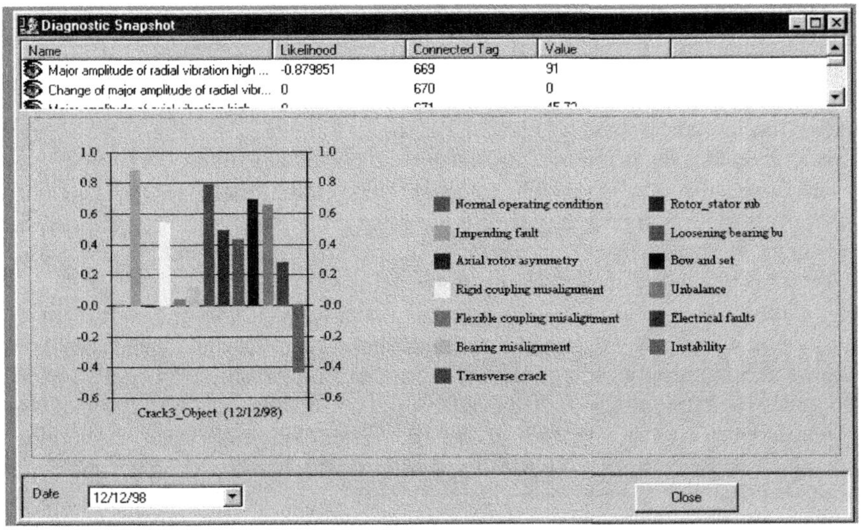

Fig. 8.10. Results for the likelihood of the faults in case of 34% crack depth at 1550 rpm.

In the third example, the same crack depth of the previous one is considered, but the rated speed is 1550 rpm. In this case the G_{limit} is equal to 60.96 μm , while the total level is 91.0 μm . Since the vibration limit for "Good state" is lower at this speed, the general state of the machine is worse (see figure 8.10). Once more the largest value of likelihood, equal to 0.79, indicates a "Transverse crack".

In the three examples presented, the qualitative fault matrix method succeeded in diagnosing the presence of a crack, but it was not able to indicate the severity nor the position of the crack relatively to the machine span. These tasks are now accomplished by introducing the model based identification of the crack.

8.4 Model Based Approach

Model based techniques in rotor dynamics consider the model of the fully assembled machine, which is composed by the sub-models of the rotor, of the bearings and of the foundation. Some identification techniques, such as the least square identification in frequency domain have proven to be quite robust even if the sub-models are not fine tuned [8.9],[8.27]-[8.29]. Anyhow, the use of a reliable model can increase the accuracy of the identification.

In general terms model based approaches can be classified in three categories [8.30],[8.31]:

1. *parameter estimation*, when the characteristic constant parameters of the process or of the system components are affected by the fault;
2. *state estimation*, when the constant parameters are unaffected by possible faults and only the state of the system, which is represented by a set of generally not measurable state variables (function of time), is affected by the faults; in this case the model acts as a state observer;
3. *parity equations*, when the faults affect some of the not measurable input variables, the parameters are constant, only output variables are measured and compared with calculated model output variables.

The last approach is the most suitable in real rotating machinery modelled by means of finite elements and will be used in the following.

Crack diagnosis is performed by means of parity equations, by introducing *external forces* that are *equivalent* to the effects of the crack, rather than with parameter estimation approach, because the system parameters are represented by the complete mass, stiffness and damping matrices of the system. This way, few unknowns of the equivalent force have to be estimated.

The estimation of external forces appears a simpler task than the estimation of system parameters, even if successful applications in rotor dynamics have been presented in literature, but limited to test-rigs or to small machines modelled using few degrees of freedom (d.o.f.s) [8.32],[8.33]. This is due mainly to the fact that the identification of the variation of the parameter with respect to its nominal value, which is proportional to the equivalent force (as shown in paragraph 8.2.1), is more accurate than the parameter estimation of the full system, being generally the change in system parameters due to faults or defects rather small. For instance, a blade loss in power plant turbine is easier represented by an equivalent fault (unbalance) than by the overall system mass and mass distribution changes. Another suitable example is a transverse crack. The crack affects only the stiffness of one small part of the rotor, roughly corresponding to one element, and its effect on the system parameter, i.e. the terms of the stiffness matrix, can be vanishing small and its effect on natural frequencies can be hard to measure in fully assembled real machines, as shown in section 3.1.1. However, this approach requires that the model of the fully assembled machine, i.e. of the parts that compose it, is reliable.

8.4.1 Definition of Equivalent External Forces to Faults

The starting point for the definition of equivalent external forces to cracks is the system of dynamic equations of the machine, obtained by considering the rotor modelled by means of finite beam elements (see figure 8.11), the bearings by equivalent damping and stiffness and the foundation by a suitable representation (pedestals, modal, transfer matrix or rigid).

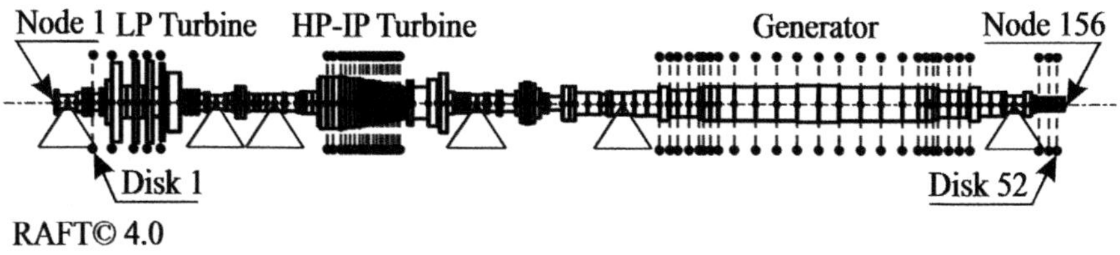

Fig. 8.11. Example of a steam turbo generator modelled by means of finite beam elements.

The system of dynamic equations of a large rotating machine, with several d.o.f.s, is typically:

$$[\mathbf{M}]\ddot{\mathbf{x}}_{total} + [\mathbf{C}]\dot{\mathbf{x}}_{total} + [\mathbf{K}]\mathbf{x}_{total} = \mathbf{F}(t) \tag{8.18}$$

where the mass matrix $[\mathbf{M}]$ takes also into account the secondary effect of the rotatory inertia, the damping matrix $[\mathbf{C}]$ includes also the speed depending gyroscopic matrix (that in this chapter is not explicated for the sake of brevity in the equations) and the stiffness matrix $[\mathbf{K}]$ takes also into account the shear effect.

Considering eq. (8.18), it seems difficult to identify the changes due to the developing fault in the matrices $[\mathbf{M}]$, $[\mathbf{C}]$ and $[\mathbf{K}]$, which are of high order, from measurement of the absolute vibration $\mathbf{x}_{total}$ in only few measuring planes along the shaft.

In real rotating machines, the measuring points of the vibration along the shaft during operation are few and the transducers are normally in correspondence of the bearings (see figure 8.12). Only lateral vibrations are normally monitored, insofar in the following only lateral behaviour of the rotor will be considered.

Fig. 8.12. Typical layout of transducers in industrial machines in correspondence of the bearings.

Fig. 8.13. Compressor and turbine of a gas turbo generator with case open. Due to the presence of the working fluid it is generally not possible to have a measuring plane inside the case.

The consideration of further measuring planes along the rotor span, due to the presence of the casings and of the possible working fluids (see figure 8.13), is practically impossible, therefore methods that reconstruct modal shapes of the rotor, like those presented in [8.34], cannot be used.

The right hand side (r.h.s.) external forces $\mathbf{F}(t)$ of eq. (8.18) are composed by the weight (which is known) and by the original unbalance and bow (which are unknown). The system parameter changes due to the fault are indicated as $[d\mathbf{M}]$, $[d\mathbf{C}]$ and $[d\mathbf{K}]$ and eq. (8.18) becomes:

$$\begin{aligned}&([\mathbf{M}]+[d\mathbf{M}])\ddot{\mathbf{x}}_{total}+([\mathbf{C}]+[d\mathbf{C}])\dot{\mathbf{x}}_{total}+([\mathbf{K}]+[d\mathbf{K}])\mathbf{x}_{total}=\\&=\mathbf{W}+\left(\bar{\mathbf{U}}+\bar{\mathbf{M}}_u\right)\ \mathrm{e}^{\mathrm{i}\Omega t}\end{aligned} \tag{8.19}$$

If the system behaviour is considered to be linear, which is acceptable for a wide class of faults [8.9],[8.35]-[8.37], then the total vibration $\mathbf{x}_{total}$ can be considered as due to two superposed effects:

$$\mathbf{x}_{total}=\mathbf{x}_{ref.}+\mathbf{x}_a \tag{8.20}$$

It can be shown that the overall behaviour of an horizontal axis heavy cracked shaft is linear, under the conditions shown in section 5.4.3. The non-linear effect of the breathing crack, which is weak in normal conditions, may influence its behaviour only in extreme operating conditions.

The first vibration vector $\mathbf{x}_{ref.}$ is the pre-fault vibration, which is due to the weight $\mathbf{W}$ and the unknown unbalance force $\bar{\mathbf{U}}\mathrm{e}^{\mathrm{i}\Omega t}$ and unbalance moment $\bar{\mathbf{M}}_u\mathrm{e}^{\mathrm{i}\Omega t}$. The second vibration $\mathbf{x}_a$ is due to the developing fault and the last is also called *additional vibration*. The vibration component $\mathbf{x}_a$ may be obtained by calculating the vector differences of the actual vibrations (due to weight, original unbalance, bow and fault) and the original vibrations measured, in the same operating conditions in a reference case (rotation speed, flow rate, power, temperature, etc.) before the fault was developing. A discussion about the possible errors introduced and their tracking is presented in [8.28]. Recalling the definition of the pre-fault vibration $\mathbf{x}_{ref.}$, the following equation holds:

$$[\mathbf{M}]\ddot{\mathbf{x}}_{ref.}+[\mathbf{C}]\dot{\mathbf{x}}_{ref.}+[\mathbf{K}]\mathbf{x}_{ref.}=\mathbf{W}+\left(\bar{\mathbf{U}}+\bar{\mathbf{M}}_u\right)\ \mathrm{e}^{\mathrm{i}\Omega t} \tag{8.21}$$

which substituted in eq. (8.19) with eq. (8.20) gives:

$$[\mathbf{M}]\ddot{\mathbf{x}}_a+[\mathbf{C}]\dot{\mathbf{x}}_a+[\mathbf{K}]\mathbf{x}_a=-[d\mathbf{M}]\ddot{\mathbf{x}}_{total}-[d\mathbf{C}]\dot{\mathbf{x}}_{total}-[d\mathbf{K}]\mathbf{x}_{total} \tag{8.22}$$

The r.h.s. of eq. (8.22) can be considered as a system of equivalent external forces which force the fault-free system to have the change in the additional vibration $\mathbf{x}_a$ that is due to the developing fault only:

$$[\mathbf{M}]\ddot{\mathbf{x}}_a+[\mathbf{C}]\dot{\mathbf{x}}_a+[\mathbf{K}]\mathbf{x}_a=\mathbf{F}_f(t) \tag{8.23}$$

A rather complete overview of the equivalent forcing systems to the most common faults in rotating machinery is presented in [8.9],[8.35]-[8.39].

Note that in eq. (8.23) system parameters $[\mathbf{M}]$, $[\mathbf{C}]$ and $[\mathbf{K}]$ are time invariant and known, but normally $[\mathbf{C}]$ and $[\mathbf{K}]$ are function of the operating speed as regards the gyroscopic effect and the bearing coefficients. Using this last approach, the problem of fault identification is reduced to an external force identification by means of parity equations. Keeping in mind that the final goal is the identification of cracks (and faults in general), this approach is preferred since the unknown fault forcing vector is sparse and only few elements are actually different from zero, because the equivalent force is applied on few nodes of the rotor model. This fact reduces sensibly the number of unknowns to be identified.

The effect of a crack on the statical and dynamical behaviour of the rotor can be simulated in the frequency domain, by applying to the rotor different sets of equivalent forces in correspondence of the cracked beam element, one set for each one of the harmonic components considered.

8.4.2 Definition of Equivalent External Forces to Cracks

The stiffness of a cracked shaft is periodical due to the breathing and the rotation of the crack, see section 5.6 where it is shown that the equivalent crack forces are given by the following expressions:

$$\widehat{\mathbf{F}}_{f_0} = -\begin{pmatrix} \frac{1}{4}[\Delta\mathbf{K}_1]\mathbf{x}_1^* + \frac{1}{4}[\Delta\mathbf{K}_1^*]\mathbf{x}_1 + \frac{1}{4}[\Delta\mathbf{K}_2]\mathbf{x}_2^* + \\ + \frac{1}{4}[\Delta\mathbf{K}_2^*]\mathbf{x}_2 + \frac{1}{4}[\Delta\mathbf{K}_3]\mathbf{x}_3^* + \frac{1}{4}[\Delta\mathbf{K}_3^*]\mathbf{x}_3 \end{pmatrix} \tag{8.24}$$

$$\widehat{\mathbf{F}}_{f_1}\mathrm{e}^{\mathrm{i}\Omega t} = -\begin{pmatrix} [\Delta\mathbf{K}_1]\mathbf{x}_s + \frac{1}{2}[\Delta\mathbf{K}_2]\mathbf{x}_1^* + \frac{1}{2}[\Delta\mathbf{K}_3]\mathbf{x}_2^* + \\ + \frac{1}{2}[\Delta\mathbf{K}_1^*]\mathbf{x}_2 + \frac{1}{2}[\Delta\mathbf{K}_2^*]\mathbf{x}_3 \end{pmatrix}\mathrm{e}^{\mathrm{i}\Omega t} \tag{8.25}$$

$$\widehat{\mathbf{F}}_{f_2}\mathrm{e}^{\mathrm{i}2\Omega t} = -\left([\Delta\mathbf{K}_2]\mathbf{x}_s + \frac{1}{2}[\Delta\mathbf{K}_3]\mathbf{x}_1^* + \frac{1}{2}[\Delta\mathbf{K}_1]\mathbf{x}_1 + \frac{1}{2}[\Delta\mathbf{K}_1^*]\mathbf{x}_3\right)\mathrm{e}^{\mathrm{i}2\Omega t} \tag{8.26}$$

$$\widehat{\mathbf{F}}_{f_3}\mathrm{e}^{\mathrm{i}3\Omega t} = -\left([\Delta\mathbf{K}_3]\mathbf{x}_s + \frac{1}{2}[\Delta\mathbf{K}_1]\mathbf{x}_2 + \frac{1}{2}[\Delta\mathbf{K}_2]\mathbf{x}_1\right)\mathrm{e}^{\mathrm{i}3\Omega t} \tag{8.27}$$

of which eqs. (8.25), (8.26) and (8.27) are respectively the first, the second and the third harmonic component of the crack force system. Note that the projections along reference axes of the harmonic component of the force system are not necessarily equal, thus the equivalent force system is not in general represented by rotating forces.

From the experimental point of view and under the previously exposed hypothesis of linearity, the difference $\mathbf{x}_a = \mathbf{x}_{total} - \mathbf{x}_{ref.}$, between the measured vibration $\mathbf{x}_{total}$ of a rotor system that has a fault and the reference case $\mathbf{x}_{ref}$, represents the vibrational behaviour due to the fault, i.e. the additional vibrations. These vibrations are used in the identification procedure, since they are due to the impending fault only. In fact, the reference case vibrations $\mathbf{x}_{ref.}$ are given by eq. (8.28):

$$[\mathbf{M}]\ddot{\mathbf{x}}_{ref.}+[\mathbf{C}]\dot{\mathbf{x}}_{ref.}+[\mathbf{K}]\mathbf{x}_{ref.}=\mathbf{F}_e \tag{8.28}$$

while those caused by the developing crack are given by:

$$[\mathbf{M}]\ddot{\mathbf{x}}_{total}+[\mathbf{C}]\dot{\mathbf{x}}_{total}+[\mathbf{K}_m]\mathbf{x}_{total}=\mathbf{F}_e+\widehat{\mathbf{F}}_{f_0}+\widehat{\mathbf{F}}_{f_1}\,\mathrm{e}^{\,\mathrm{i}\Omega t}+\widehat{\mathbf{F}}_{f_2}\,\mathrm{e}^{\,\mathrm{i}2\Omega t}+\widehat{\mathbf{F}}_{f_3}\,\mathrm{e}^{\,\mathrm{i}3\Omega t} \tag{8.29}$$

If eq. (8.29) is considered for an unknown crack, also $[\mathbf{K}_m]$ is unknown. Anyhow it can be approximated by $[\mathbf{K}]$ of the un-cracked shaft, from which it differs only very little: the crack affects the stiffness of one element only. Therefore the additional vibrations are given by:

$$[\mathbf{M}]\ddot{\mathbf{x}}_a+[\mathbf{C}]\dot{\mathbf{x}}_a+[\mathbf{K}]\mathbf{x}_a=\widehat{\mathbf{F}}_{f_0}+\widehat{\mathbf{F}}_{f_1}\,\mathrm{e}^{\,\mathrm{i}\Omega t}+\widehat{\mathbf{F}}_{f_2}\,\mathrm{e}^{\,\mathrm{i}2\Omega t}+\widehat{\mathbf{F}}_{f_3}\,\mathrm{e}^{\,\mathrm{i}3\Omega t} \tag{8.30}$$

By applying the harmonic balance criterion in the frequency domain to eq. (8.30) and considering the harmonic components of the additional vibrations $\mathbf{X}_n$, the following equations are obtained for each one of the considered harmonic components:

$$\left[-(n\Omega)^2\mathbf{M}+\mathrm{i}\,n\Omega\mathbf{C}+\mathbf{K}\right]\mathbf{X}_n=\mathbf{F}_{f_n} \qquad n=1,2,3 \tag{8.31}$$

The force vectors $\mathbf{F}_{f_n}$ in eq. (8.31) are those to be identified. For the sake of clarity and simplicity, a single fault is considered, while multiple faults can be handled using the method fully described in [8.9],[8.40].

In the r.h.s. of eq. (8.31), the equivalent force system is applied to the two nodes of the element that contains the crack. Since lateral vibrations only are considered, the model has 4 d.o.f.s per node (see figure 8.14) and the equivalent force is a vector of eight generalized forces.

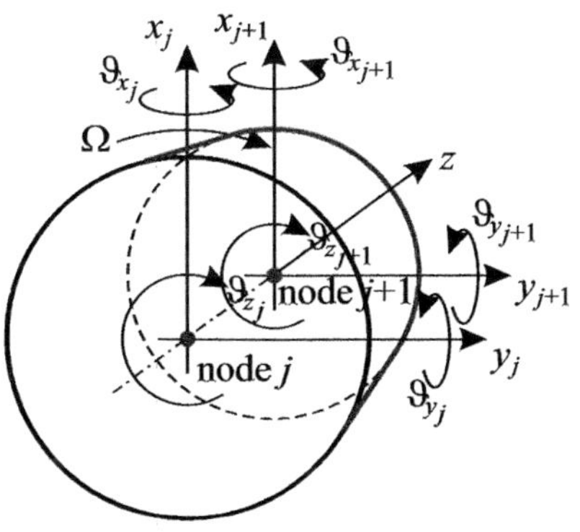

Fig. 8.14. Degrees of freedom of an element.

Due to energy considerations, it results that the most important among these forces are the bending moments that are roughly equal and opposite on the two nodes of the cracked element (see figure 8.15).

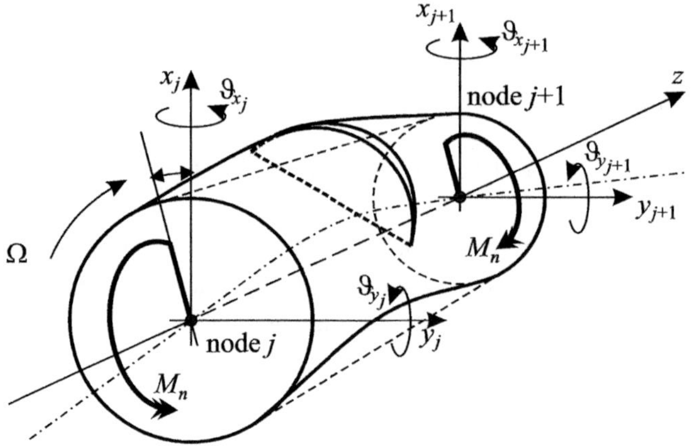

Fig. 8.15. Bending moments equivalent to crack.

In fact, cracks develop and propagate mainly to axial stresses generated by bending moments in rotating shafts; therefore generally huge bending moments are present in correspondence of cracked elements. Further on, it can be shown that in a beam element loaded by shear forces and bending moments, the elastic deflection energy associated to the shear forces is much smaller than the energy associated to the bending moments. Therefore the first one is often neglected in comparison to the second one, as it has been done also in the Bernoulli's beam in comparison to the Timoshenko's beam. For this reason, the additional deflections due to the crack can be attributed to equivalent bending moments only.

Therefore the set of equivalent forces in the case of a crack is reduced to a couple of bending moments (with their relative phase) along two orthogonal directions for each n-th harmonic component.

The equivalent force system acts on few d.o.f.s of the system, therefore vectors $\mathbf{F}_{f_n}^{(x)}$ and $\mathbf{F}_{f_n}^{(y)}$, respectively each one on the reference system directions, are not full-element vectors that are convenient to be represented by means of:

$$\mathbf{F}_{f_n}^{(x,y)} = \left\{\mathbf{L}^{(x,y)}\right\} \overline{A}^{(x,y)} \tag{8.32}$$

where $\left\{\mathbf{L}^{(x,y)}\right\}$ is the localization vector that has all null-elements except for the d.o.f. to which the forcing system is applied, and $\overline{A}^{(x,y)}$ is a complex number representing the amplitude and the phase of the fault.

In the case considered in the book, in which a crack is going to be identified, the three first harmonic components of the vibration are taken into account. For a crack located in the j-th element of the shaft line, the corresponding equivalent force systems are:

$$\mathbf{F}_{f_n}^{(x)} = \Big\{0 \;\vdots\; \underbrace{0\;\;0\;\;0\;\;\mathrm{i}}_{j\text{-th rotor node}}\;\; \underbrace{0\;\;0\;\;0\;\;-\mathrm{i}}_{j+1\text{-th rotor node}} \;\vdots\; 0\;\; \underbrace{0\;\cdots\;0}_{\text{foundation d.o.f.s}}\Big\}^{\mathrm{T}} \cdot M_n^{(x)} \mathrm{e}^{\mathrm{i}\varphi_n^{(x)}} =$$
$$= \left\{\mathbf{L}_j^{(x)}\right\} \bar{A}_n^{(x)}, \quad n = 1,2,3$$
$$\mathbf{F}_{f_n}^{(y)} = \Big\{0 \;\vdots\; \underbrace{0\;\;1\;\;0\;\;0}_{j\text{-th rotor node}}\;\; \underbrace{0\;\;-1\;\;0\;\;0}_{j+1\text{-th rotor node}} \;\vdots\; 0\;\; \underbrace{0\;\cdots\;0}_{\text{foundation d.o.f.s}}\Big\}^{\mathrm{T}} \cdot M_n^{(y)} \mathrm{e}^{\mathrm{i}\varphi_n^{(y)}} =$$
$$= \left\{\mathbf{L}_j^{(y)}\right\} \bar{A}_n^{(y)}, \quad n = 1,2,3 \tag{8.33}$$

where the only terms different from zero are those relative respectively to the rotational horizontal and vertical d.o.f.s of the extreme nodes of element *j*.

Eq. (8.31) can be now rewritten considering that, if the machine performs a speed transient, the condition monitoring systems collect data for many rotating speeds, so the additional vibrations could be available for several rotating speeds and a set of n_p rotating speeds is considered:

$$\mathbf{\Omega} = \left\{\Omega_1 \quad \Omega_2 \quad \dots \quad \Omega_{n_p}\right\}^{\mathrm{T}} \tag{8.34}$$

Then, introducing the admittance matrix $\left[\mathbf{E}(n\mathbf{\Omega})\right]$ of the system:

$$\left[\mathbf{E}(n\mathbf{\Omega})\right] = \left[-(n\mathbf{\Omega})^2\mathbf{M} + \mathrm{i}\, n\mathbf{\Omega}\mathbf{C} + \mathbf{K}\right] \quad n = 1,2,3 \tag{8.35}$$

eq. (8.31) becomes:

$$\left[\mathbf{E}(n\mathbf{\Omega})\right]\,\mathbf{X}_n = \begin{bmatrix} \mathbf{E}(n\Omega_1) & 0 & 0 & 0 \\ 0 & \mathbf{E}(n\Omega_2) & 0 & 0 \\ \vdots & \vdots & \vdots & \vdots \\ 0 & 0 & 0 & \mathbf{E}\left(n\Omega_{n_p}\right) \end{bmatrix} \begin{Bmatrix} \mathbf{X}_n \\ \mathbf{X}_n \\ \vdots \\ \mathbf{X}_n \end{Bmatrix} = \mathbf{F}_{f_n}, \quad n = 1,2,3 \tag{8.36}$$

Anyhow the equivalent force fault identification problem in eq. (8.36) is overdetermined since the number of the observations (the measured vibrations at different rotating speeds) is greater than the number of the equivalent crack forces that have to be identified. The procedure used to solve the problem is the following, by recalling that the vibration measuring planes in real machines are normally corresponding to bearing location.

Let the machine model have n_r nodes and n_b measuring planes in which the vibrations $\mathbf{X}_{m_n}$ are measured at n_p rotating speeds in two orthogonal directions, and the supporting structure is represented by means of a foundation with n_f d.o.f.s or modes [8.38]. In case of rigid foundation, obviously $n_f = 0$. The fully assembled model d.o.f.s are $(4n_r + n_f)$ while per each rotating speed are measured only $2n_b$ d.o.f.s. Note that n_e, number of elements, is $n_r - 1$ and $n_r \gg n_b$.

Since the fault has to be identified not only in its severity but also in its position, in general the procedure has to be repeated per each node of the rotor, unless the research of the fault is limited in a specified interval of the nodes.

Eq. (8.36) represents the general system of equations for all the d.o.f.s of the considered fully assembled machine. The admittance matrix $\left[\mathbf{E}(n\boldsymbol{\Omega})\right]$ has order $\left((4n_r + n_f)n_p \times (4n_r + n_f)n_p\right)$.

Now, least square identification is used here in order to evaluate the module, the phase and a residual of the fault, starting from the first element and moving along the rotor from element to element. Weighted least square technique, or more sophisticated techniques, can be employed in order to increase the robustness of the identification as described in [8.38],[8.40].

The equivalent force system, is applied in each element of the rotor model, so for all the rotating speeds the fault vector is of order $\left((4n_r + n_f)n_p \times 2\right)$. Therefore, in the first element, the localization matrix is obtained combining the localization vectors of eq. (8.33) for the i-th rotating speed is:

$$\left[\mathbf{L}_1\right]^{(i)} = \left[\underbrace{\begin{matrix} 0 & 0 & 0 & \mathrm{i} \\ 0 & 1 & 0 & 0 \end{matrix}}_{1^{\text{st}}\ \text{rotor node}} \quad \underbrace{\begin{matrix} 0 & 0 & 0 & -\mathrm{i} \\ 0 & -1 & 0 & 0 \end{matrix}}_{2^{\text{nd}}\ \text{rotor node}} \quad \vdots \quad \underbrace{\begin{matrix} 0 & 0 & \cdots & 0 \\ 0 & 0 & & 0 \end{matrix}}_{\text{foundation d.o.f.s}}\right]^{\mathrm{T}} \tag{8.37}$$

and for all the n_p rotating speeds:

$$\left[\mathbf{L}_1\right] = \begin{Bmatrix} \left[\mathbf{L}_1\right]^{(1)} \\ \vdots \\ \left[\mathbf{L}_1\right]^{(n_p)} \end{Bmatrix} \tag{8.38}$$

The effect on the measured d.o.f.s $\left[\hat{\mathbf{X}}_n\right]$, which vector is of order $(2n_b n_p \times 2)$, due to unitary force systems applied to the selected element of the model is now calculated. This is done by first substituting eq. (8.37) in the r.h.s. of eq. (8.36), and inverting matrix $\left[\mathbf{E}(n\boldsymbol{\Omega})\right]$, obtaining the matrix $\left[\mathbf{H}(n\boldsymbol{\Omega})\right]$.

$$\mathbf{X}_n = \left[\mathbf{E}(n\boldsymbol{\Omega})\right]^{-1} \mathbf{F}_{f_n}(\boldsymbol{\Omega}) = \left[\mathbf{H}(n\boldsymbol{\Omega})\right] \mathbf{F}_{f_n}(\boldsymbol{\Omega}), \quad n = 1,2,3 \tag{8.39}$$

Then, the vibrations of the d.o.f.s, which are measured, are separated from the remaining d.o.f.s of the system, by considering only the rows of $\left[\mathbf{H}(n\boldsymbol{\Omega})\right]$ corresponding to the measured d.o.f.s. The partitioned matrix is of order $\left(2n_b n_p \times (4n_r + n_f)n_p\right)$ and following eq. (8.40) is obtained:

$$\left[\hat{\mathbf{X}}_n\right] = \left[\mathbf{H}(n\boldsymbol{\Omega})\right]\Big|_{\substack{mesured \\ d.o.f.s}} \left[\mathbf{L}_1\right], \quad n = 1,2,3 \tag{8.40}$$

Now the complex vector:

$$\bar{\mathbf{A}}_n = \left\{ \bar{A}_n^{(x)} \quad \bar{A}_n^{(y)} \right\}^{\mathrm{T}} \tag{8.41}$$

(i.e. the module and the phase for each harmonic component) of the equivalent force systems, applied to the selected element, that fits best the experimental data $\mathbf{X}_{m_n}$, of order $(2n_b n_p \times 1)$, has to be estimated. The fitting is done in least square sense, since the number of the unknown (the module and the phase) is less than the equations (recalling that data are corresponding to several rotating speeds and each of the sets is composed by several measuring planes). The problem is equivalent to:

$$\min \left\| \left[\hat{\mathbf{X}}_n \right] \bar{\mathbf{A}}_n - \mathbf{X}_{m_n} \right\|, \quad n = 1,2,3 \tag{8.42}$$

whose general solution is given by means of the pseudo-inverse calculation:

$$\bar{\mathbf{A}}_n^{(1)} = \begin{Bmatrix} M_n^{(x),(1)} \, \mathrm{e}^{\mathrm{i}\varphi_n^{(x),(1)}} \\ M_n^{(y),(1)} \, \mathrm{e}^{\mathrm{i}\varphi_n^{(y),(1)}} \end{Bmatrix} = \left(\left[\hat{\mathbf{X}}_n \right]^{\mathrm{T}} \left[\hat{\mathbf{X}}_n \right] \right)^{-1} \left[\hat{\mathbf{X}}_n \right]^{\mathrm{T}} \mathbf{X}_{m_n}, \quad n = 1,2,3 \tag{8.43}$$

The module and the phase of the complex value in the rows of $\bar{\mathbf{A}}_n^{(1)}$ are the equivalent bending moments due to the crack, in the selected element, which is the first element as indicated by superscript (1) in eq. (8.43). Finally the residual in the selected element is determined, by first obtaining the calculated response due to the identified fault in the selected element:

$$\mathbf{X}_{c_n} = \left[\hat{\mathbf{X}}_n \right] \bar{\mathbf{A}}_n^{(1)}, \quad n = 1,2,3 \tag{8.44}$$

and then by normalizing it:

$$\delta_{r_n}^{(1)} = \left(\frac{ \left[\mathbf{X}_{c_n} - \mathbf{X}_{m_n} \right]^{*\mathrm{T}} \left[\mathbf{X}_{c_n} - \mathbf{X}_{m_n} \right] }{ \mathbf{X}_{m_n}^{*\mathrm{T}} \, \mathbf{X}_{m_n} } \right)^{1/2}, \quad n = 1,2,3 \tag{8.45}$$

The procedure is then iterated for all the n_e elements of the rotor (or of the subset if the research is limited to a rotor part). If a fault only is considered, a set in $\mathbb{R}$ of relative residuals given by eq. (8.45), ordered by the element number, is obtained:

$$\boldsymbol{\delta}_{r_n} = \left(\delta_{r_n}^{(1)}, \ldots, \delta_{r_n}^{(n_r - 1)} \right), \quad n = 1,2,3 \tag{8.46}$$

A graph representing the values of $\boldsymbol{\delta}_{r_n}$ along the shaft axis (from the first to the last element) can be drawn as shown for example in figure 8.16.

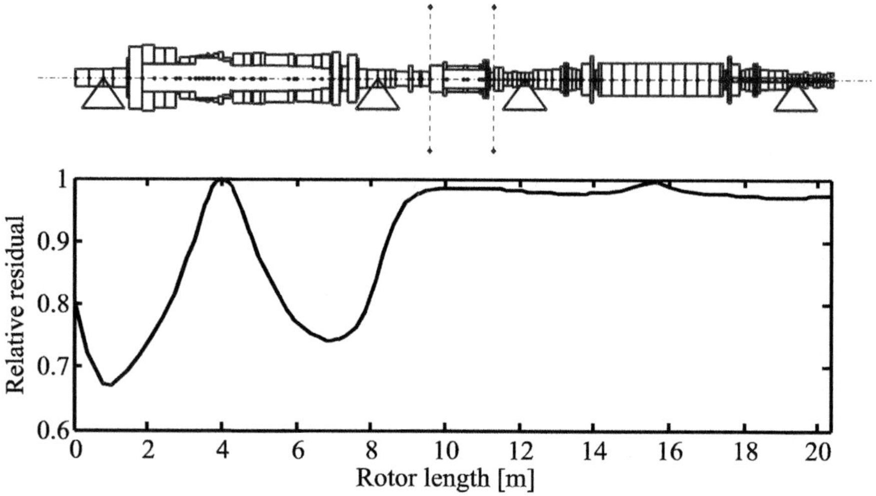

Fig. 8.16. Values of relative residual plotted along the shaft axis.

The location that corresponds to the minimum value of $\boldsymbol{\delta}_{r_n}$ indicates the most probable location of the fault (see figure 8.27, figure 8.34, figure 8.35, figure 8.54, figure 8.61 and figure 8.69), whose amplitude estimation is given by the corresponding values:

$$\mathbf{M}_n = M_n^{(x)} \mathrm{e}^{\mathrm{i}\varphi_n^{(x)}} + M_n^{(y)} \mathrm{e}^{\mathrm{i}\varphi_n^{(y)}} \tag{8.47}$$

of eq. (8.43). The closer to zero the minimum value of eq. (8.46), the better the estimation of the faults per each harmonic component.

It is worth noting that the 1X vibration components are due both to the breathing mechanism of the crack and to the local bow which generally has developed during the crack propagation. Therefore, the 1X component is useful for the localization of the crack, but not for the identification of its depth.

The 3X component is rather small and generally masked by some noise. Often this component can be recognized only when approaching the resonant condition at a rotating speed equal to ⅓ of the rotor 1st critical speed.

The 2X component is therefore the most suitable symptom for detecting position and depth of the crack; the highest values are obviously reached during a speed transient when approaching the resonant condition at ½ of the rotor 1st critical speed.

The final remark about the model based method regards the accuracy of the machine model. A full discussion would be far from the scope of this paper, but a reader potentially interested in measures of accuracy in model based identification can refer to [8.28].

8.4.3 Crack Depth Identification

The following procedure has been implemented for the identification of the crack depth [8.39]. The identification procedure, described in the previous section, identifies the crack position in a particular element of the rotor, whose length is known from the f.e.m. and equal to l (see figure 8.17). Figure 8.17 shows also the crack depth a, the d.o.f.s of the extremity nodes of the element and the bending caused by the equivalent moment components $\mathbf{M}_n$ (1X, 2X and 3X components) that are applied to this element. These equivalent bending moment components $\mathbf{M}_1$, $\mathbf{M}_2$ and $\mathbf{M}_3$ have been calculated from the corresponding 1X, 2X and 3X measured vibrational behaviour.

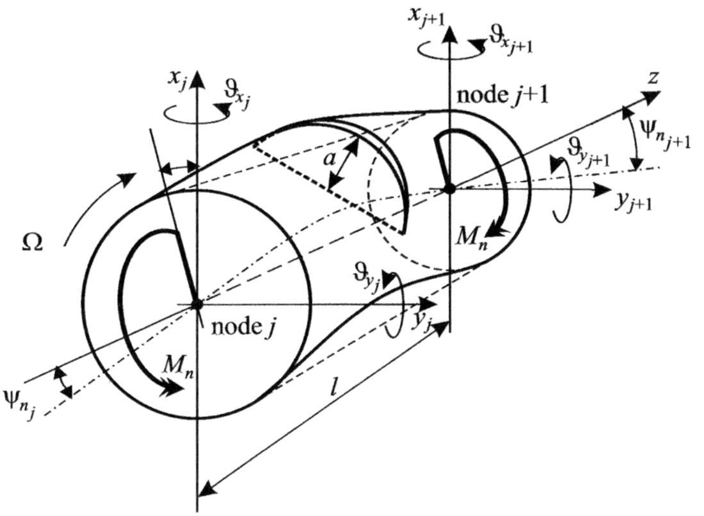

Fig. 8.17. Cracked element.

Then the static bending moment $\mathbf{M}$ in correspondence of the same element, due to the weight and to bearing alignment conditions, is calculated from model data.

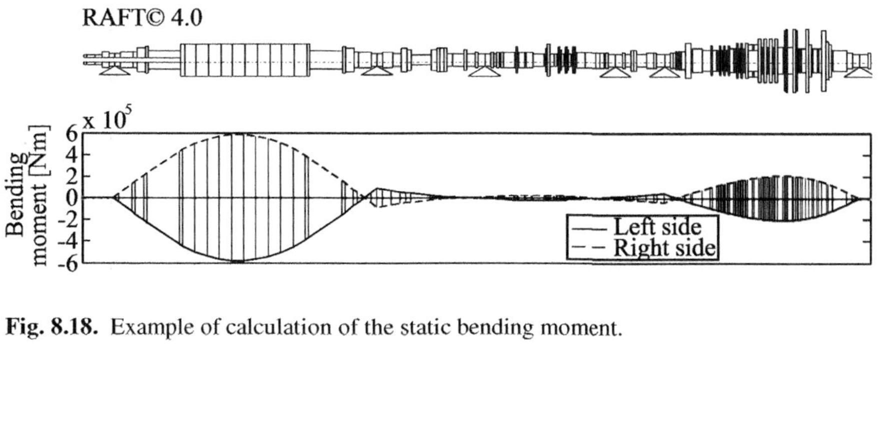

Fig. 8.18. Example of calculation of the static bending moment.

The moments $\mathbf{M}_1$, $\mathbf{M}_2$ and $\mathbf{M}_3$ represent the amplitude of the equivalent crack forces defined respectively by eqs. (8.25), (8.26) and (8.27). If we neglect all terms of eqs. (8.25), (8.26) and (8.27) except the first one, we obtain:

$$\mathbf{F}_{f_n} = -[\Delta\mathbf{K}_n]\mathbf{x}_s \tag{8.48}$$

This assumption is reasonable if heavy, horizontal axis, industrial rotating machinery are considered, since the statical deflection, in the shaft sections loaded by consistent bending moments, is much higher than the vibrations, especially when operating conditions far from resonances in critical speeds are analyzed. Moreover, it is possible to show that the projections along the reference axes are equal in case of 2X component, i.e. $M_2^{(x)} = M_2^{(y)} = M_2$ and $\varphi_2^{(x)} = \varphi_2^{(y)}$, and the equivalent force system is a rotating moment of amplitude M_2.

Even if vectors $\mathbf{F}_{f_n}^{(x,y)}$ have order $(4n_r + n_f)\times 1$, the only elements different from zero are those corresponding to the d.o.f.s of the cracked element j-th. Therefore, it is more convenient to consider only the sub-matrix $[\Delta\mathbf{K}_n]\big|_j$ of order 8×8 corresponding to those d.o.f.s and the displacement vector $\mathbf{x}_s\big|_j$ of the nodes of the cracked element (see figure 8.17):

$$\mathbf{x}_s\big|_j = \Big\{ \underbrace{x_j \quad \vartheta_{y_j} \quad y_j \quad \vartheta_{x_j}}_{j^{\text{th}}\text{ rotor node}} \quad \underbrace{x_{j+1} \quad \vartheta_{y_{j+1}} \quad y_{j+1} \quad \vartheta_{x_{j+1}}}_{j+1^{\text{th}}\text{ rotor node}} \Big\}^{\mathrm{T}} \tag{8.49}$$

By recalling eqs. (8.33) and (8.48), it follows:

$$\left[\begin{matrix} \underbrace{\begin{matrix} 0 & 0 & 0 & \mathrm{i} \\ 0 & 1 & 0 & 0 \end{matrix}}_{j\text{-th rotor node}} & \underbrace{\begin{matrix} 0 & 0 & 0 & -\mathrm{i} \\ 0 & -1 & 0 & 0 \end{matrix}}_{j+1\text{-th rotor node}} \end{matrix}\right]^{\mathrm{T}} \begin{Bmatrix} M_n^{(x)} e^{\mathrm{i}\varphi_n^{(x)}} \\ M_n^{(y)} e^{\mathrm{i}\varphi_n^{(y)}} \end{Bmatrix} = -[\Delta\mathbf{K}_n]\big|_j \mathbf{x}_s\big|_j \tag{8.50}$$

and for the modulus:

$$M_n = \|\mathbf{M}_n\| = \left\| [\Delta\mathbf{K}_n]\big|_j \mathbf{x}_s\big|_j \right\| \tag{8.51}$$

Matrix $[\Delta\mathbf{K}_n]$ is restricted to the nodes of the cracked element, therefore only the statical deflections of the nodes of the cracked element are required in eq. (8.51). Since in the crack, as in the Bernoulli's beam, the shear force can be neglected with respect to bending moment flexibility, $\mathbf{x}_s\big|_j$ is composed mainly by rotational deflections of the extremity nodes. Therefore $\mathbf{x}_s\big|_j$ can be calculated by considering the static bending moments due to static, fixed in space, force vector $\mathbf{W}$ acting on the cracked element. These bending moments do not change much from node to node of a element, for a suitably refined mesh of the rotor, and can therefore be considered constant in the cracked element and equal to M . Moment M is then proportional to the relative rotations of the extremity nodes of the cracked element, $\psi_{j+1}-\psi_j$ (see figure 8.17).

By considering the sub-matrix $\left[\mathbf{K}_m\right]\big|_j$ of order 8×8 corresponding to those d.o.f.s extracted from the matrix $\left[\mathbf{K}_m\right]$, i.e. the mean value of the stiffness matrix of the cracked element, the static deflections of the cracked element are given also by:

$$\mathbf{x}_s\big|_j = \left(\left[\mathbf{K}_m\right]\big|_j\right)^{-1}\left\{\underbrace{0 \quad \cos\varphi_s \quad 0 \quad \sin\varphi_s}_{j\text{-th rotor node}} \quad \underbrace{0 \quad -\cos\varphi_s \quad 0 \quad -\sin\varphi_s}_{j+1\text{-th rotor node}}\right\}^{\mathrm{T}} M \qquad (8.52)$$

By replacing eq. (8.52) in eq. (8.51), it finally results:

$$M_n = \left\|\left[\Delta\mathbf{K}_n\right]\big|_j\left(\left[\mathbf{K}_m\right]\big|_j\right)^{-1}\right\| M \quad\rightarrow\quad \frac{M_n}{M} = \left\|\left[\Delta\mathbf{K}_n\right]\big|_j\left(\left[\mathbf{K}_m\right]\big|_j\right)^{-1}\right\| \qquad (8.53)$$

Eq. (8.53) shows that ratio M_n/M depends only on the crack relative depth $p = a/D$, since $\left[\Delta\mathbf{K}_n\right]\big|_j$ and $\left[\mathbf{K}_m\right]\big|_j$ of the cracked element are function of p only, if regular shapes with rectilinear tips are considered.

A similar ratio M_n'/M has been calculated for different cracks depths considering the equivalent length l_c of a reference cracked beam element, instead of the length l of the cracked element in the model of the shaft. This equivalent length l_c has been tuned by means of numerical 3D analyses (see figure 8.19) as shown in [8.41] and [8.42] and in section 5.4.3.

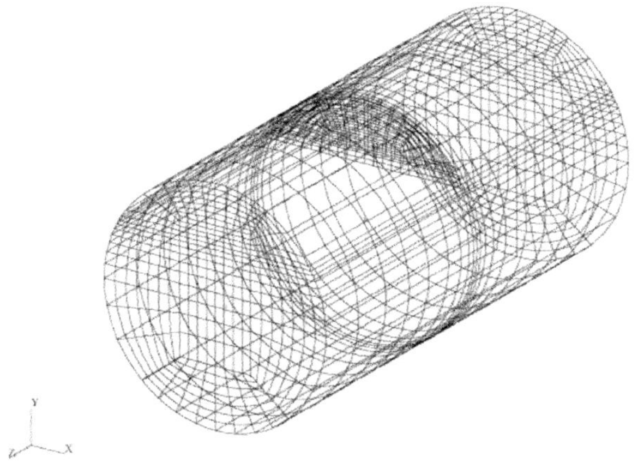

Fig. 8.19. Equivalent cracked beams models used in 3D analyses.

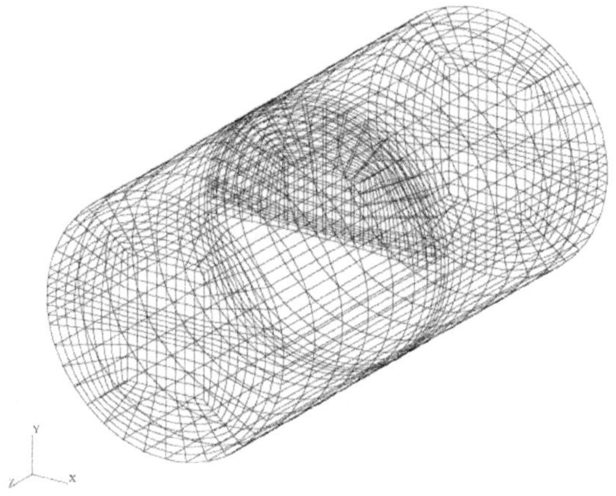

Fig. 8.19. (*continued*)

Moment M'_n gives the same relative rotation of the nodes of the beam with length l_c as M_n does with a beam with length l:

$$\Delta\psi_n = \frac{M'_n}{EJ} l_c \tag{8.54}$$

This is represented in figure 8.20 for the 1X, 2X and 3X component and expressed by the relationship in eq. (8.55).

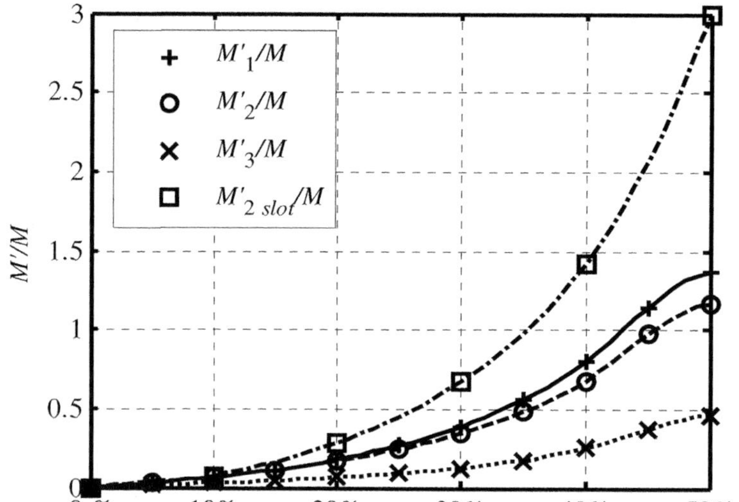

Fig. 8.20. Bending moments ratio on the equivalent cracked beam, as a function of crack relative depth $p = a/D$ for the nX component.

$$\frac{M'_n}{M} = f(p) \tag{8.55}$$

In the same figure also the curve M'_{2slot} / M for an always open crack (a slot or notch) is shown: in this case the 2X component only is present and 1X and 3X component are absent.

Eq. (8.55) can then be used for determining the crack depth. However, as shown in [8.41] and section 5.4.3, the length l_c of the equivalent, reduced stiffness, beam element that simulates the behaviour of the cracked beam, is also depending on the relative crack depth p:

$$\frac{l_c}{D} = g(p) \tag{8.56}$$

The function $g(p)$ is represented in figure 8.21. Anyhow, coming from the identification procedure, the equivalent bending moments M_n are applied to an element with an actual length l instead of l_c. It is worth noting that the nX measured displacements are the effect of the relative rotation of the cracked element extremity nodes, which is proportional to the product $M_n \cdot l$ of the identified nX bending moment component M_n applied to one element of the finite element model of the rotor, multiplied by its length l.

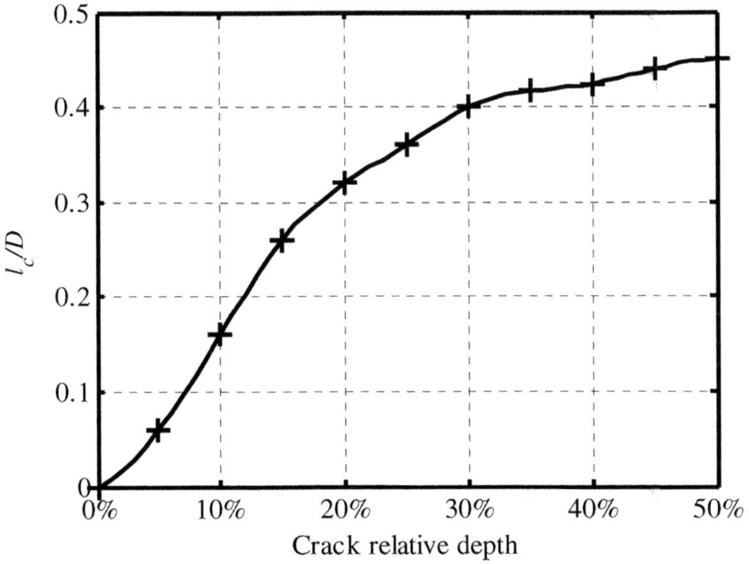

Fig. 8.21. Relationship between the crack relative depth p, the diameter D and the length l_c of equivalent beam.

The equivalent bending moment component M'_n, applied to an equivalent cracked beam element of length l_c, can therefore be calculated as:

$$M'_n \cdot l_c = M_n \cdot l \tag{8.57}$$

By assuming that the static bending moment M applied to the original element of length l does not change along the element, the same M can be considered applied to the element with equivalent length l_c. Recalling eq. (8.55) we can derive:

$$\frac{M'_n}{M} = \frac{M_n \cdot l}{M \cdot l_c} = f(p) \tag{8.58}$$

and using eq. (8.56) we get:

$$\frac{M_n \cdot l}{M \cdot D} = f(p) \cdot g(p) \tag{8.59}$$

Eq. (8.59), shown in figure 8.22 for the nX components, can then be used for determining, from the known left hand side, the relative depth of the crack.

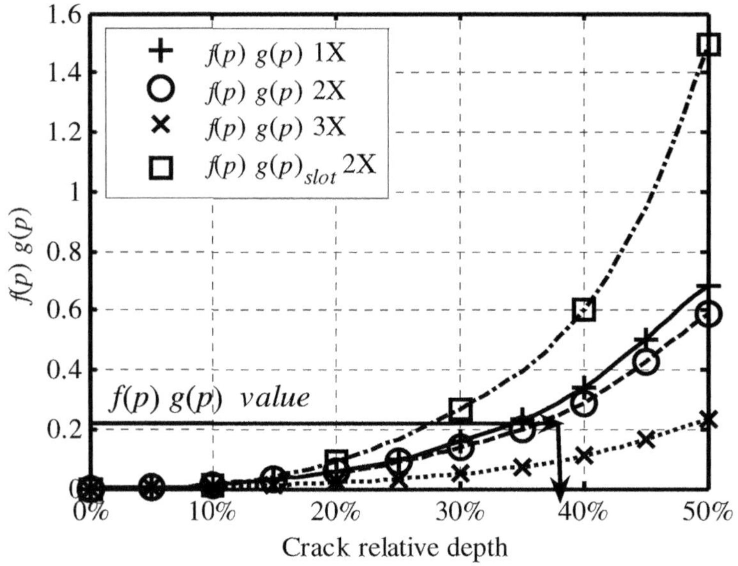

Fig. 8.22. Function for the calculation of the crack depth.

8.5 Results of Model Based Approach

The experimental results obtained by applying the model based approach are shown for two different test-rigs.

8.5.1 Test-Rig of Politcnico di Milano

The MODIAROT test-rig of Dept. of Mechanical Engineering of Politecnico di Milano, shown in figure 8.23 and figure 6.9, is composed of two rigidly coupled rotors, supported on four lemon-shaped oil-film bearings, driven through a flexible coupling by a variable speed, inverter controlled, electric motor.

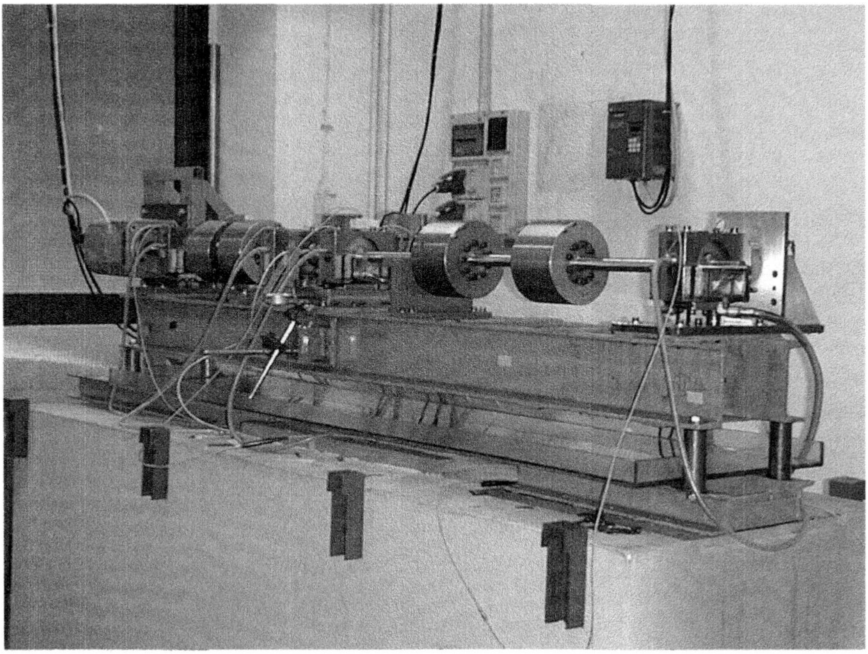

Fig. 8.23. 4-bearing 2-shaft MODIAROT test-rig on flexible foundation with fully instrumented bearing housing.

The shaft diameter is 25 mm and the total length of the two rigid coupled shafts is about 2 m, the total mass is less than 100 kg. The critical speeds are about 1200 rpm (1st) and 3400 rpm (2nd). The supporting structure is flexible and has several natural frequencies in the operating speed range (one of its resonances is about 2000 rpm). The vibrations in each bearing are measured by 2 eddy-current proximity probes fixed to the bearing housing in ±45° directions with respect to the vertical, measuring the shaft relative displacements (see figure 8.24).

Fig. 8.24. Sensor position on the bearing housing: proximity probes are ±45°, accelerometers (not shown in the pictures for clarity) are positioned on the white tapes.

Each bearing housing is equipped with 2 accelerometers measuring the absolute vibrations of the bearing housing. By rotating the reference frame and adding the absolute displacement of the bearing housing, the absolute vibrations of the shaft in correspondence of the bearing can be obtained and compared with calculated results. All signals during a run-down transient were analyzed by means of a 8 channel spectrum analyzer in tracking filter modality. This heavy processing of the experimental data introduced significant errors in the experimental data, as will be shown later.

The model of the complete rotor is composed of 47 elements and is shown in figure 8.25.

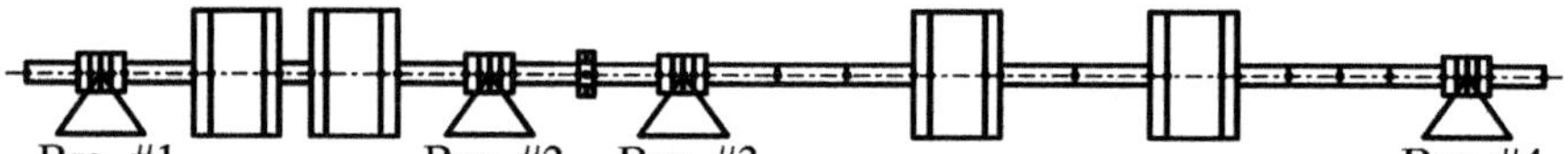

Fig. 8.25. Model of the MODIAROT test-rig.

A crack with estimated depth of about 25%-30% of the shaft diameter has been generated by means of fatigue bending solicitation in the middle of the longer shaft, starting from a very small slot, as introduced in section 3.1.1.1. The supposed crack profile is shown in figure 3.1. The tests have been carried out on both the quasi-static and the dynamic behaviour. In the first case, the longer shaft has been disassembled and positioned on supports or on a lathe. The procedure of the tests has been discussed in sections 3.1.1.1 and 3.1.2.1.

8.5.1.1 Natural Frequencies

Different values of natural frequencies should be found in cracked rotors, depending on its angular position and therefore on the "open" or "closed" situation of the crack.

The frequencies corresponding to a complete open crack and to a complete closed crack have been calculated, assuming a stiff supporting structure and measuring its natural frequencies with impact tests, as discussed in section 3.1.1.1.

8.5.1.2 Quasi-Static Behaviour

The cracked shaft, equipped with its massive disks, was mounted on a lathe and 4 proximity probes have been installed according to figure 3.9 on a movable frame. The sensor signals have been recorded during slow rotation in 11 positions along the shaft and analysed.

The sensors in opposite positions allow to separate 2X component due to geometric errors (shaft ovalization) from 2X component due to the crack: the sum indicates twice the ovalization, the difference twice the 2X displacement (see figure 8.26).

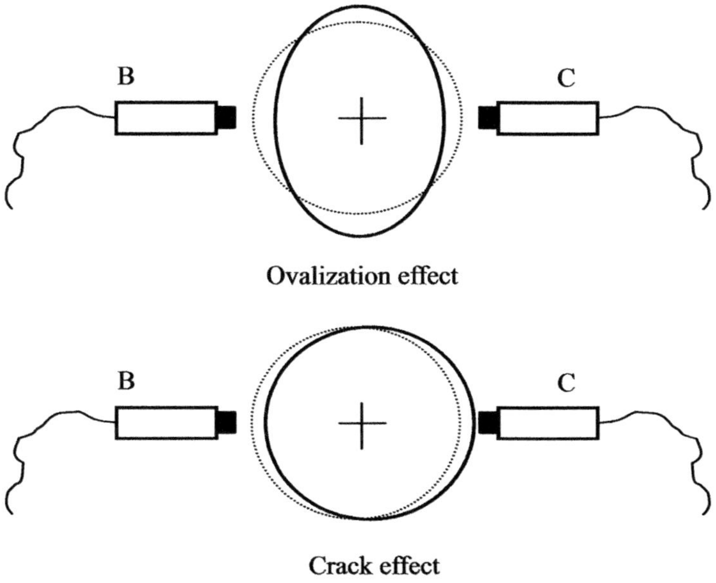

Fig. 8.26. 2X components.

The vertical and horizontal 2X should be equal, according to the results of the FLEX model as exposed in sections 5.4.3 and 5.6. The measured displacements have been compared with those calculated with the FLEX simplified model, for a depth of 24%: these comparisons have been shown in figure 3.7 and figure 3.8 for the horizontal and vertical components. The 2X measured deflection is rather linear, except for the small cracked element, and fits very well with the calculated deflection.

With regard to the other harmonic components, the 1X component is masked by high static bow, while the 3X component is so small that its presence is masked by geometric surface errors and by the measuring errors.

The displacement values of the 2X components, for all the 11 measuring points, have been used for the identification procedure. The results obtained are shown in figure 8.27.

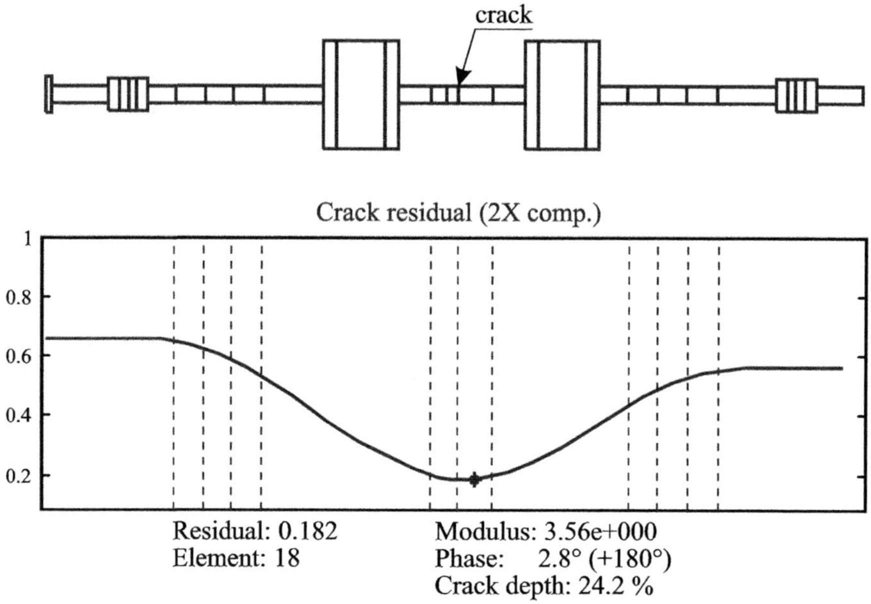

Fig. 8.27. Crack identification in quasi-static case.

The location of the suspected crack is identified with good accuracy, and also its depth is identified. The high quality of the identification is measured by the low value of the minimum relative residual ($\delta_r = 0.182$). The good identification results are probably due to the high number of measuring stations (11), which allows to define very well the deflection shape.

8.5.1.3 Dynamic Behaviour

The same cracked shaft, installed in the test-rig, shows at low speed vanishing small 2X components in the bearings, which are completely masked by 2X ovalization errors in the shaft journals.

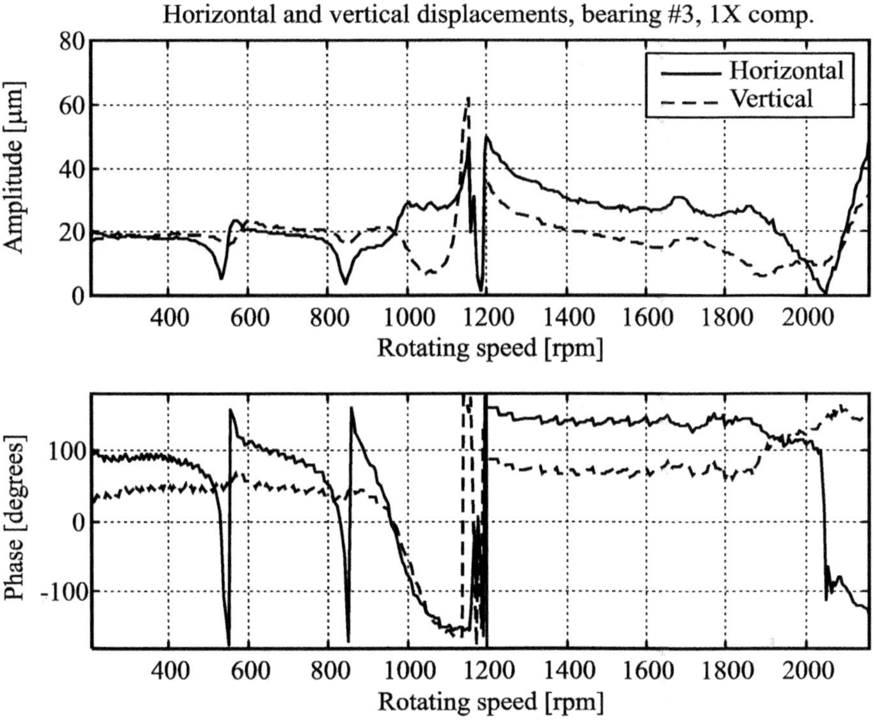

Fig. 8.28. Experimental displacements at bearing #3, 1X component of cracked shaft.

The experimental results obtained in the bearings (e.g. in bearing #3) are shown in figure 8.28 for the 1X component, where high bow is recognisable at low rotating speed, in figure 8.29 for the 2X component, where a high ovalization error is recognisable and in figure 8.30 for the 3X component, where the amplitude of 2 μm is exceeded only when passing a resonance.

The previously stored 2X components of the un-cracked rotor, i.e. the reference case, have been subtracted from the actual vibrations and the 2X additional vibrations are represented in figure 8.31. It can be seen that the crack induced 2X components in the bearings are affected by errors due to the heavy processing of the experimental data. Also the data below 550 rpm were lost.

In order to emphasise the vibrations due to the crack, a measuring station with four proximity probes, similar to that one of figure 3.9, has been positioned in correspondence of the crack. The results in terms of 2X and 3X components in vertical direction are shown in figure 8.32 and figure 8.33. The agreement with calculated results using FLEX model is much better in this measuring station with respect to the measuring planes corresponding to the bearings. In this case, the measuring station is connected to the concrete base below the flexible structure supporting the bearings. The concrete base is very stiff and the vibrations measured in the measuring station are already absolute vibrations.

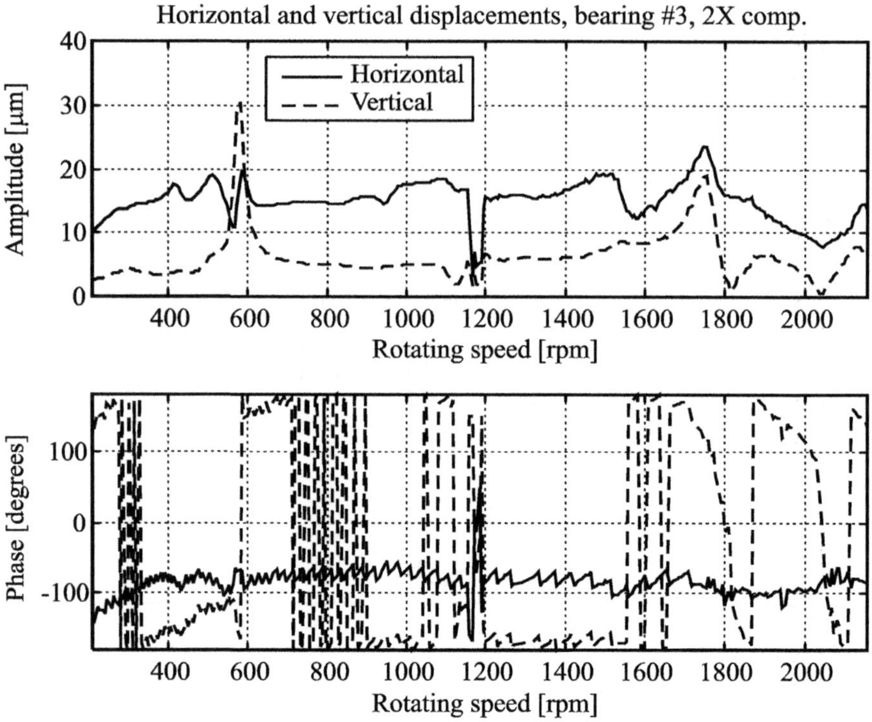

Fig. 8.29. Experimental displacements at bearing #3, 2X component of cracked shaft.

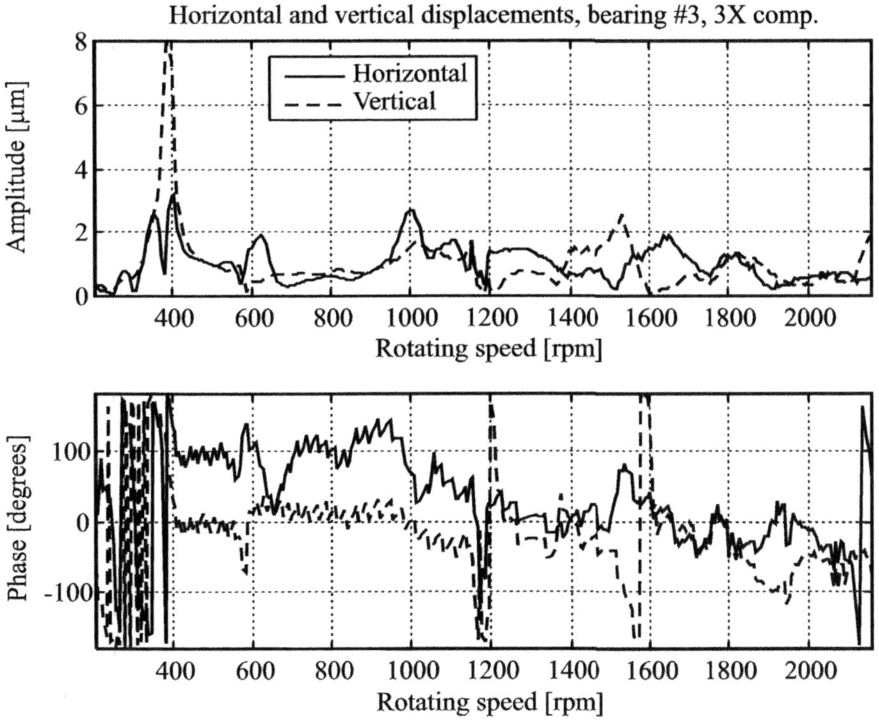

Fig. 8.30. Experimental displacements at bearing #3, 3X component of cracked shaft.

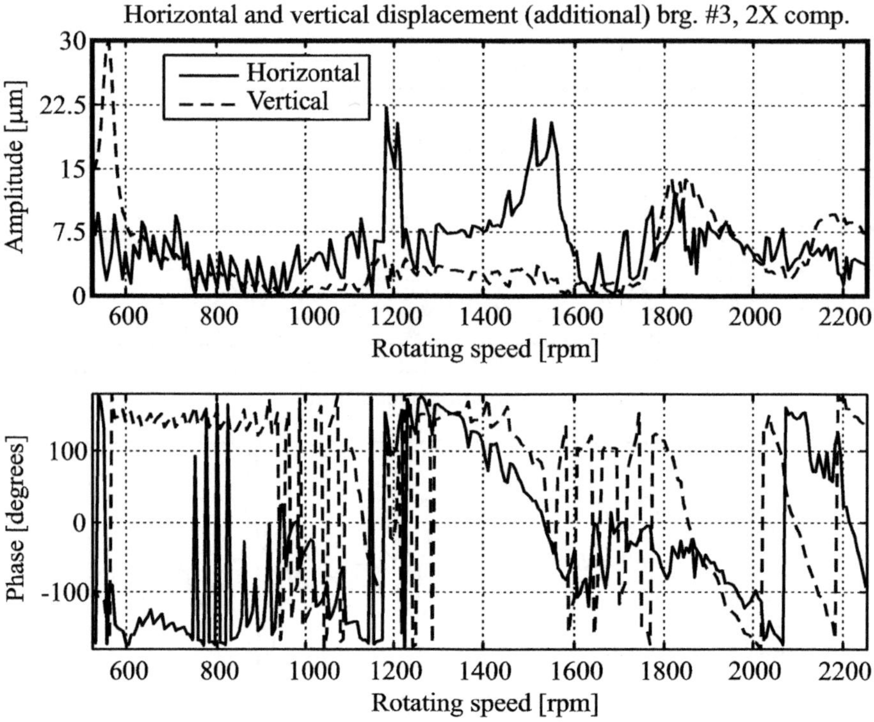

Fig. 8.31. 2X additional vibration components at bearing #3.

Very high peaks are evident in vertical direction when passing ½ of the 1st critical speed for the 2X component (see figure 8.32), and ⅓ of the 1st critical speed for the 3X component (see figure 8.33). Despite some 2X excitations, due to journal ovalization, which are not accounted for in the model, and some 2X and 3X surface errors, a rather good agreement between calculated and measured results is generally found, which confirms the validity of the model. The measured 3X component is heavily masked by 3X surface errors (of roughly 5 μm).

An attempt to identify the crack (its location and its amount) using only the "poor" bearing measurements (like those shown in figure 8.31), considering separately the 2X and 3X components, lead to the results shown in figure 8.34. The position of the identified crack is quite close to the actual position, both in 2X and 3X case, while the depth of the crack is overestimated. The poor quality of the identification is represented by the high values of the minimum relative residual (δ_r = 0.859 and 0.923).

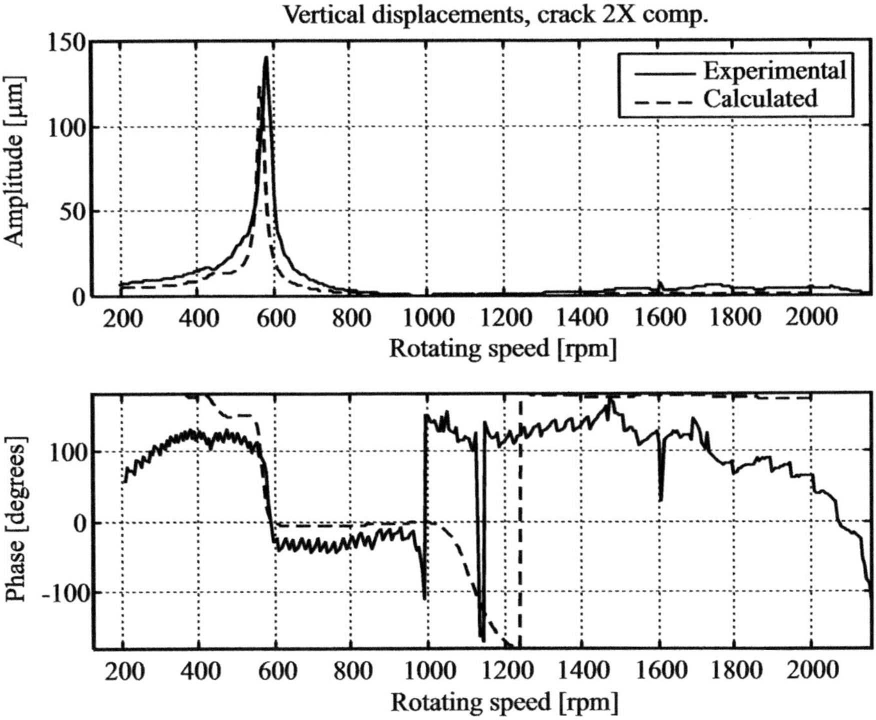

Fig. 8.32. 2X measured and calculated vertical displacements at crack location.

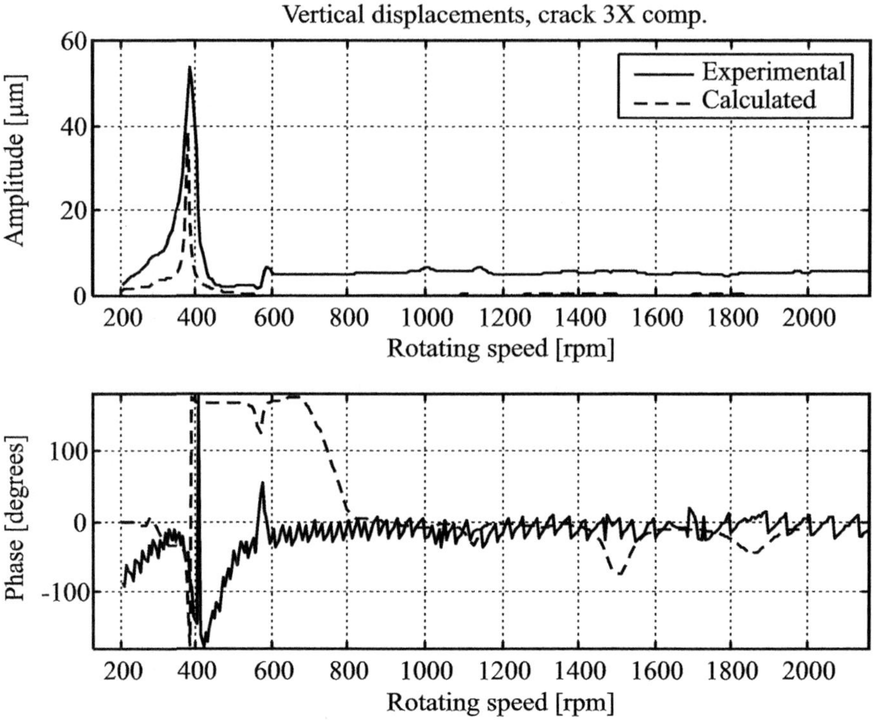

Fig. 8.33. 3X measured and calculated vertical displacements at crack location.

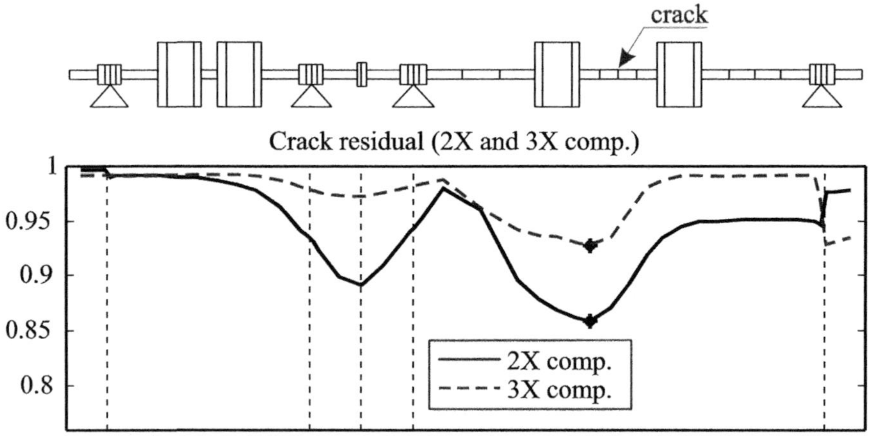

2X comp.:	Residual: 0.859	Modulus: 3.69e+000
	Element: 33	Phase: -107.3° (+180°)
		Crack depth: 39.9 %
3X comp.:	Residual: 0.923	Modulus: 6.35e-001
	Element: 33	Phase: 100.0° (+180°)

Fig. 8.34. Identification of the crack with 2X and 3X components, measured in the bearings.

Nevertheless this result can be deemed as good, considering the heavy processing of the experimental data and the lacking of experimental data in the lower speed range, where the 2X and 3X resonances amplify the corresponding vibration components. By considering also the measurements taken at crack location, which is unrealistic in plant operating machines, higher accuracy in location and depth is expected as shown for the 2X component in figure 8.35.

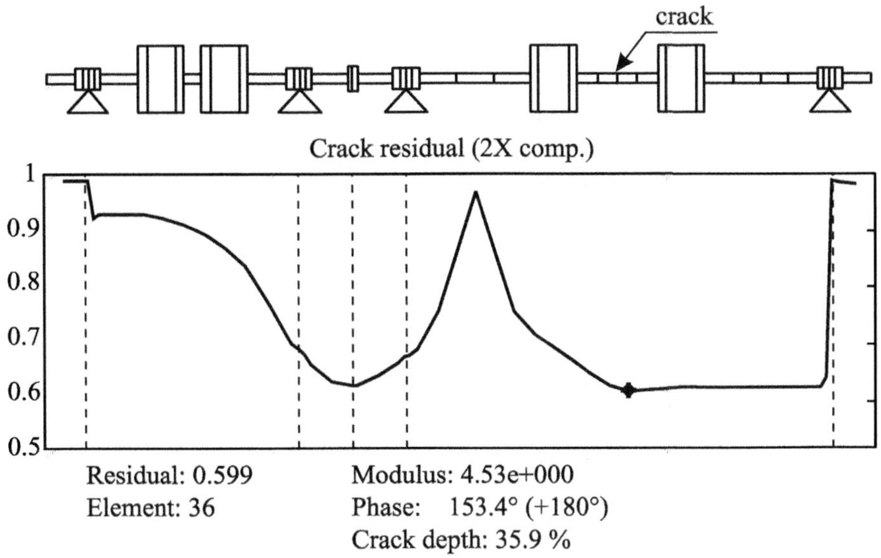

Fig. 8.35. Identification of the crack with 2X using also the measurements in the crack position.

8.5.2 Test-Rig of Électricité de France

The test-rig EURoPE, shown in figure 8.36, is composed of a shaft divided in three parts supported by two equal oil film bearings. The bearings are three-lobed shape. The nominal diameter of the shaft is 70 mm and the overall length is 3.15 m. The distance between the bearings is 1.88 m. The total mass is 450 kg and the main inertia disk has a mass of 250 kg. In this configuration the 1st critical speed is close to 1150 rpm. The supporting structure can be considered as rigid in the speed range from 0 to 1500 rpm. The proximity probes for the measurements of relative shaft journal vibrations are installed very close to the bearings, but not inside of them, as usually occurs in real machines (in figure 8.36, they are both on the bearing side towards the motor as shown in figure 8.37). The test-rig has been designed in order to be as much as possible representative of industrial machines, mainly steam turbines of power plants.

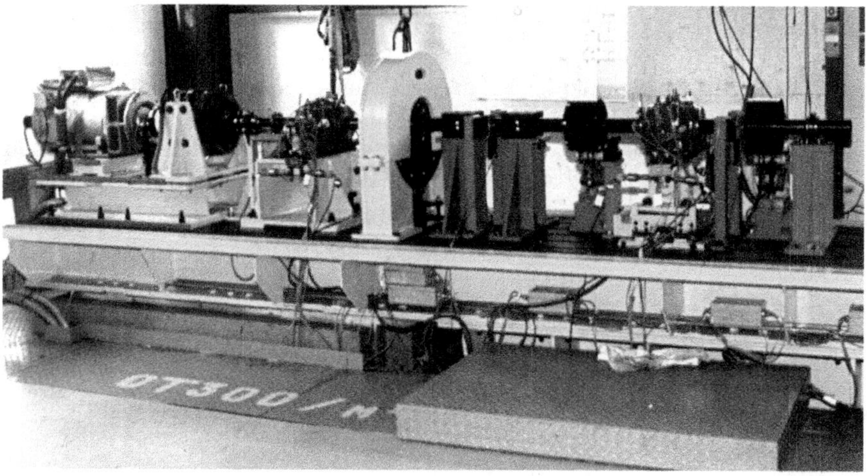

Fig. 8.36. 2-bearing 1-composed shaft test-rig on rigid foundation.

Fig. 8.37. Close-up of the position of proximity probes on the bearings (motor is on left side).

Maximum design speed is 3000 rpm. In this application for safety reasons, the maximum speed was reduced to 1500 rpm. The last is also the operating speed of turbo generators of many nuclear power plants, equipped with two pole-pairs generators.

The model of the rotor is shown in figure 8.38 and is composed of 45 finite beam elements.

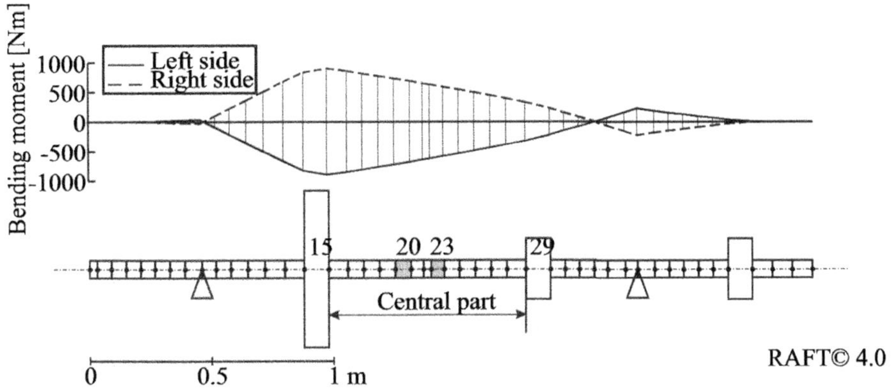

Fig. 8.38. Finite beam element model of the test-rig and statical bending moment distribution.

Since the central part of the rotor, between elements 15 and 29, can be disassembled, three types of crack were generated on three different specimen (see figure 8.39), then they were assembled in the rotor system and used for validating the identification procedure.

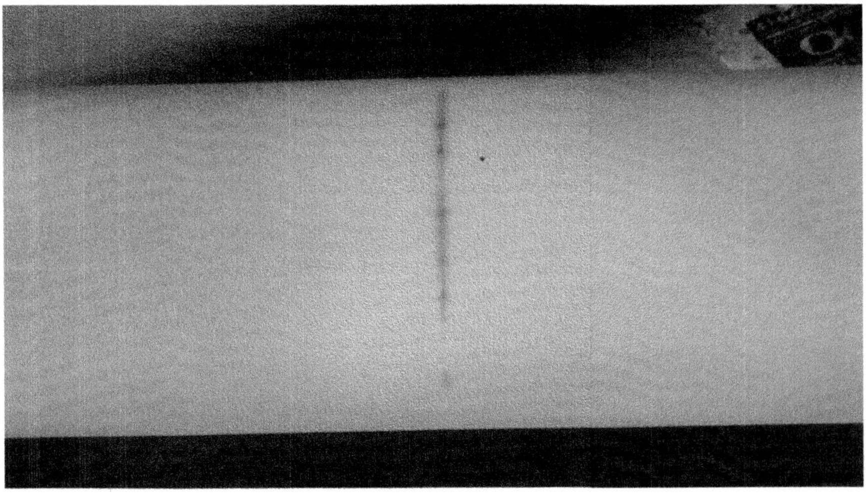

Fig. 8.39. Particular of the disassembled central part of the rotor and of the crack highlighted by dye-penetrant.

In the first case, a transverse slot of thickness 0.1 mm and of relative depth of 34% was generated by electroerosion, at a distance of 310 mm from the main inertia disk. This position corresponds roughly to element 20 in the finite beam element model of the rotor (see figure 8.38).

In the second case, starting from a notch made by electroerosion, a crack was obtained as a consequence of fatigue solicitation. The crack position along the rotor corresponds again to element 20. At the end of the tests this last shaft was broken to check the actual shape and depth of the crack. From the analysis of figure 8.40, the crack average depth results of only about 14% in average, with a maximum depth of 11.3 mm equal to 16.1%, but a relevant part of this depth is due to the notch.

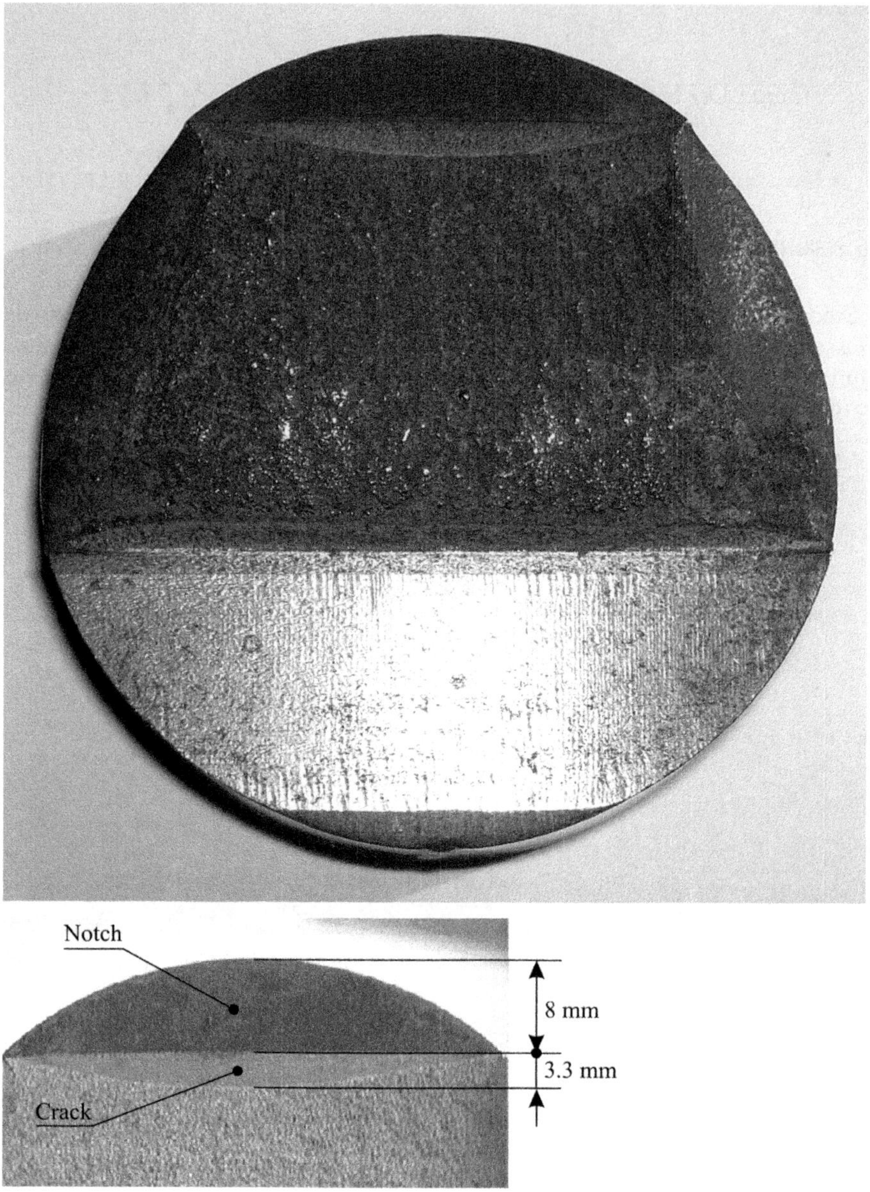

Fig. 8.40. Mean crack depth of 14% in the second experimental case.

In the third case, the crack has been started from a notch and made grown up to the depth of 33 mm, checked at the end of the tests by breaking the shaft. This depth corresponds to 47% of the shaft of 70 mm of diameter (see figure 8.41). In this case the depth of the notch is less than ⅓ of the total depth. The distance was of 430 mm from the main inertia disk, that corresponds to element 23 (see figure 8.38).

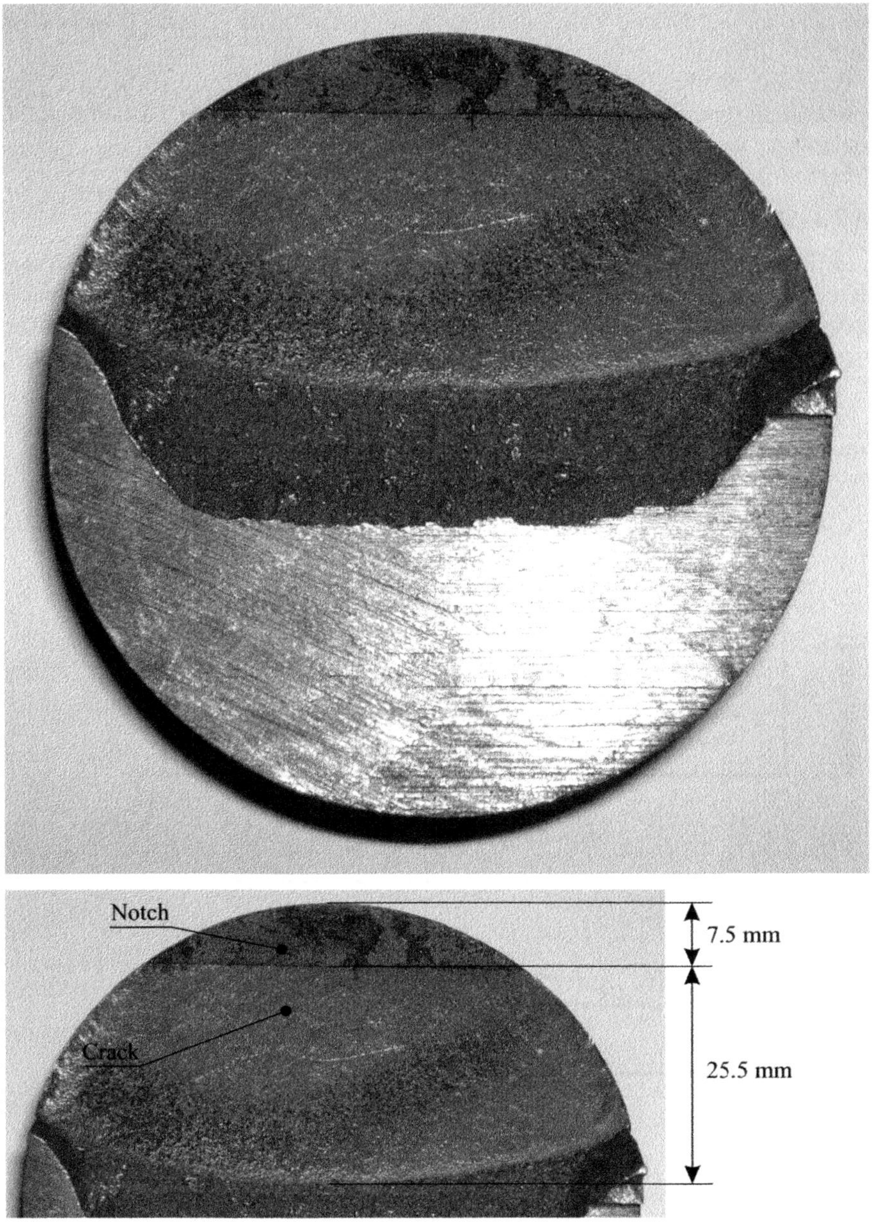

Fig. 8.41. Crack depth of 47% in the third experimental case.

The possibility of disassembling the central part of the rotor has the main advantage to not dismount the entire rotor in order to create a crack. However this leads to some difficulties to have a valid reference case. In fact, by considering a run down of the un-cracked rotor and a run-down of a cracked one, small differences in alignment might be introduced when the central part is coupled to the other extremities. Moreover, the cracked part presents usually a permanent bow due to the fatigue solicitation used to generate the crack. This bow is generally different from that present in the reference case.

The static bending moments are shown in figure 8.38. Note that the solid line represents the moments on the left side of the element, while the dashed line the moments on the right side.

8.5.2.1 Reference Situation

Even if the so called reference situation cannot actually be considered as the true reference situation of the same rotor in this test-rig, due to the reason previously expressed, nevertheless it has been used to calculate the vibration difference. The measured reference situation is reported in figure 8.42 to figure 8.47 for the 1X, 2X and 3X components.

From the analysis of the 1X component in figure 8.42 and figure 8.43, the 1st critical speed at about 1150 rpm is recognizable. Moreover, the rotor presents a bow which generates around 15 μm at very low speed in bearing #1.

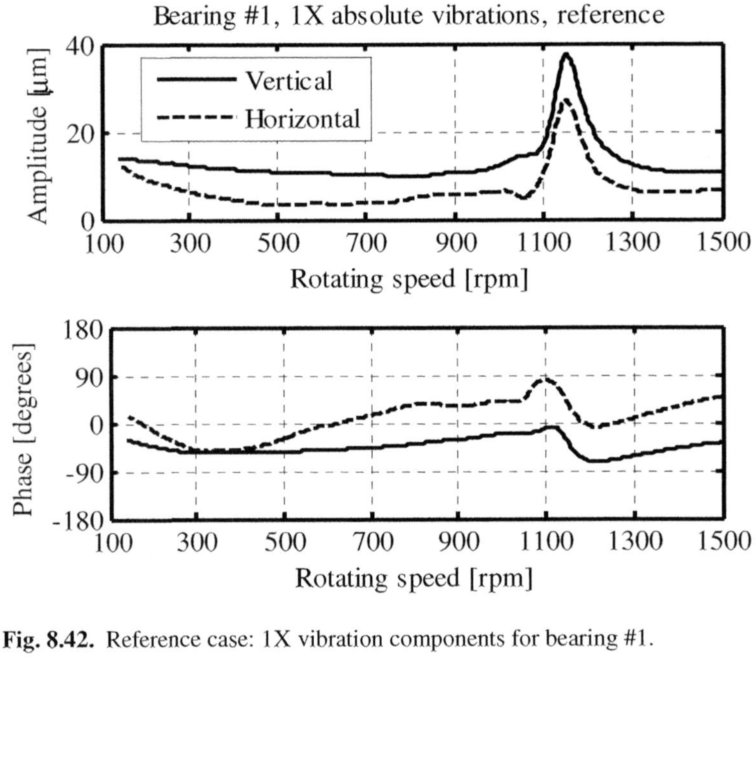

Fig. 8.42. Reference case: 1X vibration components for bearing #1.

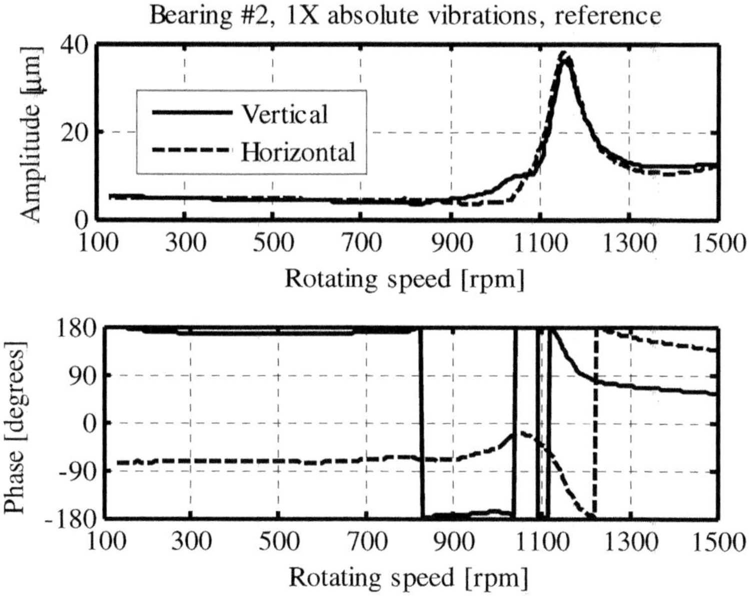

Fig. 8.43. Reference case: 1X vibration components for bearing #2.

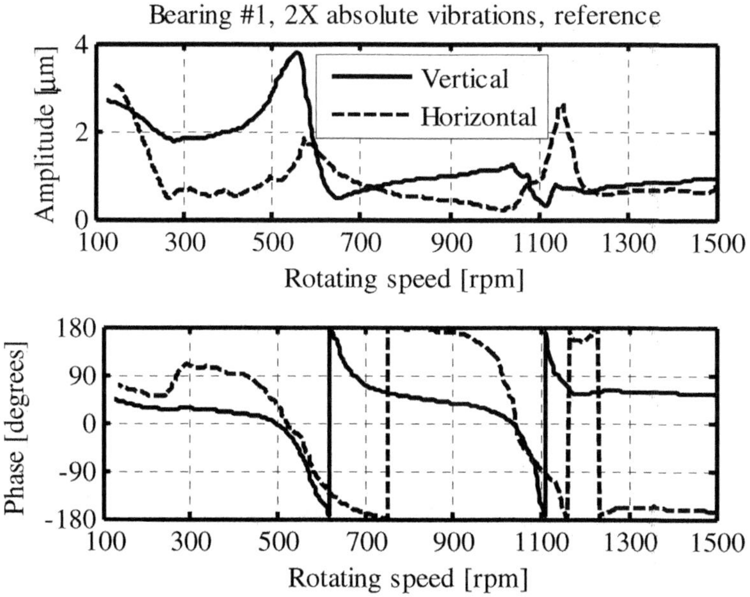

Fig. 8.44. Reference case: 2X vibration components for bearing #1.

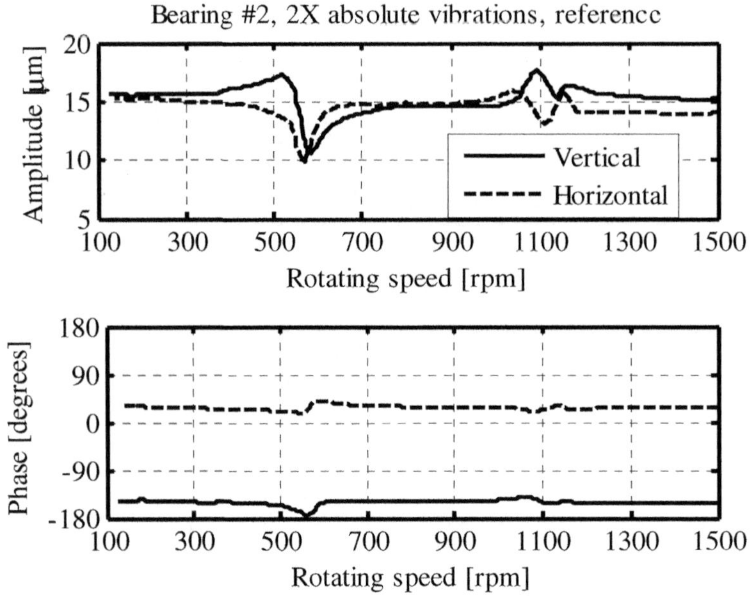

Fig. 8.45. Reference case: 2X vibration components for bearing #2.

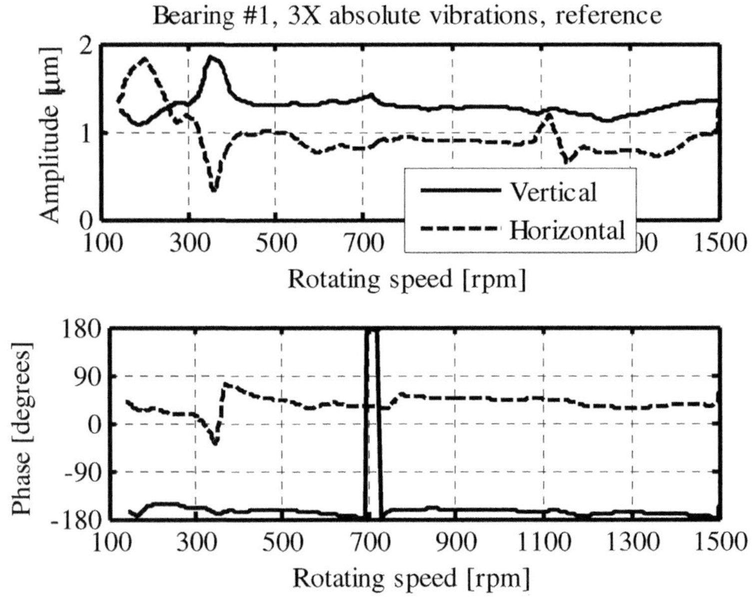

Fig. 8.46. Reference case: 3X vibration components for bearing #1.

With regard to the 2X component in figure 8.44 and figure 8.45, the 2X critical speed is rather evident at about ½ of the 1st critical speed, but also a peak at about 1100 rpm that indicates a non linear effect of the oil film can be recognized. The relatively high value of the 2X component in the second bearing (about 14 μm) at low speed, which remains the main component over all the speed range as is also shown by the phase trend, indicates a geometrical error of the shaft (journal ovalization) in the bearing. The phase difference of 180° between horizontal and vertical components is typical for ovalization errors. The 3X component has very reduced amplitude in both bearings (about 1 μm and 4 μm respectively) and is mainly due to some noise, however a smaller peak at about ⅓ of the 1st critical speed is recognizable (see figure 8.46 and figure 8.47).

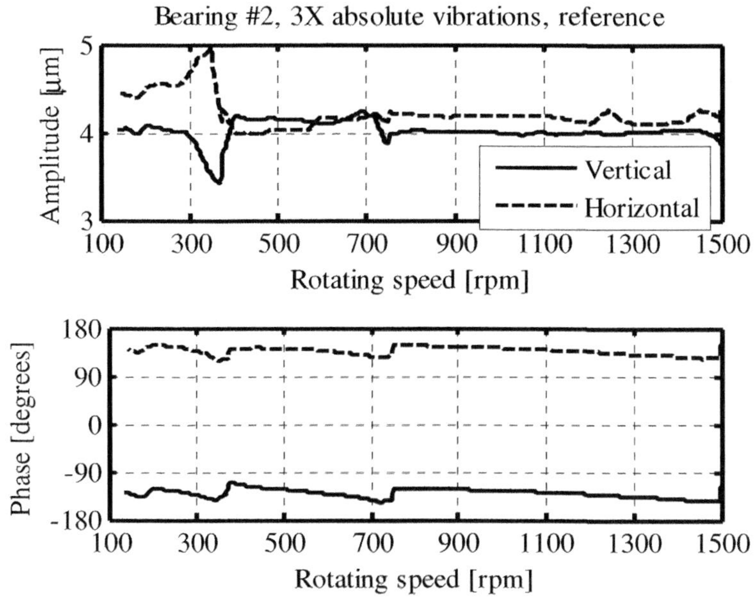

Fig. 8.47. Reference case: 3X vibration components for bearing #2.

8.5.2.2 *Case 1: Identification of a 34% Slot*

The measured and the additional vibrations in the measuring planes close to the bearings, i.e. the differences between the measured vibrations in the sensor positions in the reference case and in the case of the slot, for the first three harmonic components in the case of a 34% slot are reported in figure 8.48 to figure 8.53. Since the slot is practically an always open crack, it does not present any breathing behaviour and the only significant contribution, which is not due to the unassembly/reassembly of the rotor, is due to the rotor stiffness asymmetry which occurs twice per revolution. Therefore, only the 2X additional vibrations will be considered in the identification procedure.

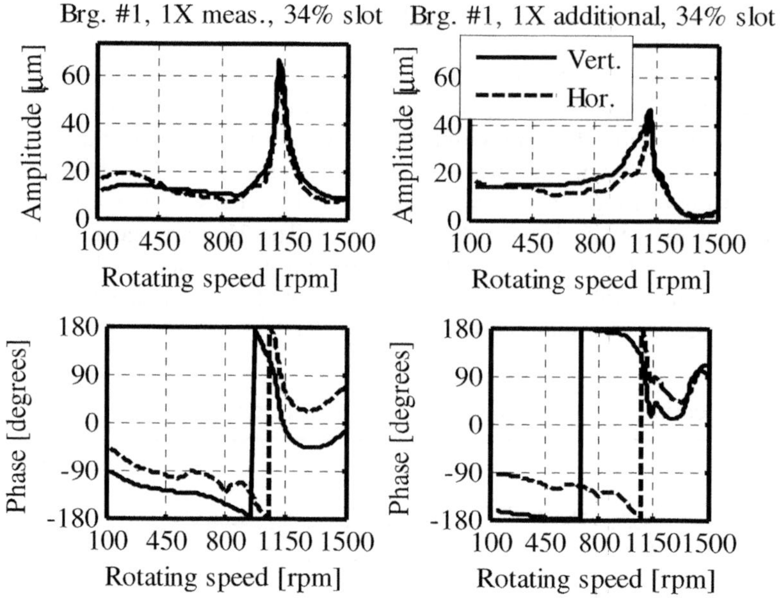

Fig. 8.48. Case 1, 34% slot (open crack): 1X vibration components for bearing #1.

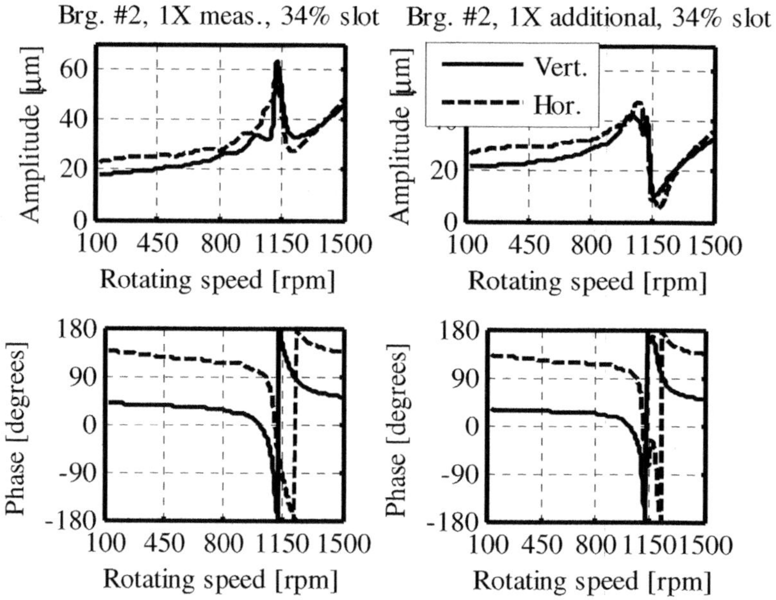

Fig. 8.49. Case 1, 34% slot (open crack): 1X vibration components for bearing #2.

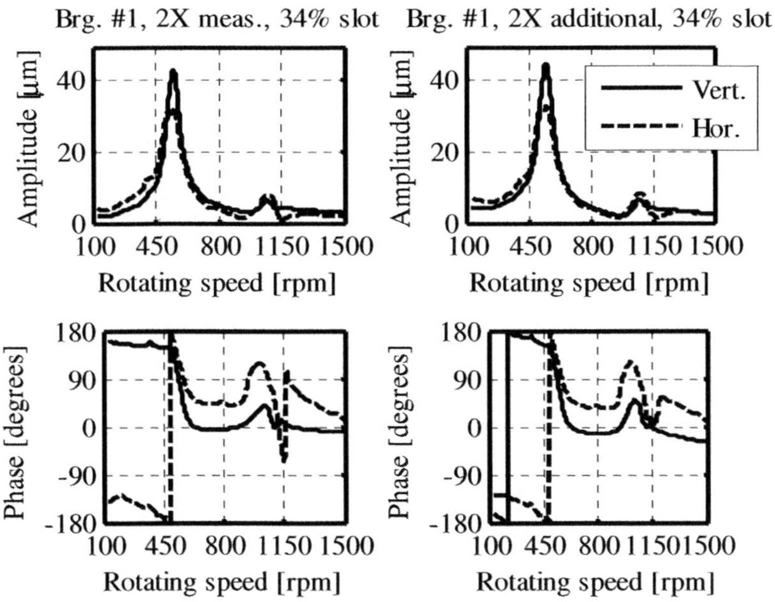

Fig. 8.50. Case 1, 34% slot (open crack): 2X vibration components for bearing #1.

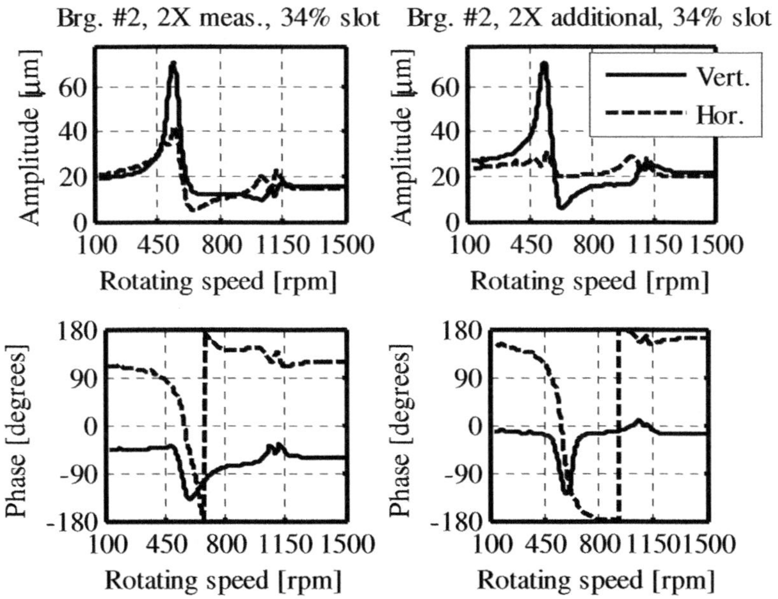

Fig. 8.51. Case 1, 34% slot (open crack): 2X vibration components for bearing #2.

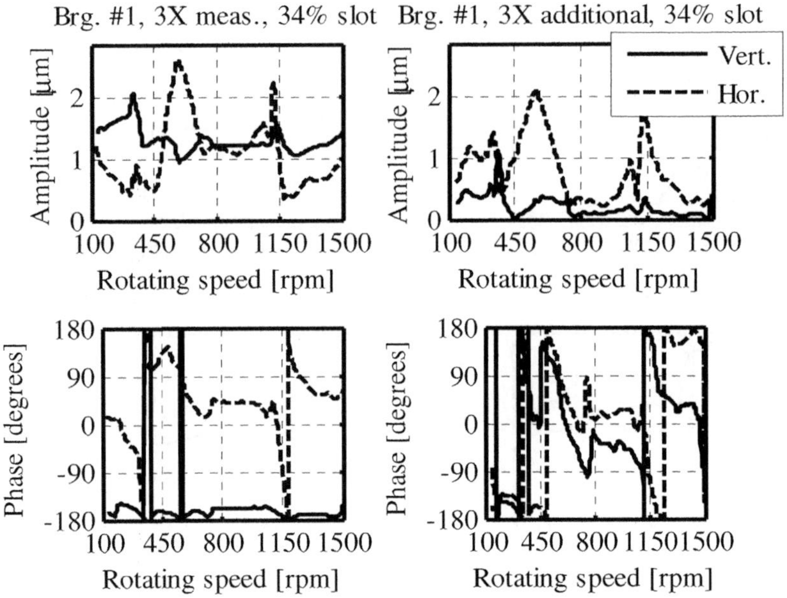

Fig. 8.52. Case 1, 34% slot (open crack): 3X vibration components for bearing #1.

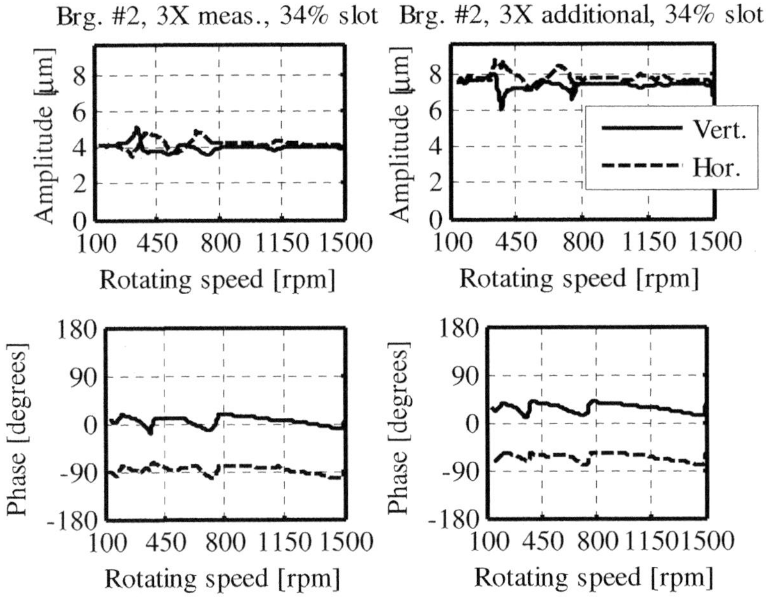

Fig. 8.53. Case 1, 34% slot (open crack): 3X vibration components for bearing #2.

With regard to the 1X components, a significant offset is present in both the bearing at very low rotating speeds, that indicates again a permanent bow in the rotor. The 3X components are rather small and the peak at ⅓ of the 1st critical speed is not present in all the measured vibrations.

The identification results (using the additional vibration at all the available rotating speeds) are shown in figure 8.54, taking into account that in this case the M'_{2slot}/M curve of figure 8.22 is used. The M_{2slot} identified moment is equal to 218 Nm, while the static bending moment M is 727 Nm (see figure 8.38) in element 19. Taking into account that shaft diameter D is 70 mm and element 19 length l is 65 mm, r.h.s. of eq. (8.59) is 0.278, that corresponds in figure 8.22 to a depth of about 32%.

The relative residual curve has its minimum in a position in the f.e. model of the rotor, which is very close to the actual crack position, with an error equal to 2.1% with respect to the overall length of the rotor. The depth is almost correctly estimated.

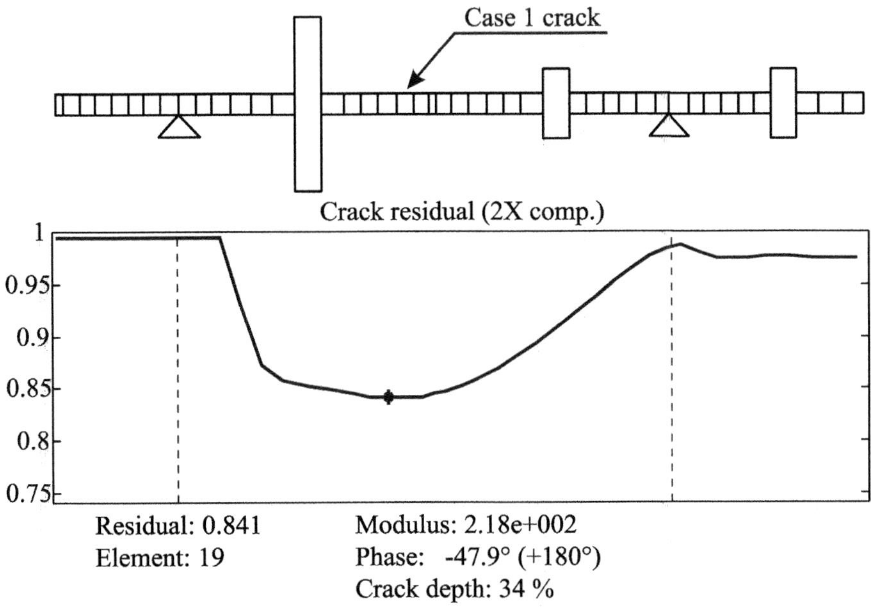

Fig. 8.54. Case 1, 34% slot (open crack): relative residual of the fault identified from 2X component.

8.5.2.3 Case 2: Identification of a 14% Crack

The additional vibrations in the bearings for the first three harmonic components in this case are reported in figure 8.55 to figure 8.60.

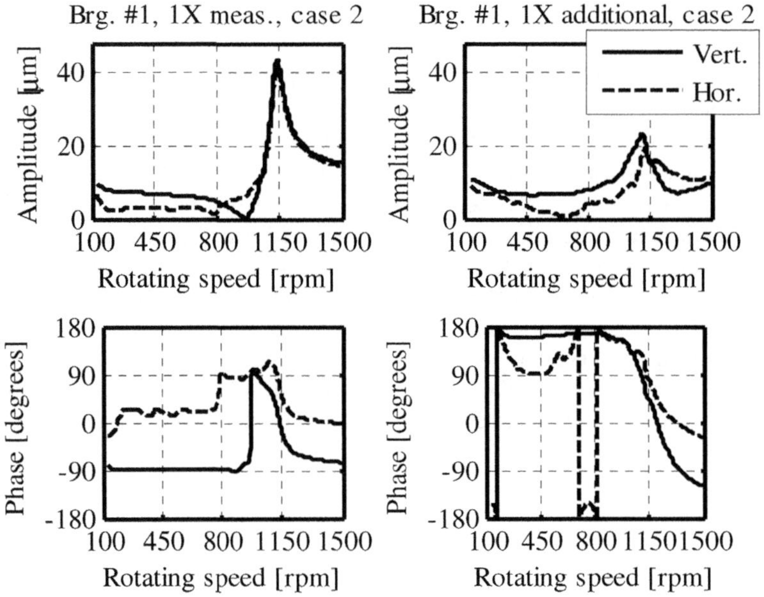

Fig. 8.55. Case 2, slot & crack: 1X vibration components for bearing #1.

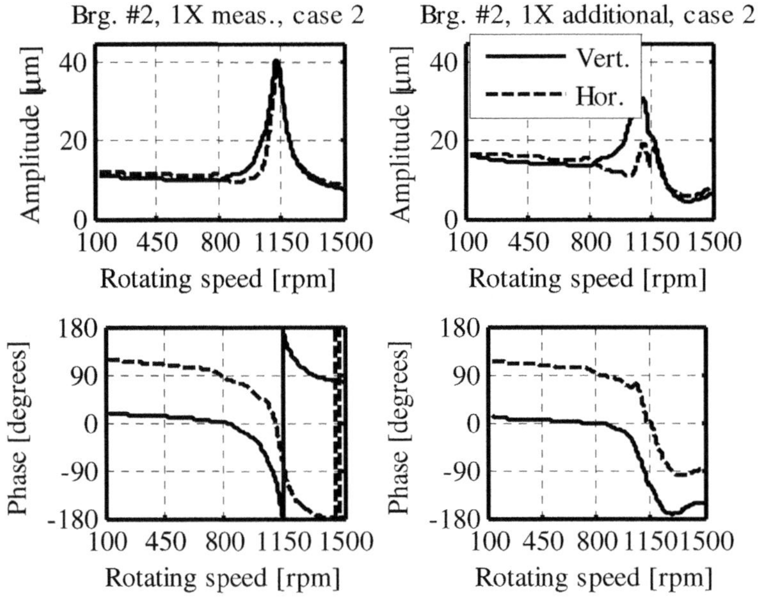

Fig. 8.56. Case 2, slot & crack: 1X vibration components for bearing #2.

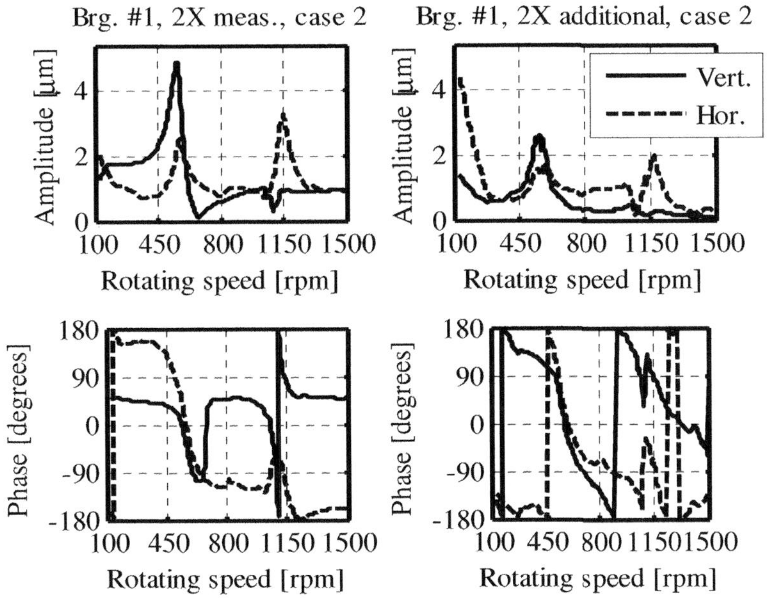

Fig. 8.57. Case 2, slot & crack: 2X vibration components for bearing 1.

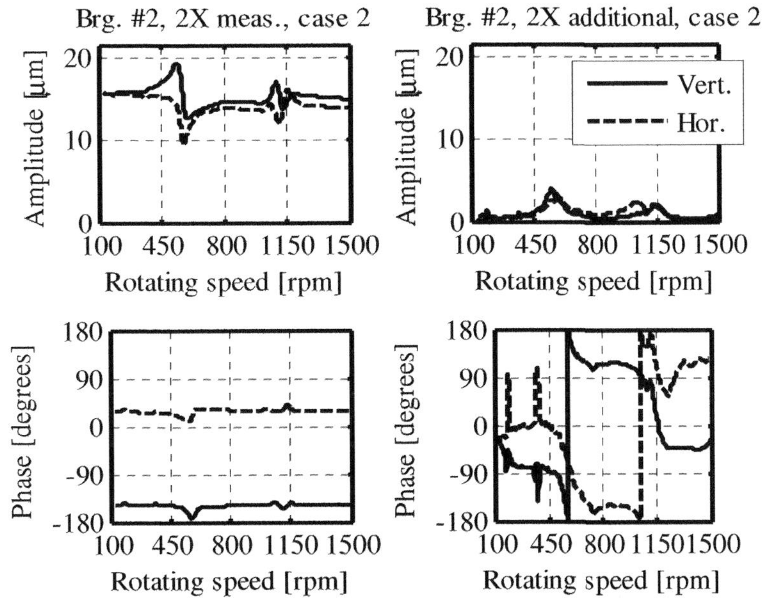

Fig. 8.58. Case 2, slot & crack: 2X vibration components for bearing #2.

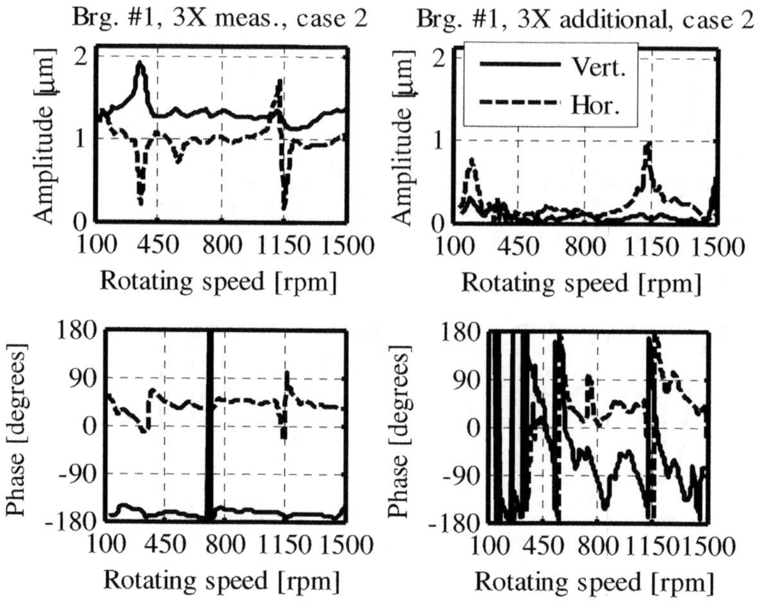

Fig. 8.59. Case 2, slot & crack: 3X vibration components for bearing #1.

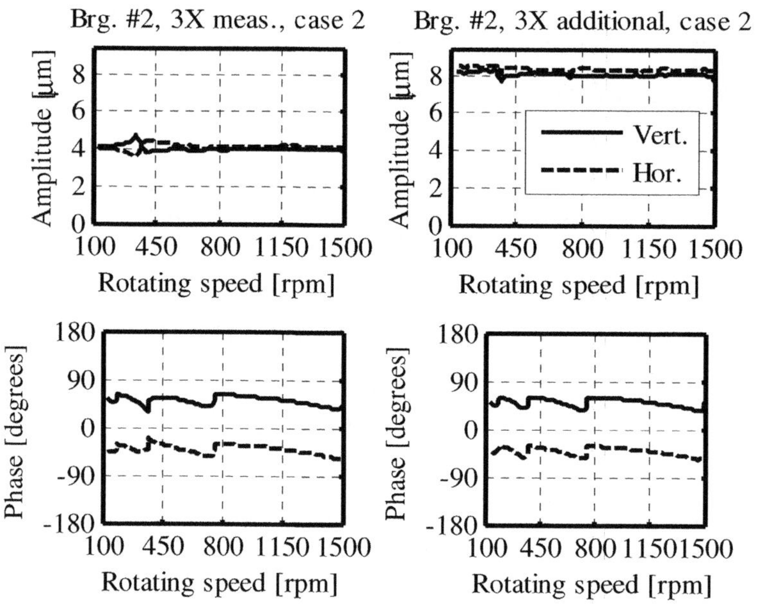

Fig. 8.60. Case 2, slot & crack: 3X vibration components for bearing #2.

The 1X behaviour shows a high run-out at low speed, due to a permanent bow of the rotor. The 3X component, instead, is rather high and constant in amplitude and phase which could indicate geometrical surface errors. As already stressed, in this case the crack is not properly a real crack, nor a slot, since it is breathing only for a smaller part of its surface (see figure 8.40), but the two curves M'_{2slot}/M and M'_2/M of figure 8.22 are very close to each other when the depth is small, so the curve relative to an actual crack is used. The results of the identification for the 2X component are reported in figure 8.61. The M_2 is equal to 19 Nm, while the static bending moment M is equal to 684 Nm in element 20 (see figure 8.38). Since the shaft diameter D is 70 mm and element 20 length l is 65 mm, the r.h.s. of eq. (8.59) is 0.026, that corresponds in figure 8.22 to a depth of 15.6%.

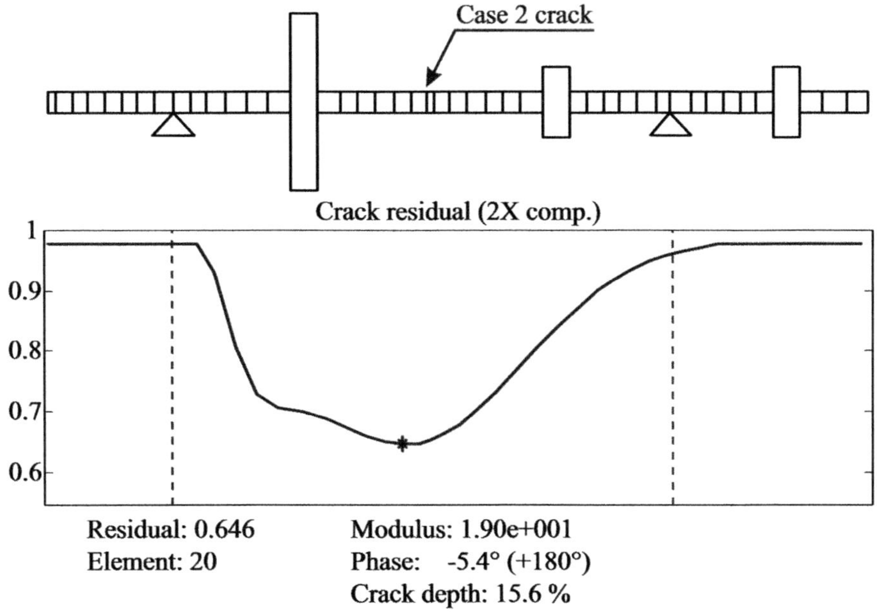

Fig. 8.61. Case 2, slot & crack: relative residual of the fault identified from 2X component.

The position of the crack, and also its relative depth is identified with good accuracy by means of the 2X component (see figure 8.61). The identification is not shown for 1X and 3X components, due to the unknown bow for the 1X and to the noise for the 3X components.

Finally the comparison between the experimental and the calculated values (with the identified equivalent bending moments) are shown in figure 8.62 for the 2X component, in vertical direction, in the measuring position close to bearing 2.

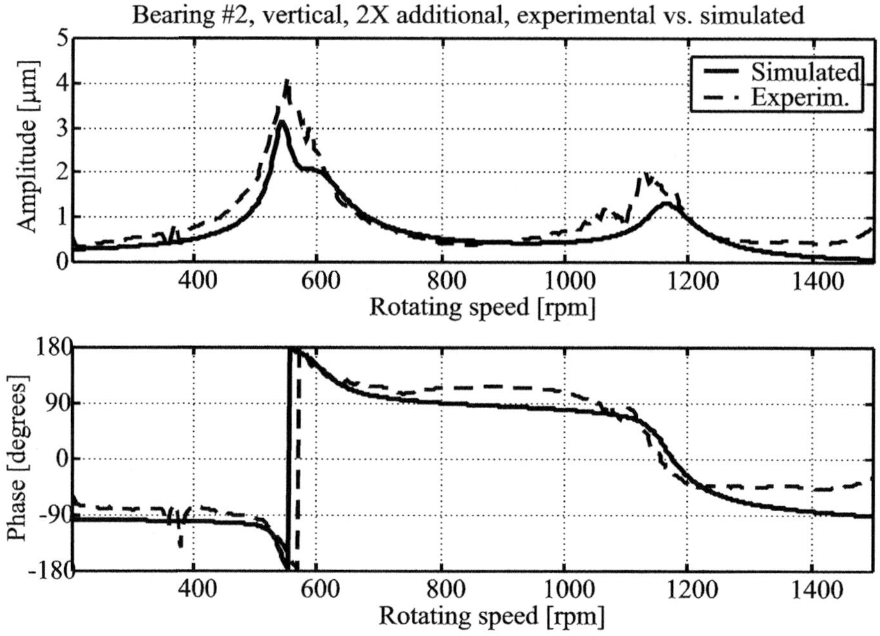

Fig. 8.62. Case 2, slot & crack: 2X measured and calculated vertical displacements close to bearing 2.

Despite the fact that the small crack (about 14%) generates only small 2X additional vibration components, less than 2 μm in bearing 1 and less than 5 μm in bearing 2, the identification can be considered successful.

8.5.2.4 Case 3: Identification of a 47% Crack

The additional vibrations obtained with a crack of about 47% depth of the diameter in the central section are shown in figure 8.63 to figure 8.68.

From the 1X component (see figure 8.63 and figure 8.64) it can be inferred that the rotor presents a permanent bow which is increased with respect to the reference rotor. In regards to the 2X component, the high amplitude (about 40÷50 μm) of the peak at ½ of the 1st critical speed (see figure 8.65 and figure 8.66) is clearly due to the crack, which produces also a high resonance amplitude (about 10 μm) of the 3X component at ⅓ of the 1st critical speed (see figure 8.67 and figure 8.68).

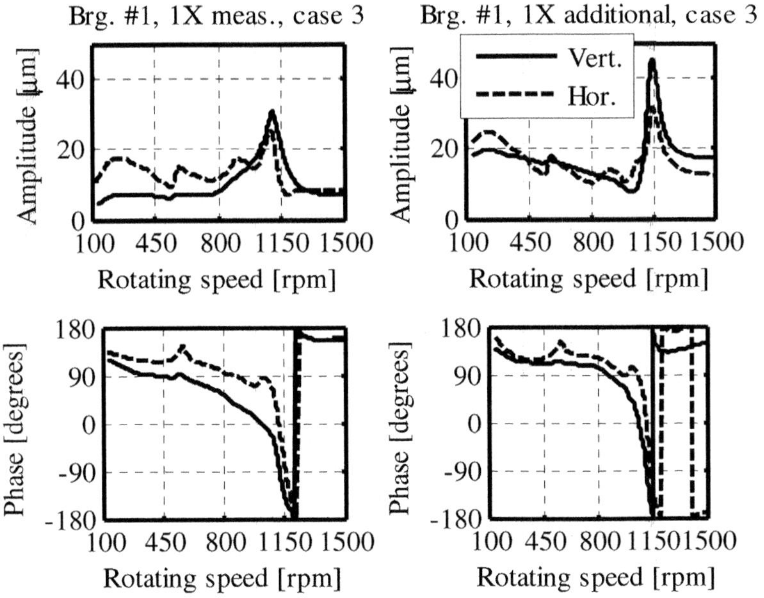

Fig. 8.63. Case 3, crack: 1X vibration components for bearing #1.

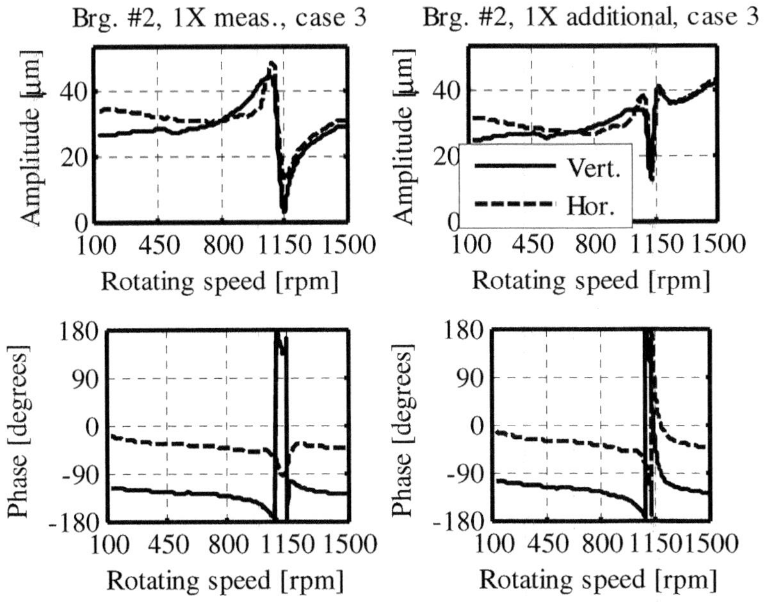

Fig. 8.64. Case 3, crack: 1X vibration components for bearing #2.

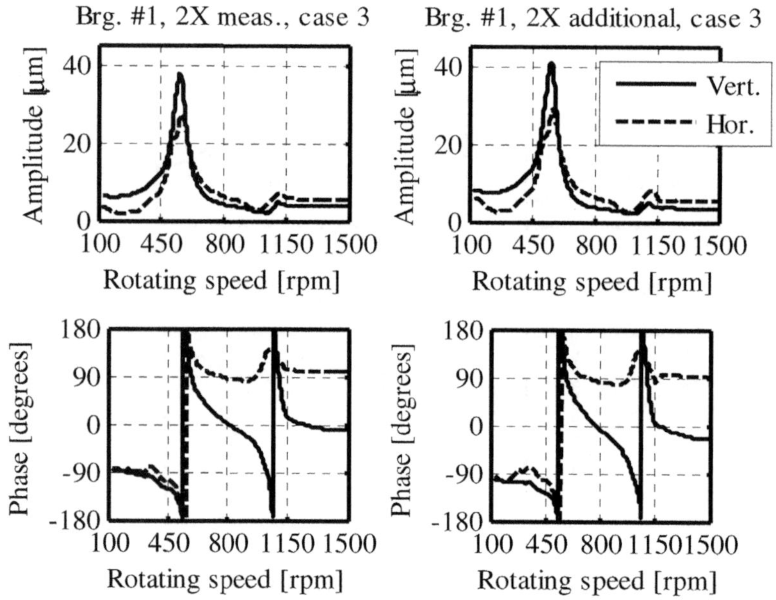

Fig. 8.65. Case 3, crack: 2X vibration components for bearing #1.

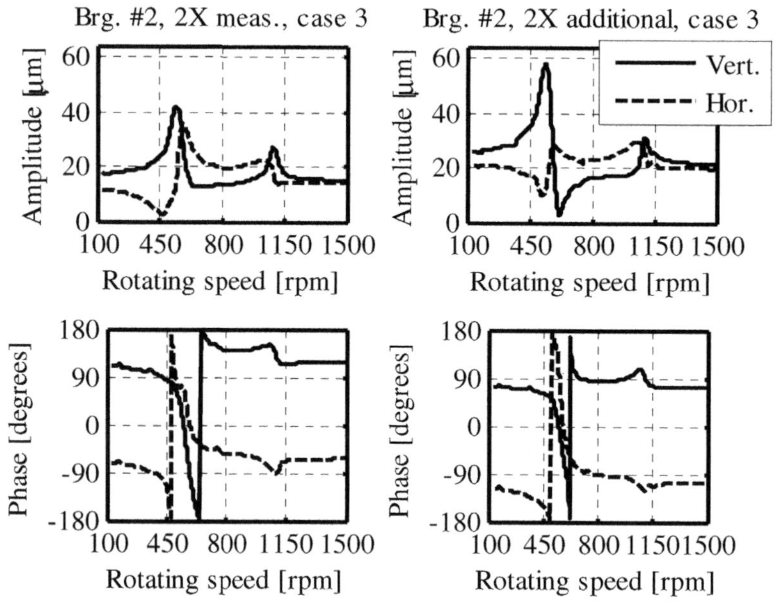

Fig. 8.66. Case 3, crack: 2X vibration components for bearing #2.

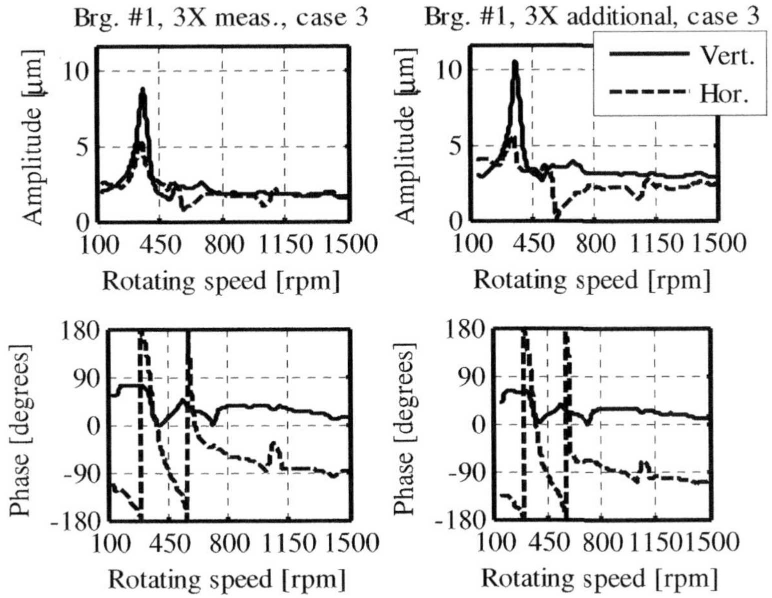

Fig. 8.67. Case 3, crack: 3X vibration components for bearing #1.

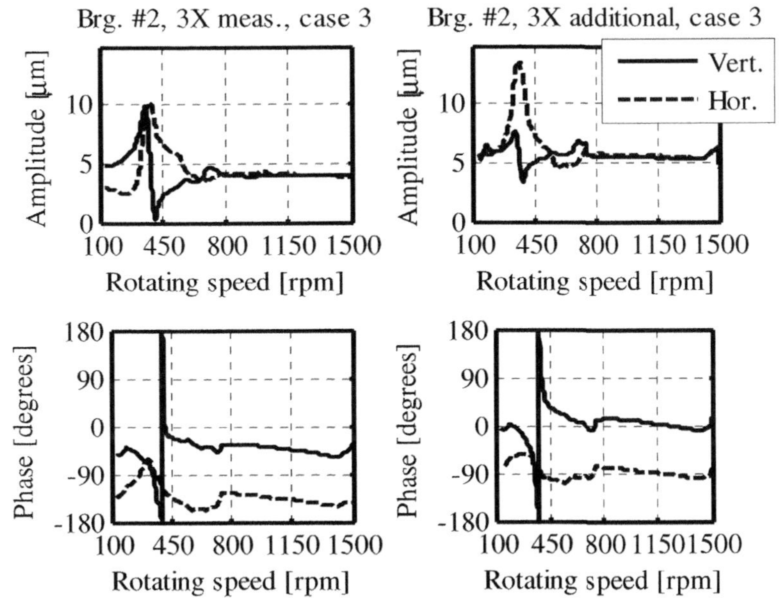

Fig. 8.68. Case 3, crack: 3X vibration components for bearing #2.

In this case the identification of the crack has been attempted not only by using the 2X component, but also by the 1X and 3X components. In fact the last one has a rather high value, while some processing were used in order to depurate the effect of the unbalance in the 1X component as explained later.

The results in figure 8.69 show that the location of the crack is precisely identified by 2X and 3X components. The 1X component identifies the crack in a beam element close to the actual one (element 24 instead of 23) with an error of 1.7% with respect to the overall length of the rotor. The crack depth is identified using only the value of the 2X equivalent bending moment M_2, which is equal to 301 Nm. The static bending moment M is equal to 548 Nm in element 23 (see figure 8.38). The shaft diameter D is 70 mm and element 23 length l is 53.5 mm, so that r.h.s. of eq. (8.59) is 0.425 that corresponds in figure 8.22 to a depth of 47%.

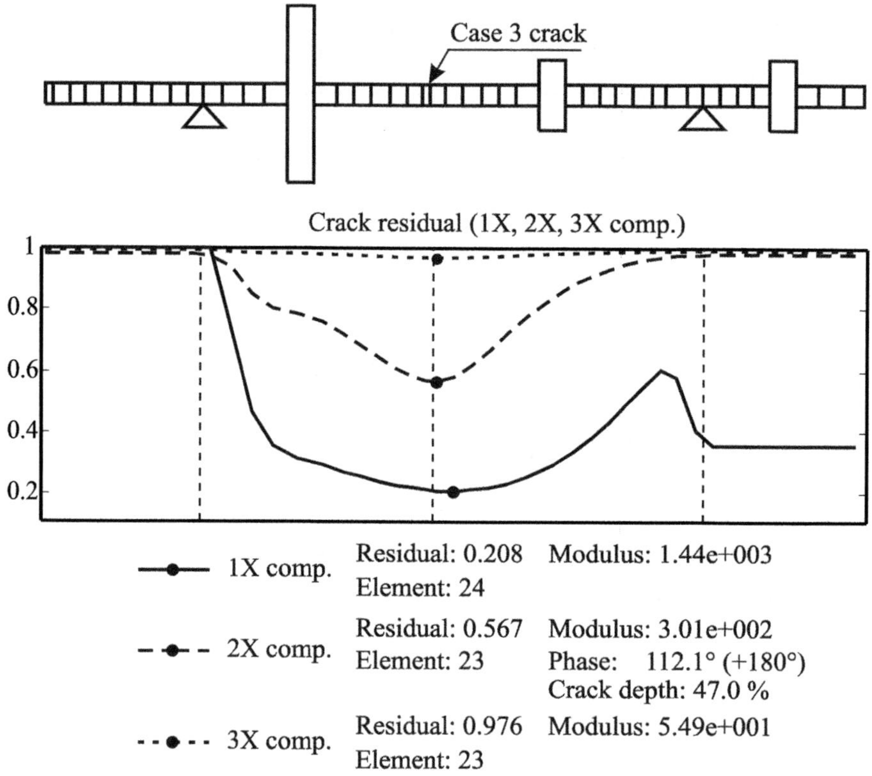

Fig. 8.69. 47% cracked shaft. Relative residuals of the crack identification.

Moreover, the 1X component identifies the position with a particularly reduced value of the relative residual. This result has been obtained by processing the experimental data in the following way: first the unbalances on the disks were identified, and then the dynamical behaviour due to the unbalances only has been subtracted from measured data in order to obtain the bow induced vibrations. This leads to a very good agreement between the experimental and the simulated behaviour for the 1X component as shown in figure 8.70 and figure 8.71, in which bow and unbalances have been superimposed. Due to this procedure, it was considered not pertinent to use M_1 to identify the crack depth.

With regard to the 2X component, the relative residual can be considered as good, but the most remarkable result is the exact identification of the depth of the crack.

Also the comparison of simulated behaviour and experimental measurements in figure 8.72 and figure 8.73 is good.

As concerning the 3X component, the relative residual is quite high, but this can be explained by considering that this component is normally masked by noise. This notwithstanding, if M_3 were deemed useful to identify the crack depth, it resulted equal to 54.9 Nm that made r.h.s. of eq. (8.59) to be equal to 0.077 that corresponds in figure 8.22 to a depth of 30%. However, the residual curve presents a well defined minimum in correspondence of the crack even this is not so evident in figure 8.69 due to the scale.

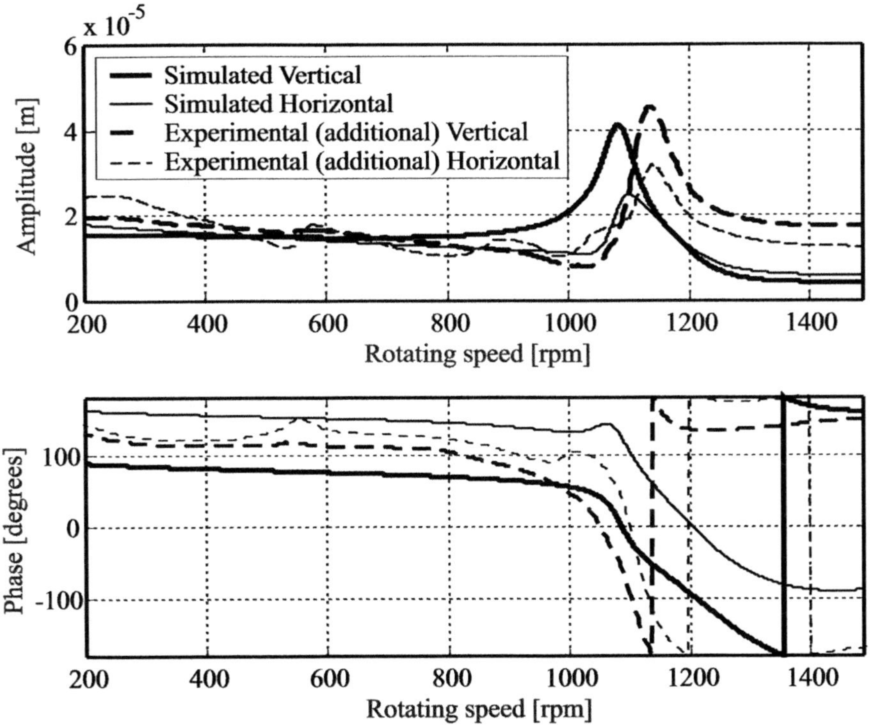

Fig. 8.70. Case 3, crack: comparison between simulated and experimental 1X vibration components for bearing 1.

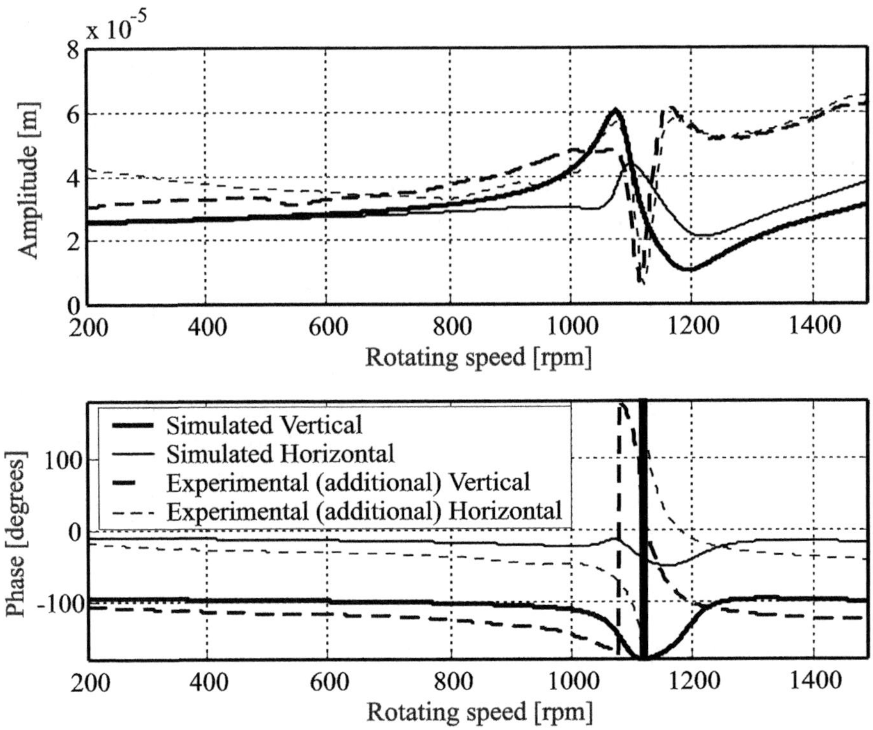

Fig. 8.71. Case 3, crack: comparison between simulated and experimental 1X vibration components for bearing 2.

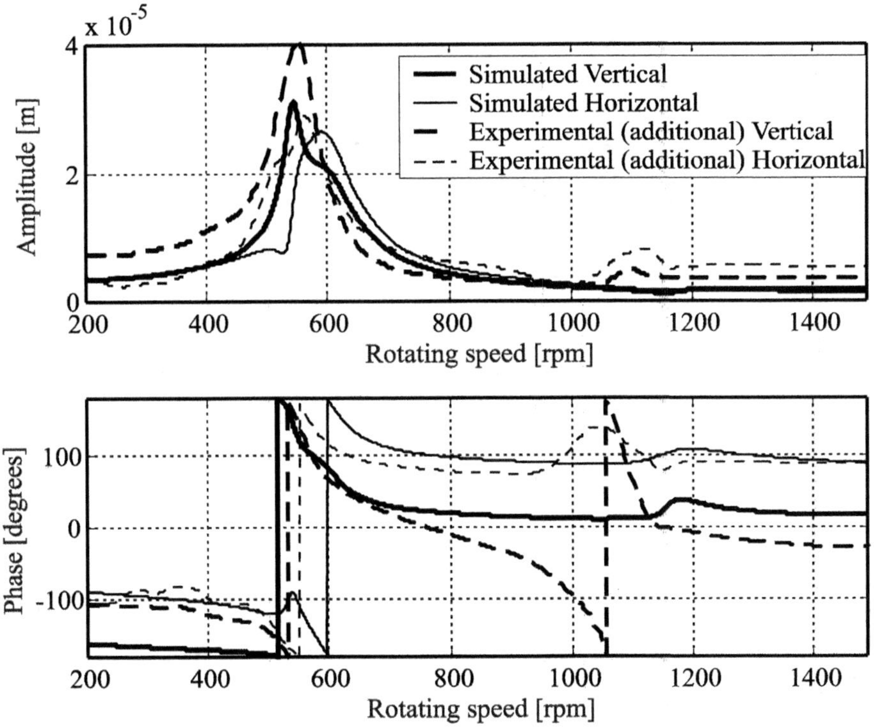

Fig. 8.72. Case 3, crack: comparison between simulated and experimental 2X vibration components for bearing 1.

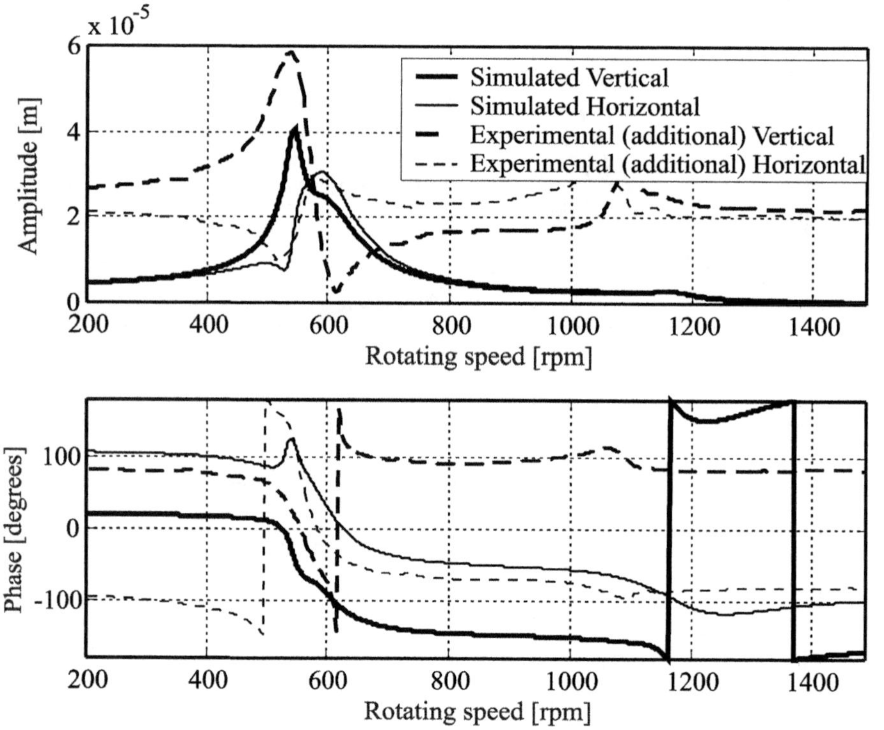

Fig. 8.73. Case 3, crack: comparison between simulated and experimental 2X vibration components for bearing 2.

References

[8.1] White MF, Jecmenica M (1999) Fault diagnosis using a fault matrix incorporating fuzzy logic. 12th International Congress on Condition Monitoring and Diagnostic Engineering Management-COMADEM 99, July 1999, Sunderland, U.K.

[8.2] Sohre JS (1968) Operating problems with high speed turbomachinery, causes and corrections. ASME Petroleum Mech. Eng. Conf., September 1968, Dallas, Texas.

[8.3] Sohre JS (1991) Turbomachinery problems and their correction. Standardization and Condition Monitoring workshop, chapter 7, March 1991, Houston, Texas.

[8.4] Rao JS (1996) Rotor Dynamics. 3rd Edition, New Age Publishers, new Delhi, India.

[8.5] Taylor JI (1994) The Vibration Analysis Handbook. Bayonet Point, FL, USA.

[8.6] Bloch HP, Geitner FK (1986) Practical machinery management for process plants, vol. 2: machinery failure analysis and troubleshooting. Gulf Professional Publishing, USA.

[8.7] Ehrich FH (1992) Handbook of rotordynamics. McGraw-Hill Inc., USA.

[8.8] Muszyńska A (2005) Rotordynamics. CRC Press, Boca Raton, FL, USA.

[8.9] Bachschmid N, Pennacchi P, Vania A (2002) Identification of multiple faults in rotor systems. Journal of Sound and Vibration, 254(2), 327-366.

[8.10] Pennacchi P, Vania A (2005) Diagnosis and model based identification of a coupling misalignment. Shock and Vibration, 12(4), 293-308.

[8.11] Bachschmid N, Pennacchi P, Tanzi E (2006) Effect of oil film bearing misalignment on rigidly coupled shafts. Proc. of Гидродинамическая теория смазки – 120 лет; Hydrodynamic theory of lubrication – 120th anniversary, Vol. 1, May 18-20, Orel, Russia, 457-467.

[8.12] Kiciński J (2006) Rotor dynamics. IFFM Publishers, Gdańsk, Poland.

[8.13] Pennacchi P, Vania A (2007) Analysis of rotor-to-stator rub in a large steam turbogenerator. International Journal of Rotating Machinery, vol. 2007, Article ID 90631, 1-8.

[8.14] Bachschmid N, Pennacchi P, Vania A, Zanetta GA, Gregori L (2003) Identification of rub and unbalance in a 320MW turbogenerator. International Journal of Rotating Machinery, 9(2), 97-112.

[8.15] Bachschmid N, Pennacchi P, Vania A (2004) Diagnostic significance of orbit shape analysis and its application to improve machine faults detection. Journal of the Brazilian Society of Mechanical Sciences and Engineering, XXVI(2), 200-208.

[8.16] Bachschmid N, Pennacchi P, Vania A (2007) Thermally induced vibrations due to rub in real rotors. Journal of Sound and Vibration, 299(4-5), 683-719.

[8.17] Adams ML Jr (2001) Rotating machinery vibration. Marcel Dekker Inc., New York, NY, USA.

[8.18] Pennacchi P, Vania A (2004) Accuracy in the identification of a generator thermal bow. Journal of Sound and Vibration, 274(1-2), 273-295.

[8.19] Bachschmid N, Pennacchi P (2006) Faults identification and corrective actions in rotating machinery at rated speed. Shock and Vibration, 13(4-5), 485-503.

[8.20] Pennacchi P, Bachschmid N, Vania A, Zanetta GA, Gregori L (2006) Use of modal representation for the supporting structure in model based fault identification of large rotating machinery: part 2 – application to a real machine. Mechanical Systems and Signal Processing, 20(3), 682-701.

[8.21] Vania A (2001) Analysis of the vibrations of a generator rotor caused by a short-circuit of the rotor windings. Proc. of Survelliance 4 – Acoustical and Vibratory Surveillance Methods and Diagnostic Techniques, 16-18 October 2001, Compiègne, France, 507-516.

[8.22] Pennacchi P, Frosini L (2005) Dynamical Behaviour of a Three-Phase Generator Due to Unbalanced Magnetic Pull. IEE Proc. Electric Power Applications, 152(6), 1389-1400.

[8.23] Frosini L, Pennacchi P (2004) Detection and modelling of rotor eccentricity in electrical machines: an overview. IMechE paper C623/060/2004, Proc. of 8th International Conference on Vibrations in Rotating Machinery, 7-9 September 2004, Swansea, Wales, 501-510.

[8.24] Childs D (1993) Turbomachinery rotordynamics. John Wiley & Sons Inc., New York, NY, USA.

[8.25] Robert CE Sr, Robert CE Jr (1998) Machinery malfunction diagnosis and correction. Hewlett-Packard Company, New Jersey, USA.

[8.26] Bachschmid N, Pennacchi P, Vania A (2008) Steam whirl analysis in a high pressure cylinder of a turbo generator. Mechanical Systems and Signal Processing, 22(1), 121-132.

[8.27] Bachschmid N, Pennacchi P (2002) Multiple fault identification method in the frequency domain for rotor systems. Shock and Vibration, 9(4-5), 203-215.

[8.28] Vania A, Pennacchi P (2004) Experimental and theoretical application of fault identification measures of accuracy in rotating machine diagnostics. Mechanical Systems and Signal Processing, 18(2), 329-352.

[8.29] Bachschmid N, Pennacchi P (2003) Accuracy of fault detection in real rotating machinery using model based diagnostic techniques. JSME International Journal Series C, 46(3), 1026-1034.

[8.30] Isermann R (1995) Fault detection and diagnosis – Methods and applications – . 2nd International Symposium on Acoustical and Vibratory Surveillance Methods and Diagnostic Techniques, October 1995, Senlis, France, 777-797.

[8.31] Isermann R (1997) Supervision, fault-detection and fault-diagnosis methods – An introduction. Control Engineering Practice, 5(5), 639-652.

[8.32] Sinha JK, Friswell MI, Lees AW (2002) The identification of the unbalance and the foundation model of a flexible rotating machine from a single run-down. Mechanical Systems and Signal Processing, 16(2-3), 255-271.

[8.33] Aenis M, Knopf E, Nordmann R. (2002) Active magnetic bearings for the identification and fault diagnosis in turbomachinery. Mechatronics, 12, 1011-1021.

[8.34] Kreuzinger-Janik T, Irretier H (2000) Unbalance identification of flexible rotors based on experimental modal analysis. IMechE paper C576-029, 7th Int. Conf. on Vibrations in Rotating Machinery, September 2000, University of Nottingham, 335-346.

[8.35] Platz R, Markert R, Seidler M (2000) Validation of online diagnostics of malfunctions in rotor systems. IMechE paper C576-025, 7th Int. Conf. on Vibrations in Rotating Machinery, September 2000, University of Nottingham, 581-590.

[8.36] Markert R, Platz R, Siedler M (2000) Model Based Fault Identification in Rotor Systems by Least Squares Fitting. ISROMAC-8 Conference, March 2000, Honolulu, Hawaii, 901-915.

[8.37] Platz R, Markert R, (2001) Fault models for on-line identification of malfunctions in rotor systems. Survelliance 4 – Acoustical and Vibratory Surveillance Methods and Diagnostic Techniques, October 2001, Compiègne, France, 435-446.

[8.38] Pennacchi P, Bachschmid N, Vania A, Zanetta GA, Gregori L (2006) Use of modal representation for the supporting structure in model based fault identification of large rotating machinery: part 1 – theoretical remarks. Mechanical Systems and Signal Processing, 20(3), 662-681.

[8.39] Pennacchi P, Bachschmid N, Vania A (2006) A model-based identification method of transverse cracks in rotating shafts suitable for industrial machines.. Mechanical Systems and Signal Processing, 20(8), 2112-2147.

[8.40] Pennacchi P, Vania A, Bachschmid N, (2007) Increasing the robustness of fault identification in rotor dynamics by means of M-estimators. Mechanical Systems and Signal Processing, 21(8), 3003-3029.

[8.41] Bachschmid N, Vania A, Audebert S (2000) A Comparison of Different Methods for Transverse Crack Modelling in Rotor Systems. ISROMAC-8 Conference, March 2000, Honolulu, Hawaii, 1057-1064.

[8.42] Bachschmid N, Pennacchi P, Tanzi E, Audebert S (2003) Transverse Crack Modeling and Validation in Rotor Systems, Including Thermal Effects. International Journal of Rotating Machinery, 9(2), 113-126.

Index

Batch number: 10755284

Printed by Printforce, the Netherlands